Nuclear Terrorism & The Dirty Bomb

Edited by Paul F. Kisak

Contents

Chapter 1

Nuclear terrorism

Nuclear terrorism refers to an act of terrorism in which a person or persons belonging to a terrorist organization detonates a nuclear device.[1] Some definitions of nuclear terrorism include the sabotage of a nuclear facility and/or the detonation of a radiological device, colloquially termed a dirty bomb, but consensus is lacking. In legal terms, nuclear terrorism is an offense committed if a person unlawfully and intentionally "uses in any way radioactive material ···with the intent to cause death or serious bodily injury; or with the intent to cause substantial damage to property or to the environment; or with the intent to compel a natural or legal person, an international organization or a State to do or refrain from doing an act", according to the 2005 United Nations International Convention for the Suppression of Acts of Nuclear Terrorism.[2]

The possibility of terrorist organizations using nuclear weapons (including those of a relatively smaller size, such as those contained within suitcases (suitcase nuclear device), is something which is known of within U.S. culture, and at times previously discussed within the political settings of the U.S. It is considered plausible that terrorists could acquire a nuclear weapon.[3] However, despite thefts and trafficking of small amounts of fissile material, all low-concern and less than Category III Special nuclear material (SNM), there is no credible evidence that any terrorist group has succeeded in obtaining Category I SNM, the necessary multi-kilogram critical mass amounts of weapons grade plutonium required to make a nuclear weapon.[4][5]

1.1 Scope

Main article: Vulnerability of nuclear plants to attack

Nuclear terrorism could include:

- Acquiring or fabricating a nuclear weapon
- Fabricating a dirty bomb

- Attacking a nuclear reactor, e.g., by disrupting critical inputs (e.g. water supply)
- Attacking or taking over a nuclear-armed submarine, plane or base.[6]

Nuclear terrorism, according to a 2011 report published by the Belfer Center for Science and International Affairs at Harvard University, can be executed and distinguished via four pathways:[7]

- The use of a nuclear weapon that has been stolen or purchased on the black market
- The use of a crude explosive device built by terrorists or by nuclear scientists who the terrorist organization has furtively recruited
- The use of an explosive device constructed by terrorists and their accomplices using their own fissile material
- The acquisition of fissile material from a nation-state.

Former U.S. President Barack Obama called nuclear terrorism "the single most important national security threat that we face". In his first speech to the U.N. Security Council, President Obama said that "Just one nuclear weapon exploded in a city -- be it New York or Moscow, Tokyo or Beijing, London or Paris -- could kill hundreds of thousands of people". It would "destabilize our security, our economies, and our very way of life".[8]

1.2 History

As early as December 1945, politicians worried about the possibility of smuggling nuclear weapons into the United States, though this was still in the context of a battle between the superpowers of the Cold War. Congressmen quizzed the "father of the atomic bomb," J. Robert Oppenheimer, about the possibility of detecting a smuggled atomic bomb:

1

Sen. Millikin: We... have mine-detecting devices, which are rather effective... I was wondering if anything of that kind might be available to use as a defense against that particular type of use of atomic bombs.

Dr. Oppenheimer: If you hired me to walk through the cellars of Washington to see whether there were atomic bombs, I think my most important tool would be a screwdriver to open the crates and look. I think that just walking by, swinging a little gadget would not give me the information.[9]

This sparked further work on the question of smuggled atomic devices during the 1950s.

Discussions of non-state nuclear terrorism among experts go back at least to the 1970s. In 1975 The Economist warned that "You can make a bomb with a few pounds of plutonium. By the mid-1980s the power stations may easily be turning out 200,000 lb of the stuff each year. And each year, unless present methods are drastically changed, many thousands of pounds of it will be transferred from one plant to another as it proceeds through the fuel cycle. The dangers of robbery in transit are evident.... Vigorous co-operation between governments and the International Atomic Energy Agency could, even at this late stage, make the looming perils loom a good deal smaller." [10] And the New York Times commented in 1981 that The Nuclear Emergency Search Team's "origins go back to the aftershocks of the Munich Olympic massacre in mid-1972. Until that time, no one in the United States Government had thought seriously about the menace of organized, international terrorism, much less nuclear terrorism. There was a perception in Washington that the value of what is called 'special nuclear material' - plutonium or highly enriched uranium (HEU) - was so enormous that the strict financial accountability of the private contractors who dealt with it would be enough to protect it from falling into the wrong hands. But it has since been revealed that the physical safeguarding of bomb-grade material against theft was almost scandalously neglected." [11]

This discussion took on a larger public character in the 1980s after NBC aired Special Bulletin, a television dramatization of a nuclear terrorist attack on the United States.[12] In 1986 a private panel of experts known as the International Task Force on the Prevention of Terrorism released a report urging all nuclear-armed states to beware the dangers of terrorism and work on equipping their nuclear arsenals with permissive action links. "The probability of nuclear terrorism," the experts warned, "is increasing and the consequences for urban and industrial societies could be catastrophic." [13]

The World Institute for Nuclear Security is an organization which seeks to prevent nuclear terrorism and improve world nuclear security. It works alongside the International Atomic Energy Agency. WINS was formed in 2008, less than a year after a break-in at the Pelindaba nuclear facility in South Africa, which contained enough enriched uranium to make several nuclear bombs.

The Global Initiative to Combat Nuclear Terrorism (GICNT) is an international partnership of 86 nations and 4 official observers working to improve capacity on a national and international level for prevention, detection, and response to a nuclear terrorist event. Partners join the GICNT by endorsing the Statement of Principles, a set of broad nuclear security objectives. GICNT partner nations organize and host workshops, conferences, and exercises to share best practices for implementing the Statement of Principles. The GICNT also holds Plenary meetings to discuss improvements and changes to the partnership.

1.3   Militant groups

Nuclear weapons materials on the black market are a global concern.[14][15] and there is concern about the possible detonation of a small, crude nuclear weapon by a militant group in a major city, with significant loss of life and property.[16][17]

It is feared that a terrorist group could detonate a dirty bomb, a type of radiological weapon. A dirty bomb is made of any radioactive source and a conventional explosive. There would be no nuclear blast and likely no fatalities, but the radioactive material is dispersed and can cause extensive fallout depending on the material used. A foot-long stick of radioactive cobalt could be taken from a food irradiation plant and combined with ten pounds of explosives to contaminate 1,000 square kilometers and make some areas uninhabitable for decades.[17] There are other radiological weapons called radiological exposure devices where an explosive is not necessary. A radiological weapon may be very appealing to terrorist groups as it is highly successful in instilling fear and panic among a population (particularly because of the threat of radiation poisoning) and would contaminate the immediate area for some period of time, disrupting attempts to repair the damage and subsequently inflicting significant economic losses.

1.3.1   al-Qaeda

According to *Bunn & Wier*, Osama bin Laden requested a ruling (a *fatwa*), and was subsequently informed via a cleric of Saudi Arabia during 2003, of it being in accordance with Islamic law for him to use a nuclear device against civilians if it were the only course of action available to him in a situation of defending Muslims against the actions of the U.S. military.[18]

According to leaked diplomatic documents, al-Qaeda can produce radiological weapons, after sourcing nuclear material and recruiting rogue scientists to build "dirty bombs".[19] Al-Qaeda, along with some North Caucasus terrorist groups that seek to establish an Islamic Caliphate in Russia, have consistently stated they seek nuclear weapons and have tried to acquire them.[7] Al-Qaeda has sought nuclear weapons for almost two decades by attempting to purchase stolen nuclear material and weapons and has sought nuclear expertise on numerous occasions. Osama bin Laden stated that the acquisition of nuclear weapons or other weapons of mass destruction is a "religious duty."[20] While pressure from a wide range of counter-terrorist activity has hampered Al-Qaeda's ability to manage such a complex project, there is no sign that it has jettisoned its goals of acquiring fissile material. Statements made as recently as 2008 indicate that Al-Qaeda's nuclear ambitions are still very strong.[7]

1.3.2 ISIS

ISIS has demonstrated ambition to use weapons of mass destruction,[21] although the chances of them obtaining a nuclear bomb are small, the group have been trying/suspected to be trying to obtain a nuclear dirty bomb.[22] In July 2014, ISIS militants captured nuclear materials from Mosul University. In a letter to UN Secretary-General Ban Ki-moon, Iraq's UN Ambassador Mohamed Ali Alhakim said that the materials had been kept at the university and "can be used in manufacturing weapons of mass destruction", however Nuclear experts regarded the threat as insignificant. International Atomic Energy Agency spokeswoman Gill Tudor said that the seized materials were "low grade and would not present a significant safety, security or nuclear proliferation risk".[23][24]

In October 2015 it was reported that Moldovan authorities working with the FBI have stopped four attempts from 2010 to 2015 by gangs with suspected connections to Russia's intelligence services that sought to sell radioactive material to ISIS and other Middle Eastern extremists. The last reported case came in February 2015-a smuggler with a large amount of radioactive caesium specifically sought a buyer from ISIS. The Criminal organizations are thriving on black market nuclear materials in Moldova, since relations between Russia and the West deteriorated, it is difficult to know whether smugglers are succeeding in selling radioactive material originating from Russia to Islamist terrorists and elsewhere.[21][25][26]

In March 2016, it was reported that a senior Belgian nuclear

official was being monitored by ISIS suspects linked to the November 2015 Paris attacks leading Belgium authorities to suspect that ISIS was planning on abducting the official to obtain nuclear materials for a dirty bomb.[27]

In April 2016, EU and NATO security chiefs warned that ISIS are plotting to carry out nuclear attacks on the UK and Europe.[28]

1.3.3 North Caucasus terrorists

North Caucasus terrorists have attempted to seize a nuclear submarine armed with nuclear weapons. They have also engaged in reconnaissance activities on nuclear storage facilities and have repeatedly threatened to sabotage nuclear facilities. Similar to Al-Qaeda, these groups' activities have been hampered by counter-terrorism activity; nevertheless they remain committed to launching such a devastating attack within Russia.[7]

1.3.4 Aum Shinrikyo

The Japanese terror cult Aum Shinrikyo, which used nerve gas to attack a Tokyo subway in 1995, has also tried to acquire nuclear weapons. However, according to nuclear terrorism researchers at Harvard University's Belfer Center for Science and International Affairs, there is no evidence that they continue to do so.[7]

1.4 Incidents involving nuclear material

Information reported to the International Atomic Energy Agency (IAEA) shows "a persistent problem with the illicit trafficking in nuclear and other radioactive materials, thefts, losses and other unauthorized activities".[29] The IAEA Illicit Nuclear Trafficking Database notes 1,266 incidents reported by 99 countries over the last 12 years, including 18 incidents involving HEU or plutonium trafficking:[30]

- There have been 18 incidents of theft or loss of highly enriched uranium (HEU) and plutonium confirmed by the IAEA.[20]

- Security specialist Shaun Gregory argued in an article that terrorists have attacked Pakistani nuclear facilities three times in the recent past; twice in 2007 and once in 2008.[31]

- In November 2007, burglars with unknown intentions infiltrated the Pelindaba nuclear research facility near

Pretoria, South Africa. The burglars escaped without acquiring any of the uranium held at the facility.[32][33]

- In June 2007, the Federal Bureau of Investigation released to the press the name of Adnan Gulshair el Shukrijumah, allegedly the operations leader for developing tactical plans for detonating nuclear bombs in several American cities simultaneously.[34]

- In November 2006, MI5 warned that al-Qaida were planning on using nuclear weapons against cities in the United Kingdom by obtaining the bombs via clandestine means.[35]

- In February 2006, Oleg Khinsagov of Russia was arrested in Georgia, along with three Georgian accomplices, with 79.5 grams of 89 percent HEU.[20]

- The Alexander Litvinenko poisoning with radioactive polonium "represents an ominous landmark: the beginning of an era of nuclear terrorism," according to Andrew J. Patterson.[36]

- In June 2002, U.S. citizen José Padilla was arrested for allegedly planning a radiological attack on the city of Chicago; however, he was never charged with such conduct. He was instead convicted of charges that he conspired to "murder, kidnap and maim" people overseas.

1.5   Pakistan

After several incidents in Pakistan in which terrorists attacked three of its military nuclear facilities, it became clear that there emerged a serious danger that they would gain access to the country's nuclear arsenal, according to a journal published by the US Military Academy at West Point.[37] In January 2010, it was revealed that the US army was training a specialised unit "to seal off and snatch back" Pakistani nuclear weapons in the event that militants would obtain a nuclear device or materials that could make one. Pakistan supposedly possesses about 80 nuclear warheads. US officials refused to speak on the record about the American safety plans.[38]

A study by the Belfer Center for Science and International Affairs at Harvard University titled "Securing the Bomb 2010," found that Pakistan's stockpile "faces a greater threat from Islamic extremists seeking nuclear weapons than any other nuclear stockpile on earth." [39]

According to Rolf Mowatt-Larssen, a former investigator with the CIA and the US Department of Energy, there is "a greater possibility of a nuclear meltdown in Pakistan than anywhere else in the world. The region has more violent extremists than any other, the country is unstable, and its arsenal of nuclear weapons is expanding." [40]

Nuclear weapons expert David Albright and author of "Peddling Peril" has also expressed concerns that Pakistan's stockpile may not be secure despite assurances by both Pakistan, U.S. and Southeast Asia government. He stated that Pakistan "has had many leaks from its program of classified information and sensitive nuclear equipment, and so you have to worry that it could be acquired in Pakistan." [41]

A 2010 study by the Congressional Research Service titled 'Pakistan's Nuclear Weapons: Proliferation and Security Issues' noted that even though Pakistan had taken several steps to enhance nuclear security in recent years, "instability in Pakistan has called the extent and durability of these reforms into question." [42]

1.6   United States

President Barack Obama has reviewed Homeland Security policy and concluded that "attacks using improvised nuclear devices ... pose a serious and increasing national security risk". [43] In their presidential contest, President George W. Bush and Senator John Kerry both agreed that the most serious danger facing the United States is the possibility that terrorists could obtain a nuclear bomb.[4] Most nuclear-weapon analysts agree that "building such a device would pose few technological challenges to reasonably competent terrorists". The main barrier is acquiring highly enriched uranium.[44]

In 2004, Graham Allison, U.S. Assistant Secretary of Defense during the Clinton administration, wrote that "on the current path, a nuclear terrorist attack on America in the decade ahead is more likely than not". [45] In 2004, Bruce Blair, president of the Center for Defense Information stated: "I wouldn't be at all surprised if nuclear weapons are used over the next 15 or 20 years, first and foremost by a terrorist group that gets its hands on a Russian nuclear weapon or a Pakistani nuclear weapon". [17] In 2006, Robert Gallucci, Dean of the Georgetown University School of Foreign Service, estimated that, "it is more likely than not that al-Qaeda or one of its affiliates will detonate a nuclear weapon in a U.S. city within the next five to ten years." [45] Despite a number of claims,[46][47] there is no credible evidence that any terrorist group has yet succeeded in obtaining a nuclear bomb or the materials needed to make one.[4][5]

Detonation of a nuclear weapon in a major U.S. city could kill more than 500,000 people and cause more than a trillion dollars in damage.[16][17] Hundreds of thousands could die from fallout, the resulting fires and collapsing buildings. In this scenario, uncontrolled fires would burn for days and emergency services and hospitals would be completely overwhelmed.*[4]*[48]*[49] The likely socio-economic consequences in the United States outside the immediate vicinity of an attack, and possibly in other countries, would also likely be far-reaching. A Rand Corporation report speculates that there may be an exodus from other urban centers by populations fearful of another nuclear attack.*[50]

The Obama administration will focus on reducing the risk of high-consequence, non-traditional nuclear threats. Nuclear security is to be strengthened by enhancing "nuclear detection architecture and ensuring that our own nuclear materials are secure," and by "establishing well-planned, well-rehearsed, plans for co-ordinated response." *[43] According to senior Pentagon officials, the United States will make "thwarting nuclear-armed terrorists a central aim of American strategic nuclear planning." *[51] Nuclear attribution is another strategy being pursued to counter terrorism. Led by the National Technical Nuclear Forensics Center, attribution would allow the government to determine the likely source of nuclear material used in the event of a nuclear attack. This would prevent terrorist groups, and any states willing to help them, from being able to pull off a covert attack without assurance of retaliation.*[52]

In July 2010 medical personnel from the U.S. Army practiced the techniques they would use to treat people injured by an atomic blast. The exercises were carried out at a training center in Indiana, and were set up to "simulate the aftermath of a small nuclear bomb blast, set off in a U.S. city by terrorists." *[53]

Stuxnet is a computer worm discovered in June 2010 that is believed to have been created by the United States and Israel to attack Iran's nuclear facilities.*[54]

1.6.1 Nuclear power plants

After 9/11, nuclear power plants were to be prepared for an attack by a large, well-armed terrorist group. But the Nuclear Regulatory Commission, in revising its security rules, decided not to require that plants be able to defend themselves against groups carrying sophisticated weapons. According to a study by the Government Accountability Office, the N.R.C. appeared to have based its revised rules "on what the industry considered reasonable and feasible to defend against rather than on an assessment of the terrorist threat itself" .*[55]*[56] If terrorist groups could sufficiently damage safety systems to cause a core meltdown at a nuclear power plant, and/or sufficiently damage spent fuel pools, such an attack could lead to widespread radioactive contamination. The Federation of American Scientists have said that if nuclear power use is to expand sig-

nificantly, nuclear facilities will have to be made extremely safe from attacks that could release massive quantities of radioactivity into the community. New reactor designs have features of passive safety, which may help. In the United States, the NRC carries out "Force on Force" (FOF) exercises at all Nuclear Power Plant (NPP) sites at least once every three years.*[57]

The peace group Plowshares have shown how nuclear weapons facilities can be penetrated, and the groups actions represent extraordinary breaches of security at nuclear weapons plants in the United States. The National Nuclear Security Administration has acknowledged the seriousness of the 2012 Plowshares action. Non-proliferation policy experts have questioned "the use of private contractors to provide security at facilities that manufacture and store the government's most dangerous military material" .*[58]

1.6.2 Hoaxes

In late 1974, President Gerald Ford was warned that the FBI received a communication from an extortionist wanting $200,000 ($1,000,000 today) after claiming that a nuclear weapon had been placed somewhere in Boston. A team of experts rushed in from the United States Atomic Energy Commission but their radiation detection gear arrived at a different airport. Federal officials then rented a fleet of vans to carry concealed radiation detectors around the city but forgot to bring the tools they needed to install the equipment. The incident was later found to be a hoax. However, the government's response made clear the need for an agency capable of effectively responding to such threats in the future. Later that year, President Ford created the Nuclear Emergency Search Team (NEST), which under the Atomic Energy Act is tasked with investigating the "illegal use of nuclear materials within the United States, including terrorist threats involving the use of special nuclear materials" .*[59]

One of its first responses by the Nuclear Emergency Search/Support Team was in Spokane, Washington on November 23, 1976. An unknown group called the "Days of Omega" had mailed an extortion threat claiming it would explode radioactive containers of water all over the city unless paid $500,000 ($2,100,000 today). Presumably, the radioactive containers had been stolen from the Hanford Site, less than 150 miles to the southwest. Immediately, NEST flew in a support aircraft from Las Vegas and began searching for non-natural radiation, but found nothing. No one ever responded despite the elaborate instructions given, or made any attempt to claim the (fake) money which was kept under surveillance. Within days, the incident was deemed a hoax, though the case was never solved. To avoid panic, the public was not notified until a few years later.*[60]*[61]

1.7 Policy landscape

1.7.1 Recovery

The Cooperative Threat Reduction Program (CTR), which is also known as the Nunn–Lugar Cooperative Threat Reduction, is a 1992 law sponsored by Senators Sam Nunn and Richard Lugar. The CTR established a program that gave the U.S. Department of Defense a direct stake in securing loose fissile material inside the since-dissolved USSR. According to Graham Allison, director of Harvard University's Belfer Center for Science and International Affairs, this law is a major reason why not a single nuclear weapon has been discovered outside the control of Russia's nuclear custodians.[62] The Belfer Center is itself running the *Project on Managing the Atom*, Matthew Bunn is a co-principal investigator of the project, Martin B. Malin is its executive director (circa. 2014).[63]

In August 2002, the United States launched a program to track and secure enriched uranium from 24 Soviet-style reactors in 16 countries, in order to reduce the risk of the materials falling into the hands of terrorists or "rogue states". The first such operation was *Project Vinca*, "a multinational, public-private effort to remove nuclear material from a poorly-secured Yugoslav research institute." The project has been hailed as "a nonproliferation success story" with the "potential to inform broader 'global cleanout' efforts to address one of the weakest links in the nuclear nonproliferation chain: insufficiently secured civilian nuclear research facilities." [64]

In 2004, the U.S. Global Threat Reduction Initiative (GTRI) was established in order to consolidate nuclear stockpiles of highly enriched uranium (HEU), plutonium, and assemble nuclear weapons at fewer locations.[65] Additionally, the GTRI converted HEU fuels to low-enriched uranium (LEU) fuels, which has prevented their use in making a nuclear bomb within a short amount of time. HEU that has not been converted to LEU has been shipped back to secure sites, while amplified security measures have taken hold around vulnerable nuclear facilities.[66]

1.7.2 Options

Robert Gallucci, President of the John D. and Catherine T. MacArthur Foundation, argues that traditional deterrence is not an effective approach toward terrorist groups bent on causing a nuclear catastrophe.[67] Henry Kissinger, stating the wide availability of nuclear weapons makes deterrence "decreasingly effective and increasingly hazardous." [68] Preventive strategies, which advocate the elimination of an enemy before it is able to mount an attack, are risky and controversial, therefore difficult to implement. Gallucci believes that "the United States should instead consider a policy of expanded deterrence, which focuses not on the would-be nuclear terrorists but on those states that may deliberately transfer or inadvertently lead nuclear weapons and materials to them. By threatening retaliation against those states, the United States may be able to deter that which it cannot physically prevent." [67]

Graham Allison makes a similar case, arguing that the key to expanded deterrence is coming up with ways of tracing nuclear material to the country that forged the fissile material. "After a nuclear bomb detonates, nuclear forensic cops would collect debris samples and send them to a laboratory for radiological analysis. By identifying unique attributes of the fissile material, including its impurities and contaminants, one could trace the path back to its origin." [69] The process is analogous to identifying a criminal by fingerprints. "The goal would be twofold: first, to deter leaders of nuclear states from selling weapons to terrorists by holding them accountable for any use of their own weapons; second, to give every leader the incentive to tightly secure their nuclear weapons and materials." [69]

1.7.3 Nuclear skeptics

John Mueller, a scholar of international relations at the Ohio State University, is a prominent nuclear skeptic. He makes three claims: (1) the nuclear intent and capability of terrorist groups such as Al Qaeda has been "fundamentally exaggerated;" (2) "the likelihood a terrorist group will come up with an atomic bomb seems to be vanishingly small;" and (3) policymakers are guilty of an "atomic obsession" that has led to "substantively counterproductive" policies premised on "worst case fantasies." [70] In his book *Atomic Obsession: Nuclear Alarmism from Hiroshima to Al-Qaeda* he argues that: "anxieties about terrorists obtaining nuclear weapons are essentially baseless: a host of practical and organizational difficulties make their likelihood of success almost vanishingly small". [71]

Intelligence officials have pushed back, testifying before Congress that the inability to recognize the shifting modus oparandi of terrorist groups was part of the reason why members of Aum Shinrikyo, for example, were "not on anybody's radar screen." [72] Matthew Bunn, associate professor at Harvard University's John F. Kennedy School of Government, argues that "Theft of HEU and plutonium is not a hypothetical worry, it is an ongoing reality." [30] Almost all of the stolen HEU and plutonium that has been seized over the years had never been missed before it was seized. The IAEA Illicit Nuclear Trafficking Database notes 1,266 incidents reported by 99 countries over the last 12 years, including 18 incidents involving HEU or plutonium trafficking.[30]

Keir Lieber and Daryl Press argue that despite the prominent U.S. focus on nuclear terrorism, "the fear of terrorist transfer [of nuclear weapons] seems greatly exaggerated... [and] the dangers of a state giving nuclear weapons to terrorists have been overstated." A decade of terrorism statistics show a strong correlation between attack fatalities and the attribution of the attack, and Lieber and Press assert that "neither a terror group nor a state sponsor would remain anonymous after a nuclear terror attack." About 75 percent of attacks with 100 or more fatalities were traced to the culprits; also, 97 percent of attacks on U.S. soil or that of a major ally (resulting in 10 or more deaths) were attributed to the guilty party. Lieber and Press conclude that the lack of anonymity would deter a state from providing terrorist groups with nuclear weapons.[73]

The use of HEU and plutonium in satellites has raised the concern that a sufficiently motivated rogue state could retrieve materials from a satellite crash (notably on land as occurred with Kosmos-954, Mars-96 and Fobos-Grunt) and then use these to supplement the yield of an already working nuclear device. This has been discussed recently in the UN and the Nuclear Emergency Search Team regularly consults with Roscosmos and NASA about satellite re-entries that may have contained such materials. As yet no parts were verifiably recovered from Mars 96 but recent Wikileaks releases suggest that one of the "cells" may have been recovered by mountain climbers in Chile.

1.7.4 Security summits

On April 12–13, 2010, President of the United States Barack Obama initiated and hosted the first-ever nuclear security summit in Washington D.C., commonly known as the Washington Nuclear Security Summit. The goal was to strengthen international cooperation to prevent nuclear terrorism. President Obama, along with nearly fifty world leaders, discussed the threat of nuclear terrorism, what steps needed to be taken to mitigate illicit nuclear trafficking, and how to secure nuclear material. The Summit was successful in that it produced a consensus delineating nuclear terrorism as a serious threat to all nations. Finally, the Summit produced over four-dozen specific actions embodied in commitments by individual countries and the Joint Work Plan.[74] However, world leaders at the Summit failed to agree on baseline protections for weapons-usable material, and no agreement was reached on ending the use of highly enriched uranium (HEU) in civil nuclear functions. Many of the shortcomings of the Washington Nuclear Security Summit were addressed at the Seoul Nuclear Security Summit in March 2012.

According to Graham Allison, director of Harvard University's Belfer Center for Science and International Affairs,

the objectives of the Nuclear Security Summit in Seoul are to continue to, "assess the progress made since the Washington Summit and propose additional cooperation measures to (1) Combat the threat of nuclear terrorism, (2) protect nuclear materials and related facilities, and (3) prevent illicit trafficking in nuclear materials."[75]

1.8 Media coverage

In 2011, the British news agency, the *Telegraph*, received leaked documents regarding the Guantanamo Bay interrogations of Khalid Sheikh Mohammed. The documents cited Khalid saying that, if Osama bin Laden is captured or killed by the Coalition of the Willing, an al-Qaeda sleeper cell will detonate a "weapon of mass destruction" in a "secret location" in Europe, and promised it would be "a nuclear hellstorm".[76][77][78][79][80] No such attack occurred.

1.9 See also

- The Apollo Affair - allegations of theft of HEU from the US' NUMEC facility by Israel, losses later recovered from pipes in facility and additional large amounts were lost after alleged theft was discovered and security enhanced.

- Atomic spies

- Crimes involving radioactive substances

- Guantanamo Bay files leak

- International Project - Forum «Nuclear Security – Counteraction Measures to Acts of Nuclear Terrorism»

- Lists of nuclear disasters and radioactive incidents

- Nuclear espionage

- Nuclear warfare

- Pelindaba

- Superphénix

- Terrorism

- 2014 Nuclear Security Summit

- Vulnerability of nuclear plants to attack

- War on Terror

- Weapons of mass destruction

- World Institute for Nuclear Security

1.10 References

[1] "Nuclear Security Dossier: Nuclear Terrorism Fact Sheet". *Harvard Kennedy School, Belfer Center for Science and International Affairs*. Retrieved 28 January 2013.

[2] "International Convention for the Suppression of Acts of Nuclear Terrorism - Article 1" (PDF). United Nations. 2005. Retrieved 13 April 2012.

[3] *Nuclear Terrorism: Frequently Asked Questions*, Belfer Center for Science and International Affairs. September 26, 2007

[4] Matthew Bunn. Preventing a Nuclear 9/11 *Issues in Science and Technology*, Winter 2005, p. v.

[5] Ajay Singh. Nuclear terrorism —Is it real or the stuff of 9/11 nightmares? *UCLA Today*, February 11, 2009.

[6] Ruff, Tilman (November 2006), *Nuclear terrorism* (PDF), energyscience.org.au

[7] Bunn, Matthew, Colonel Yuri Morozov, Rolf Mowatt-Larssen, Simon Saradzhyan, William Tobey, Colonel General (ret.) Viktor I. Yesin, and Major General (ret.) Pavel S. Zolotarev (2011). "The U.S.-Russia Joint Threat Assessment on Nuclear Terrorism" (PDF). Belfer Center for Science and International Affairs, Harvard University. Retrieved July 26, 2012.

[8] Graham Allison (January 26, 2010). "A Failure to Imagine the Worst". *Foreign Policy*.

[9] Alex Kingsbury, "History's Troubling Lessons", *U.S. News and World Report* (February 18, 2007).

[10] "Nuclear Terrorism." *The Economist* (January 25, 1975) p. 38.

[11] Larry Collins, "Combating Nuclear Terrorism." *New York Times* (December 14, 1980) Sec. 6 pg. 37.

[12] Sally Bedell, "A Realistic Film Stirs NBC Debate," *New York Times* (March 17, 1983) B13; Sally Bedell. "NBC Nuclear Terror Show Criticized," *New York Times* (March 22, 1983) C15; Aljean Harmetz, "NBC Film on Terror Wins Prize," New York Times *(July 8, 1983) C19.

[13] D. Costello. "Experts Warn on Nuclear Terror." *Courier-Mail* (June 26, 1986).

[14] Jay Davis. After A Nuclear 9/11 *The Washington Post.* March 25, 2008.

[15] Brian Michael Jenkins. A Nuclear 9/11? *CNN.com*, September 11, 2008.

[16] Orde Kittrie. Averting Catastrophe: Why the Nuclear Nonproliferation Treaty is Losing its Deterrence Capacity and How to Restore It May 22, 2007, p. 338.

[17] Nicholas D. Kristof. A Nuclear 9/11 *The New York Times*, March 10, 2004.

[18] M. BUNN & A. WIER. *The Seven Myths of Nuclear Terrorism* (PDF). Belfer Centre for Science and International Affairs. Retrieved 2015-08-08. External link in |publisher= (help)

[19] "al-Qaeda moving world toward 'nuclear 9/11'". *The Age*. Melbourne. February 3, 2011.

[20] Bunn, Matthew & Col-Gen. E.P. Maslin (2010). "All Stocks of Weapons-Usable Nuclear Materials Worldwide Must be Protected Against Global Terrorist Threats" (PDF). Belfer Center for Science and International Affairs, Harvard University. Retrieved July 26, 2012.

[21] "Smugglers Tried to Sell Nuclear Material to ISIS". NBC News. 7 October 2015.

[22] "The Risk of a Nuclear ISIS Grows". the Huffington post. 8 October 2015.

[23] Cowell, Alan (10 July 2014). "Low-Grade Nuclear Material Is Seized by Rebels in Iraq, U.N. Says". *The New York Times*. Retrieved 15 July 2014.

[24] Sherlock, Ruth (10 July 2014). "Iraq jihadists seize 'nuclear material', says ambassador to UN". *The Telegraph*. London. Retrieved 15 July 2014.

[25] "Nuclear smuggling deals 'thwarted' in Moldova". BBC News. 7 October 2015.

[26] "FBI foils smugglers' plot to sell nuclear material to Isis". the independent. 7 October 2015.

[27] "Brussels attacks: Belgium fears Isis seeking to make 'dirty' nuclear bomb". the independent. 25 March 2016.

[28] "Nato raises 'justified concern' that Isil is plotting nuclear attack on Britain". MSN. 19 April 2016.

[29] IAEA Illicit Trafficking Database (ITDB) p. 3.

[30] Bunn, Matthew. "Securing the Bomb 2010: Securing All Nuclear Materials in Four Years" (PDF). President and Fellows of Harvard College. Retrieved 28 January 2013.

[31] Rhys Blakeley. "Terrorists 'have attacked Pakistan nuclear sites three times'," *Times Online* (August 11, 2009).

[32] "IOL - Pretoria News". *IOL*.

[33] Washington Post. December 20, 2007. Op-Ed by Micah Zenko

[34] "Feds Hoped to Snag Bin Laden Nuke Expert in JFK Bomb Plot". *Fox News*. June 4, 2007.

[35] Teather, David; Younge, Gary (January 5, 2005). "Briton accused of trying to sell missiles". *The Guardian*. London.

[36] "Ushering in the era of nuclear terrorism," by Patterson, Andrew J. MD, PhD, *Critical Care Medicine*, v. 35, p.953-954, 2007.

[37] Blakely, Rhys (August 11, 2009). "Terrorists 'have attacked Pakistan nuclear sites three times'". *Times Online*, London

[38] "Login".

[39] Pakistan nuclear weapons at risk of theft by terrorists, US study warns, The Guardian, 2010-04-12

[40] Could terrorists get hold of a nuclear bomb?, BBC, 2010-04-12

[41] Official: Terrorists seek nuclear material, but lack ability to use it, CNN, 2010-04-13

[42] Pakistan's Nuclear Weapons: Proliferation and Security Issues, Congressional Research Service, 2010-02-23

[43] The White House. Homeland Security

[44] Charles D. Ferguson. Preventing a nuclear 9/11 : First, secure the highly enriched uranium *The New York Times*, September 24, 2004.

[45] Orde Kittrie. Averting Catastrophe: Why the Nuclear Nonproliferation Treaty is Losing its Deterrence Capacity and How to Restore It May 22, 2007, p. 342.

[46] Paul Williams (2005). *The Al Qaeda Connection : International Terrorism, Organized Crime, and the Coming Apocalypse*, Prometheus Books, pp. 192–194.

[47] Nuclear 9/11: Interview with Dr. Paul L. Williams *Global Politician*, September 11, 2007.

[48] Controlling Nuclear Warheads and Materials p. 16.

[49] Bleek, Philipp, Anders Corr, and Micah Zenko. Nuclear 9/11: What if Port is Ground Zero? *The Houston Chronicle*, May 1, 2005.

[50] *Considering the Effects of a Catastrophic Terrorist Attack* by Charles Meade & Roger C. Molander p 9, Retrieved March 11, 2013 - this report uses smuggled nuclear weapons in container ships at a US port as an example, so speculates an exodus from coastal cities

[51] Thom Shanker and Eric Schmitt. U.S. to Make Stopping Nuclear Terror Key Aim *The New York Times*, December 18, 2009.

[52] Richelson, Jeffrey. "U.S. Nuclear Detection and Counterterrorism, 1998-2009". George Washington University.

[53] Deborah Block. US Military Practices Medical Response to Nuclear Attack *Voice of America*, 26 July 2010.

[54] Zetter, Kim (25 March 2013). "Legal Experts: Stuxnet Attack on Iran Was Illegal 'Act of Force'". Wired.

[55] Elizabeth Kolbert (28 March 2011). "The Nuclear Risk". *The New Yorker*.

[56] Daniel Hirsch et al. The NRC's Dirty Little Secret, *Bulletin of the Atomic Scientists*, May 1, 2003, vol. 59 no. 3, pp. 44-51.

[57] Charles D. Ferguson & Frank A. Settle (2012). "The Future of Nuclear Power in the United States" (PDF). *Federation of American Scientists*.

[58] Kennette Benedict (9 August 2012). "Civil disobedience". *Bulletin of the Atomic Scientists*.

[59] "Nuclear Emergency Support Team (NEST)" (PDF). U.S. Department of Energy. Retrieved 2012-10-21.

[60] Peck, Chris (1981-02-08). "The day they said they'd nuke Spokane-Part 1" (scan). *The Spokesman-Review*. p. 17. Retrieved 2012-10-21.

[61] Peck, Chris (1981-02-08). "The day they said they'd nuke Spokane-Part 2" (scan). *The Spokesman-Review*. p. 24. Retrieved 2012-10-21.

[62] Allison, Graham (December 29, 2011). "Washington Can Work: Celebrating Twenty Years With Zero Nuclear Terrorism". *The Huffington Post*. Retrieved July 26, 2012.

[63] "Managing the Atom - Harvard - Belfer Center for Science and International Affairs".

[64] Philipp C. Bleek. "Project Vinca: Lessons for Securing Civil Nuclear Material Stockpiles." *The Nonproliferation Review* (Fall-Winter 2003) p. 1.

[65] Bunn, Matthew & Eben Harrell (2012). "Consolidation: Thwarting Nuclear Theft" (PDF). Belfer Center for Science and International Affairs, Harvard University. Retrieved July 26, 2012.

[66] Wier, Anthony and Matthew Bunn (November 19, 2006). "Bombs That Won't Go Off". *The Washington Post*. Retrieved July 26, 2012.

[67] Gallucci, Robert (September 2006). "Averting Nuclear Catastrophe: Contemplating Extreme Responses to U.S. Vulnerability". *Annals of the American Academy of Political and Social Science*. **607**: 51–58. doi:10.1177/0002716206290457. Retrieved 28 January 2013.

[68] Kissinger, Henry (15 January 2008). "Toward a Nuclear-Free World". *NTI*. Retrieved 28 January 2013.

[69] Allison, Graham (13 March 2009). "How to Keep the Bomb From Terrorists". *Newsweek*. Retrieved 28 January 2013.

[70] Mueller, John (15 January 2008). *The Atomic Terrorist: Assessing the Likelihood*, prepared for presentation at the Program on International Security Policy (PDF). University of Chicago.

[71] http://www.oup.com/us/catalog/general/subject/HistoryWorld/?view=usa&ci=9780195381368 Atomic Obsession: Nuclear Alarmism from Hiroshima to Al-Qaeda: Oxford University Press

[72] Allison, Graham (2004). *Nuclear Terrorism: The Ultimate Preventable Catastrophe*. New York: Macmillan. p. 15. ISBN 9781429945516.

[73] Lieber, Keir; Press, Daryl (Summer 2013). "Why States Won't Give Nuclear Weapons to Terrorists". *International Security*. **38** (1): 80–84, 104. doi:10.1162/isec_a_00127.

[74] Tobey, William (2011). "Planning for Success at the 2012 Seoul Nuclear Security Summit" (PDF). The Stanley Foundation. Retrieved July 26, 2012.

[75] "2012 Seoul Nuclear Security Summit Q&A with Professor Graham Allison" (PDF). Belfer Center for Science and International Affairs, Harvard University. 2012. Retrieved July 26, 2012.

[76] Hope, Christopher (April 25, 2011). "WikiLeaks: Guantanamo Bay terrorist secrets revealed". London: Telegraph.co.uk. Retrieved April 27, 2011.

[77] Gould, Martin. "WikiLeaks: Al-Qaida Already Has Nuclear Capacity". NewsMax. Retrieved April 27, 2011.

[78] "'Nuclear hellstorm' if bin Laden caught - 9/11 mastermind". News.com.au. April 25, 2011. Retrieved April 27, 2011.

[79] "'Nuclear hellstorm' if bin Laden caught: 9/11 mastermind". News.Yahoo.com. 2011-04-25. Retrieved April 27, 2011.

[80] http://newstabulous.com/al-qaeda-hid-bomb-in-europe-wikileaks-releases-secret-files/9722/

1.11   Further reading

- Allison, Graham (9 August 2004). *Nuclear Terrorism: The Ultimate Preventable Catastrophe*. New York, New York: Times Books. ISBN 978-0-8050-7651-6.

- Byrne, John and Steven M. Hoffman (1996). *Governing the Atom: The Politics of Risk*, Transaction Publishers.

- Cooke, Stephanie (2009). *In Mortal Hands: A Cautionary History of the Nuclear Age*, Black Inc.

- Ferguson, Charles D., and William C. Potter, with Amy Sands, Leonard S. Spector and Fred L. Wehling (2004). *The Four Faces of Nuclear Terrorism*. Monterey, California: Center for Nonproliferation Studies. ISBN 1-885350-09-0.

- Jones, Ishmael (2010) [2008]. *The Human Factor: Inside the CIA's Dysfunctional Intelligence Culture*. Encounter Books. ISBN 978-1-59403-382-7.

- Levi, Michael (2007). *On Nuclear Terrorism*. Cambridge, Massachusetts: Harvard University Press. ISBN 978-0-674-02649-0.

- Lovins, Amory B. and John H. Price (1975). *Non-Nuclear Futures: The Case for an Ethical Energy Strategy*. Ballinger Publishing Company, 1975, ISBN 0-88410-602-0

- Schell, Jonathan (2007). *The Seventh Decade: The New Shape of Nuclear Danger*. New York, New York: Metropolitan Books.

1.12   External links

- Nuclear Terrorism publications from Harvard Kennedy School faculty and fellows

- What if the terrorists go nuclear?, Center for Defense Information

- Preventing Catastrophic Nuclear Terrorism, Council on Foreign Relations

- Use of nuclear and radiological weapons by terrorists?, International Review of the Red Cross

- "Can Terrorists Build Nuclear Weapons?", Nuclear Control Institute

- Annotated bibliography, Alsos Digital Library for Nuclear Issues

- Fallout: After a Nuclear Attack - slideshow by *Life magazine*

- Nuclear Emergency and Radiation Resources

Chapter 2

Radiological weapon

A **radiological weapon** or **radiological dispersion device** (**RDD**) is any weapon that is designed to spread radioactive material with the intent to kill and cause disruption. According to the U.S. Department of Defense, an RDD is "any device, including any weapon or equipment, other than a nuclear explosive device, specifically designed to employ radioactive material by disseminating it to cause destruction, damage, or injury by means of the radiation produced by the decay of such material" .[1][2]

One type of RDD is a "conventional explosive combined with some type of radiological material" , also known as a dirty bomb. It is not a true nuclear weapon and does not yield the same explosive power. It uses conventional explosives to spread radioactive material, most commonly the spent fuels from nuclear power plants or radioactive medical waste. It is not a Weapon of Mass Destruction (WMD), but rather, as researcher Peter Probst calls it, a "weapon of mass disruption" (Hughes, 2002). In fact, effective dispersal ranges are rather limited. Most deaths (if any) would come from the initial explosion (non-nuclear), but it does depend on the type of radiological material used. (Department of Homeland Security [DHS], 2003)." [1][3][4]

Another version is the salted bomb, a true nuclear weapon designed to produce larger amounts of nuclear fallout than a regular nuclear weapon.

2.1 Explanation

Radiological weapons of mass destruction have been suggested as a possible weapon of terrorism used to create panic and casualties in densely populated areas. They could also render a great deal of property uninhabitable for an extended period, unless costly remediation were undertaken. The radiological source and quality greatly impacts the effectiveness of a radiological weapon.

Factors such as: energy and type of radiation, half-life, longevity, availability, shielding, portability, and the role of the environment will determine the effect of the radiolog-

ical weapon. Radioisotopes that pose the greatest security risk include: 137
Cs
, used in radiological medical equipment, 60
Co
, 241
Am
, 252
Cf
, 192
Ir
, 238
Pu
, 90
Sr
, 226
Ra
, and 238
U
.

All of these isotopes, except for the final one, are created in nuclear power plants. While the amount of radiation dispersed from the event will likely be minimal, the fact of any radiation may be enough to cause panic and disruption.

2.2 History

The professional history of radioactive weaponry may be traced to a 1940 science fiction story, "Solution Unsatisfactory"[5] by Robert A. Heinlein and a 1943 memo from James Bryant Conant, Arthur Holly Compton and Harold Urey to Brigadier General Leslie Groves, head of the Manhattan Project.

Transmitting a report entitled, "Use of Radioactive Materials as a Military Weapon," the Groves memo states:

> As a gas warfare instrument the material would ... be inhaled by personnel. The amount

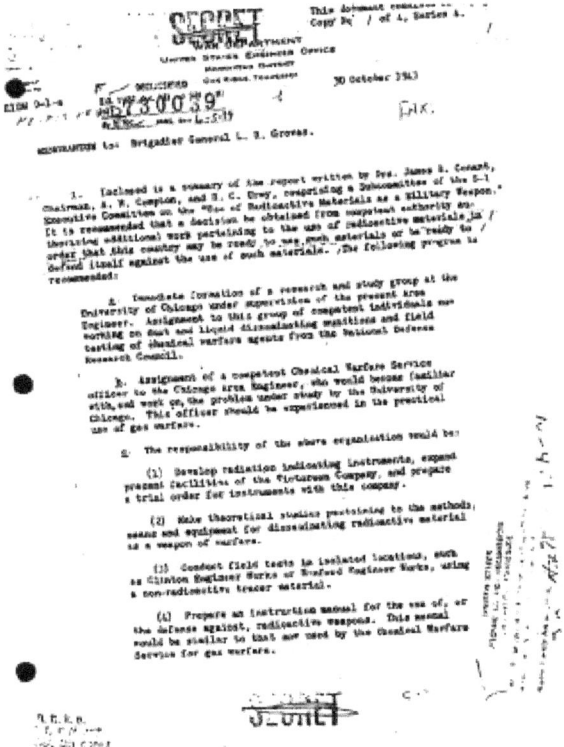

October 30, 1943 memo from Drs. Conant, Compton, and Urey to Brigadier General L. R. Groves, Manhattan District, Oak Ridge, Tennessee; declassified June 5, 1974.

necessary to cause death to a person inhaling the material is extremely small. It has been estimated that one millionth of a gram accumulating in a person's body would be fatal. There are no known methods of treatment for such a casualty.... It cannot be detected by the senses; It can be distributed in a dust or smoke form so finely powdered that it will permeate a standard gas mask filter in quantities large enough to be extremely damaging....

Radioactive warfare can be used [...] To make evacuated areas uninhabitable; To contaminate small critical areas such as rail-road yards and airports; As a radioactive poison gas to create casualties among troops; Against large cities, to promote panic, and create casualties among civilian populations.

Areas so contaminated by radioactive dusts and smokes, would be dangerous as long as a high enough concentration of material could be maintained.... they can be stirred up as a fine dust from the terrain by winds, movement of vehicles or troops, etc., and would remain a potential hazard for a long time.

These materials may also be so disposed as

to be taken into the body by ingestion instead of inhalation. Reservoirs or wells would be contaminated or food poisoned with an effect similar to that resulting from inhalation of dust or smoke. For days production could contaminate a million gallons of water to an extent that a quart drunk in one day would probably result in complete incapacitation or death in about a month's time.

The United States, however, chose not to pursue radiological weapons during World War II, though early on in the project considered it as a backup plan in case nuclear fission proved impossible to tame. Some US policymakers and scientists involved in the project felt that radiological weapons would qualify as chemical weapons and thus violate international law.

2.3   Deployment

One possible way of dispersing the material is by using a dirty bomb, a conventional explosive which disperses radioactive material. Dirty bombs are not a type of nuclear weapon, which requires a nuclear chain reaction and the creation of a critical mass. Whereas a nuclear weapon will usually create mass casualties immediately following the blast, a dirty bomb scenario would initially cause only minimal casualties from the conventional explosion.

Means of radiological warfare that do not rely on any specific weapon, but rather on spreading radioactive contamination via a food chain or water table, seem to be more effective in some ways, but share many of the same problems as chemical warfare.

2.4   Military uses

Radiological weapons are considered to be militarily useless for a state-sponsored army and are initially not hoped to be used by any military forces. Firstly, the use of such a weapon is of no use to an occupying force, as the target area becomes uninhabitable (due to the fallout caused by radioactive poisoning of the involved environment).

Furthermore, area-denial weapons are generally of limited use to an attacking army, as it slows the rate of advance.

2.5   Dirty bombs

A dirty bomb is a radiological weapon dispersed with conventional explosives.

There is currently (as of 2007) an ongoing debate about the damage that terrorists using such a weapon might inflict. Many experts believe that a dirty bomb such that terrorists might reasonably be able to construct would be unlikely to harm more than a few people and hence it would be no more deadly than a conventional bomb. Furthermore, the casualties would be a result of the initial explosion, because alpha and beta emitting material needs to be inhaled to do damage to the human body. Gamma radiation emitting material is so radioactive that it can't be deployed without wrapping an amount of shielding material around the bomb that would make transport by car or plane impossible without risking detection. Because of this a dirty bomb with radioactive material around an explosive device would be almost useless, unless said shielding was removed shortly before detonation. This is not only because of the effectiveness but also because this material would be easy to clean up. Furthermore, the possibility of terrorists making a gas or aerosol that is radioactive is very unlikely because of the complex chemical work to achieve this goal.*[6]

Hence, this line of argument goes, the objectively dominant effect would be the moral and economic damage due to the massive fear and panic such an incident would spur. On the other hand, some believe that the fatalities and injuries might be in fact much more severe. This point was made by physicist Peter D. Zimmerman (King's College London) who reexamined the Goiânia accident which is arguably comparable.*[7] and popularized in a subsequent fictionalized account produced by the BBC and broadcast in the United States by PBS.*[8] The latter program showed how shielding might be used to minimize the detection risk.

2.6 Salted bomb

Main article: Salted bomb

A salted bomb is a theoretical nuclear weapon designed to produce enhanced quantities of radioactive fallout, rendering a large area uninhabitable. As far as is publicly known none have ever been built.

2.7 See also

- Biological warfare
- Chemical warfare
- Cobalt bomb
- Lists of nuclear disasters and radioactive incidents
- Nuclear fallout

- Nuclear weapon
- Radioactive contamination
- Weapon of mass destruction
- Nuclear terrorism

2.8 References

[1] Rickert, Paul (2005-12-31). "The Likely Effect of a Radiological Dispersion Device". Liberty University. pp. 2, 3. Retrieved 21 October 2014.

[2] Ford, J. (March 1998). "Radiological Dispersion Devices: Assessing the transnational threat". National Defense University - Institute for National Strategic Studies - Strategic Forum. Archived from the original on December 12, 2005. Retrieved December 31, 2005.

[3] Hughes, D. (4 March 2002). "When terrorists go nuclear". *Popular Mechanics*. Archived from the original on September 19, 2005. Retrieved December 31, 2005.

[4] "Radiological Dispersion Devices Fact Sheet". Department of Homeland Security. 10 February 2003. Archived from the original on December 29, 2005. Retrieved December 31, 2005.

[5] Full story at publisher's web site

[6]

[7] Dirty Bombs: The Threat Revisited in Defense Horizons, Feb. 2004, a publication of the National Defense University

[8] Dirty Bomb

2.9 External links

- Annotated bibliography for radiological dispersal devices (RDD) from the Alsos Digital Library for Nuclear Issues. This page has no results.

Chapter 3

Radiological warfare

Radiological warfare is any form of warfare involving deliberate radiation poisoning or contamination of an area with radiological sources.

Radiological weapons are normally considered weapons of mass destruction (WMDs), although radiological weapons can be specific in who they target, such as the radiation poisoning of Alexander Litvinenko.

In the 1964 edition of the DOD/AEC book *The Effects of Nuclear Weapons*, a section titled Radiological Warfare details some of the most common WMDs.[1] The Fission products from a conventional nuclear explosive weapon are as much a radiological weapon as weapons solely designed for the purpose of mass radiological warfare. The standard high-fission thermonuclear weapon is automatically a weapon of radiological warfare, as dirty as a cobalt bomb.

Initially, gamma radiation from the fission products of an equivalent size fission-fusion-fission bomb are much more intense than Co-60: 15,000 times more intense at 1 hour; 35 times more intense at 1 week; 5 times more intense at 1 month; and about equal at 6 months. Thereafter fission drops off rapidly so that Co-60 fallout is 8 times more intense than fission at 1 year and 150 times more intense at 5 years. The very long-lived isotopes produced by fission would overtake the ^{60}Co again after about 75 years.[2] Other salted bomb variants that don't use cobalt have also been theorized.

A far lower-tech radiological weapon than those discussed above is a "dirty bomb" / Radiological dispersal device, which refers to a conventional explosive bomb with a radiological side effect due to strapping radiation sources to it, is a very inefficient way to spread radiation, and all such "weapons" have problems that render them likely impractical for military uses.

Rather, radiological warfare with dirty bombs would be of vastly more use to terrorists spreading or intensifying fear, uncertainty and doubt. The release of radioactive material may involve no special "weapon" and include no direct killing of people from its radiation source, but rather could make whole areas or structures unusable or unfavorable for the support of human life. The elevated radiation levels in the targeted areas would make these areas dangerous to humans. An area, once contaminated with radiation, is often expensive to clean up. Decontamination of the built environment would take time.

Like land mines, radiological weapons can be used as an area denial weapons method.

3.1 See also

- Nuclear detection
- Operation Peppermint
- Alexander Litvinenko

3.2 References

[1] Samuel Glasstone, *The Effects of Nuclear Weapons*, 1962, Revised 1964, U.S. Dept of Defense and U.S. Dept of Energy, pp.464–5. This section was removed from later editions, but, according to Glasstone in 1978, not because it was inaccurate or because the weapons had changed.

[2] Sublette, Carey. "Nuclear Weapons Frequently Asked Questions (Section 1)". Retrieved 25 July 2014.

Chapter 4

Nuclear weapon

"Atom bomb" redirects here. For other uses, see Atom bomb (disambiguation).

The mushroom cloud of the atomic bombing of the Japanese city of Nagasaki on August 9, 1945 rose some 11 miles (18 km) above the bomb's hypocenter.

A **nuclear weapon** is an explosive device that derives its destructive force from nuclear reactions, either fission (fission bomb) or a combination of fission and fusion (thermonuclear weapon). Both reactions release vast quantities of energy from relatively small amounts of matter. The first test of a fission ("atomic") bomb released the same amount of energy as approximately 20,000 tons of TNT (84 TJ). The first thermonuclear ("hydrogen") bomb test released the same amount of energy as approximately 10 million tons of TNT (42 PJ).

A thermonuclear weapon weighing little more than 2,400 pounds (1,100 kg) can produce an explosive force comparable to the detonation of more than 1.2 million tons of TNT (5.0 PJ).[1] A nuclear device no larger than traditional bombs can devastate an entire city by blast, fire, and radiation. Nuclear weapons are considered weapons of mass destruction, and their use and control have been a major focus of international relations policy since their debut.

Nuclear weapons have been used twice in nuclear warfare, both times by the United States against Japan near the end of World War II. On August 6, 1945, the U.S. Army Air Forces detonated a uranium gun-type fission bomb nicknamed "Little Boy" over the Japanese city of Hiroshima; three days later, on August 9, the U.S. Army Air Forces detonated a plutonium implosion-type fission bomb codenamed "Fat Man" over the Japanese city of Nagasaki. The bombings resulted in the deaths of approximately 200,000 civilians and military personnel from acute injuries sustained from the explosions.[2] The ethics of the bombings and their role in Japan's surrender remain the subject of scholarly and popular debate.

Since the atomic bombings of Hiroshima and Nagasaki, nuclear weapons have been detonated on over two thousand occasions for the purposes of testing and demonstration. Only a few nations possess such weapons or are suspected of seeking them. The only countries known to have detonated nuclear weapons—and acknowledge possessing them—are (chronologically by date of first test) the United States, the Soviet Union (succeeded as a nuclear power by Russia), the United Kingdom, France, the People's Republic of China, India, Pakistan, and North Korea. Israel is also believed to possess nuclear weapons, though in a policy of deliberate ambiguity, it does not acknowledge having them. Germany, Italy, Turkey, Belgium and the Netherlands are nuclear weapons sharing states.[3][4][5] South Africa is the only country to have independently developed and then renounced and dismantled its nuclear weapons.[6]

The Treaty on the Non-Proliferation of Nuclear Weapons aimed to reduce the spread of nuclear weapons, but its effectiveness has been questioned, and political tensions remained high in the 1970s and 1980s. Modernisation of weapons continues to occur.[7]

4.1 Types

Main article: Nuclear weapon design

There are two basic types of nuclear weapons: those that

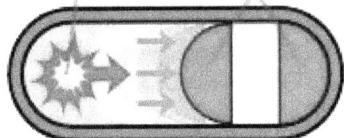

Gun-type assembly method

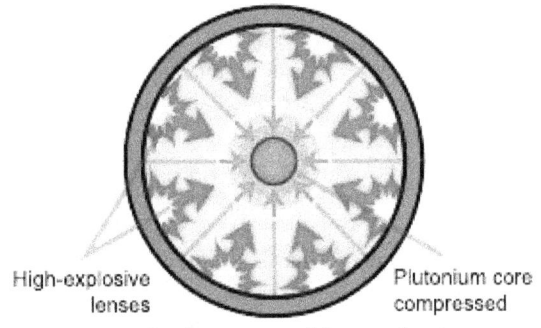

Implosion assembly method

The two basic fission weapon designs

derive the majority of their energy from nuclear fission reactions alone, and those that use fission reactions to begin nuclear fusion reactions that produce a large amount of the total energy output.

4.1.1 Fission weapons

All existing nuclear weapons derive some of their explosive energy from nuclear fission reactions. Weapons whose explosive output is exclusively from fission reactions are commonly referred to as **atomic bombs** or **atom bombs** (abbreviated as **A-bombs**). This has long been noted as something of a misnomer, as their energy comes from the nucleus of the atom, just as it does with fusion weapons.

In fission weapons, a mass of fissile material (enriched uranium or plutonium) is assembled into a supercritical mass—the amount of material needed to start an exponentially growing nuclear chain reaction—either by shooting one

piece of sub-critical material into another (the "gun" method) or by compressing using explosive lenses a sub-critical sphere of material using chemical explosives to many times its original density (the "implosion" method). The latter approach is considered more sophisticated than the former and only the latter approach can be used if the fissile material is plutonium.

A major challenge in all nuclear weapon designs is to ensure that a significant fraction of the fuel is consumed before the weapon destroys itself. The amount of energy released by fission bombs can range from the equivalent of just under a ton to upwards of 500,000 tons (500 kilotons) of TNT (4.2 to 2.1×10^6 GJ).[8]

All fission reactions necessarily generate fission products, the radioactive remains of the atomic nuclei split by the fission reactions. Many fission products are either highly radioactive (but short-lived) or moderately radioactive (but long-lived), and as such are a serious form of radioactive contamination if not fully contained. Fission products are the principal radioactive component of nuclear fallout.

The most commonly used fissile materials for nuclear weapons applications have been uranium-235 and plutonium-239. Less commonly used has been uranium-233. Neptunium-237 and some isotopes of americium may be usable for nuclear explosives as well, but it is not clear that this has ever been implemented, and even their plausible use in nuclear weapons is a matter of scientific dispute.[9]

4.1.2 Fusion weapons

Main article: Thermonuclear weapon

The other basic type of nuclear weapon produces a large proportion of its energy in nuclear fusion reactions. Such fusion weapons are generally referred to as **thermonuclear weapons** or more colloquially as **hydrogen bombs** (abbreviated as **H-bombs**), as they rely on fusion reactions between isotopes of hydrogen (deuterium and tritium). All such weapons derive a significant portion, and sometimes a majority, of their energy from fission. This is because a fission reaction is required as a "trigger" for the fusion reactions, and the fusion reactions can themselves trigger additional fission reactions.[10]

Only six countries—United States, Russia, United Kingdom, People's Republic of China, France and India—have conducted thermonuclear weapon tests. (Whether India has detonated a "true", multi-staged thermonuclear weapon is controversial.)[11] North Korea claims to have tested a fusion weapon as of January 2016, though this claim is disputed.[12] Thermonuclear weapons are considered much more difficult to successfully design and execute than prim-

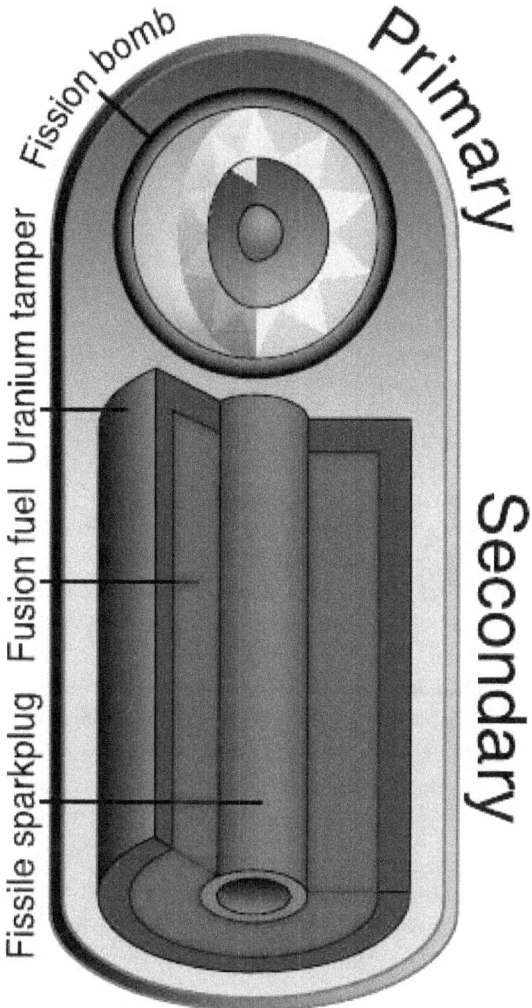

Fission bomb

Primary

Uranium tamper

Fusion fuel

Fissile sparkplug

Secondary

The basics of the Teller–Ulam design for a hydrogen bomb: a fission bomb uses radiation to compress and heat a separate section of fusion fuel.

itive fission weapons. Almost all of the nuclear weapons deployed today use the thermonuclear design because it is more efficient.

Thermonuclear bombs work by using the energy of a fission bomb to compress and heat fusion fuel. In the Teller-Ulam design, which accounts for all multi-megaton yield hydrogen bombs, this is accomplished by placing a fission bomb and fusion fuel (tritium, deuterium, or lithium deuteride) in proximity within a special, radiation-reflecting container. When the fission bomb is detonated, gamma rays and X-rays emitted first compress the fusion fuel, then heat it to thermonuclear temperatures. The ensuing fusion reaction creates enormous numbers of high-speed neutrons, which can then induce fission in materials not normally prone to it, such as depleted uranium. Each of these components is

known as a "stage", with the fission bomb as the "primary" and the fusion capsule as the "secondary". In large, megaton-range hydrogen bombs, about half of the yield comes from the final fissioning of depleted uranium.[8]

Virtually all thermonuclear weapons deployed today use the "two-stage" design described above, but it is possible to add additional fusion stages—each stage igniting a larger amount of fusion fuel in the next stage. This technique can be used to construct thermonuclear weapons of arbitrarily large yield, in contrast to fission bombs, which are limited in their explosive force. The largest nuclear weapon ever detonated, the Tsar Bomba of the USSR, which released an energy equivalent of over 50 megatons of TNT (210 PJ), was a three-stage weapon. Most thermonuclear weapons are considerably smaller than this, due to practical constraints from missile warhead space and weight requirements.[13]

Edward Teller, often referred to as the "father of the hydrogen bomb"

Fusion reactions do not create fission products, and thus contribute far less to the creation of nuclear fallout than fission reactions, but because all thermonuclear weapons contain at least one fission stage, and many high-yield thermonuclear devices have a final fission stage, thermonuclear weapons can generate at least as much nuclear fallout as fission-only weapons.

4.1.3 Other types

Main articles: boosted fission weapon, neutron bomb, and radiological bomb

There are other types of nuclear weapons as well. For example, a boosted fission weapon is a fission bomb that increases its explosive yield through a small amount of fusion reactions, but it is not a fusion bomb. In the boosted bomb, the neutrons produced by the fusion reactions serve primarily to increase the efficiency of the fission bomb. There are two types of boosted fission bomb: internally boosted, in which a deuterium-tritium mixture is injected into the bomb core, and externally boosted, in which concentric shells of lithium-deuteride and depleted uranium are layered on the outside of the fission bomb core.

Some weapons are designed for special purposes; a neutron bomb is a thermonuclear weapon that yields a relatively small explosion but a relatively large amount of neutron radiation; such a device could theoretically be used to cause massive casualties while leaving infrastructure mostly intact and creating a minimal amount of fallout. The detonation of any nuclear weapon is accompanied by a blast of neutron radiation. Surrounding a nuclear weapon with suitable materials (such as cobalt or gold) creates a weapon known as a salted bomb. This device can produce exceptionally large quantities of long-lived radioactive contamination. It has been conjectured that such a device could serve as a "doomsday weapon" because such a large quantity of radioactivities with half-lives of decades, lifted into the stratosphere where wind currents would distribute it around the globe, would make all life on the planet extinct.

In connection with the Strategic Defense Initiative, research into the Nuclear pumped laser was conducted under the DOD program Project Excalibur but this did not result in a working weapon. The concept involves the tapping of the energy of an exploding nuclear bomb to power a single-shot laser which is directed at a distant target.

During the Starfish Prime high-altitude nuclear test in 1962, an unexpected effect was produced which is called a Nuclear electromagnetic pulse. This is an intense flash of electromagnetic energy produced by a rain of high energy electrons which in turn are produced by a nuclear bomb's gamma rays. This flash of energy can permanently destroy or disrupt electronic equipment if insufficiently shielded. It has been proposed to use this effect to disable an enemy's military and civilian infrastructure as an adjunct to other nuclear or conventional military operations against that enemy. Because the effect is produced by very high altitude nuclear detonations, it can produce damage to electronics over a very wide, even continental, geographical area.

Research has been done into the possibility of pure fu-sion bombs: nuclear weapons that consist of fusion reactions without requiring a fission bomb to initiate them. Such a device might provide a simpler path to thermonuclear weapons than one that required development of fission weapons first, and pure fusion weapons would create significantly less nuclear fallout than other thermonuclear weapons, because they would not disperse fission products. In 1998, the United States Department of Energy divulged that the United States had, "...made a substantial investment" in the past to develop pure fusion weapons, but that, "The U.S. does not have and is not developing a pure fusion weapon", and that, "No credible design for a pure fusion weapon resulted from the DOE investment".[14]

Antimatter, which consists of particles resembling ordinary matter particles in most of their properties but having opposite electric charge, has been considered as a trigger mechanism for nuclear weapons.[15] A major obstacle is the difficulty of producing antimatter in large enough quantities, and there is no evidence that it is feasible beyond the military domain.[16] However, the U.S. Air Force funded studies of the physics of antimatter in the Cold War, and began considering its possible use in weapons, not just as a trigger, but as the explosive itself.[17] A fourth generation nuclear weapon design is related to, and relies upon, the same principle as Antimatter-catalyzed nuclear pulse propulsion.[18]

Most variation in nuclear weapon design is for the purpose of achieving different yields for different situations, and in manipulating design elements to attempt to minimize weapon size.[8]

4.2 Weapons delivery

See also: Nuclear weapons delivery, Nuclear triad, Strategic bomber, Intercontinental ballistic missile, and Submarine-launched ballistic missile

Nuclear weapons delivery—the technology and systems used to bring a nuclear weapon to its target—is an important aspect of nuclear weapons relating both to nuclear weapon design and nuclear strategy. Additionally, development and maintenance of delivery options is among the most resource-intensive aspects of a nuclear weapons program: according to one estimate, deployment costs accounted for 57% of the total financial resources spent by the United States in relation to nuclear weapons since 1940.[19]

Historically the first method of delivery, and the method used in the two nuclear weapons used in warfare, was as a gravity bomb, dropped from bomber aircraft. This is usually the first method that countries developed, as it does not place many restrictions on the size of the weapon and *weapon miniaturization* requires considerable weapons de-

The first nuclear weapons were gravity bombs, such as this "Fat Man" weapon dropped on Nagasaki, Japan. They were very large and could only be delivered by heavy bomber aircraft

A demilitarized and commercial launch of the Russian Strategic Rocket Forces R-36 ICBM; also known by the NATO reporting name: SS-18 Satan. Upon its first fielding in the late 1960s, the SS-18 remains the single highest throw weight missile delivery system ever built.

sign knowledge. It does, however, limit attack range, response time to an impending attack, and the number of weapons that a country can field at the same time.

With the advent of miniaturization, nuclear bombs can be delivered by both strategic bombers and tactical fighter-

bombers, allowing an air force to use its current fleet with little or no modification. This method may still be considered the primary means of nuclear weapons delivery; the majority of U.S. nuclear warheads, for example, are free-fall gravity bombs, namely the B61.*[8]

Montage of an inert test of a United States Trident SLBM (submarine launched ballistic missile), from submerged to the terminal, or re-entry phase, of the multiple independently targetable reentry vehicles

More preferable from a strategic point of view is a nuclear weapon mounted onto a missile, which can use a ballistic trajectory to deliver the warhead over the horizon. Although even short-range missiles allow for a faster and less vulnerable attack, the development of long-range intercontinental ballistic missiles (ICBMs) and submarine-launched ballistic missiles (SLBMs) has given some nations the ability to plausibly deliver missiles anywhere on the globe with a high likelihood of success.

More advanced systems, such as multiple independently targetable reentry vehicles (MIRVs), can launch multiple warheads at different targets from one missile, reducing the chance of a successful missile defense. Today, missiles are most common among systems designed for delivery of nuclear weapons. Making a warhead small enough to fit onto a missile, though, can be difficult.*[8]

Tactical weapons have involved the most variety of delivery types, including not only gravity bombs and missiles but also artillery shells, land mines, and nuclear depth charges and torpedoes for anti-submarine warfare. An atomic mortar was also tested at one time by the United States. Small, two-man portable tactical weapons (somewhat misleadingly referred to as suitcase bombs), such as the Special Atomic Demolition Munition, have been developed, although the difficulty of combining sufficient yield with portability limits their military utility.*[8]

4.3 Nuclear strategy

Main articles: Nuclear strategy and Deterrence theory
See also: Nuclear peace, Essentials of Post–Cold War
Deterrence, Single Integrated Operational Plan, Nuclear
warfare, and On Thermonuclear War

Nuclear warfare strategy is a set of policies that deal with
preventing or fighting a nuclear war. The policy of trying to
prevent an attack by a nuclear weapon from another country by threatening nuclear retaliation is known as the strategy of nuclear deterrence. The goal in deterrence is to always maintain a second strike capability (the ability of a
country to respond to a nuclear attack with one of its own)
and potentially to strive for first strike status (the ability to
completely destroy an enemy's nuclear forces before they
could retaliate). During the Cold War, policy and military
theorists in nuclear-enabled countries worked out models
of what sorts of policies could prevent one from ever being attacked by a nuclear weapon, and developed weapon
game theory models that create the greatest and most stable
deterrence conditions.

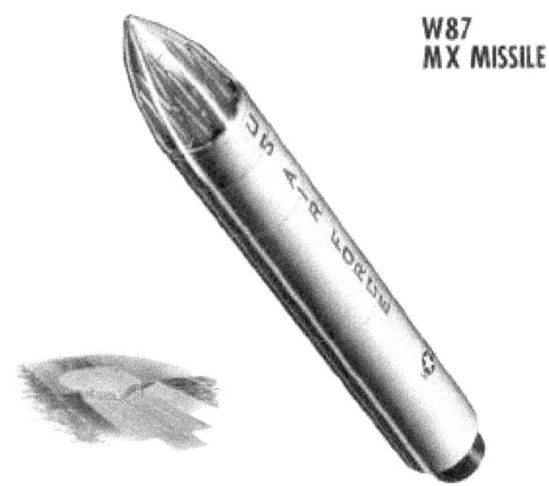

**W87
MX MISSILE**

*The now decommissioned United States' Peacekeeper missile was an
ICBM developed to entirely replace the minuteman missile in the late
1980s. Each missile, like the heavier lift Russian SS-18 Satan, could
contain up to ten nuclear warheads (shown in red), each of which
could be aimed at a different target. A factor in the development of
MIRVs was to make complete missile defense very difficult for an
enemy country.*

Different forms of nuclear weapons delivery (see above) allow for different types of nuclear strategies. The goals of
any strategy are generally to make it difficult for an enemy
to launch a pre-emptive strike against the weapon system
and difficult to defend against the delivery of the weapon
during a potential conflict. Sometimes this has meant keeping the weapon locations hidden, such as deploying them

on submarines or land mobile transporter erector launchers
whose locations are very hard for an enemy to track, and
other times, this means protecting them by burying them in
hardened missile silo bunkers.

Other components of nuclear strategies have included using
missile defense (to destroy the missiles before they land)
or implementation of civil defense measures (using early-
warning systems to evacuate citizens to safe areas before an
attack).

Weapons designed to threaten large populations, or to
generally deter attacks are known as *strategic weapons*.
Weapons designed for use on a battlefield in military situations are called *tactical weapons*.

There are critics of the very idea of nuclear strategy for
waging nuclear war who have suggested that a nuclear war
between two nuclear powers would result in mutual anni-
hilation. From this point of view, the significance of nuclear weapons is purely to deter war because any nuclear
war would immediately escalate out of mutual distrust and
fear, resulting in mutually assured destruction. This threat
of national, if not global, destruction has been a strong motivation for anti-nuclear weapons activism.

Critics from the peace movement and within the military establishment have questioned the usefulness of such weapons
in the current military climate. According to an advisory
opinion issued by the International Court of Justice in 1996,
the use of (or threat of use of) such weapons would generally be contrary to the rules of international law applicable
in armed conflict, but the court did not reach an opinion as
to whether or not the threat or use would be lawful in specific extreme circumstances such as if the survival of the
state were at stake.

Another deterrence position in nuclear strategy is that
nuclear proliferation can be desirable. This view argues
that, unlike conventional weapons, nuclear weapons successfully deter all-out war between states, and they succeeded in doing this during the Cold War between the
U.S. and the Soviet Union.[20] In the late 1950s and early
1960s, Gen. Pierre Marie Gallois of France, an adviser to
Charles DeGaulle, argued in books like *The Balance of Terror: Strategy for the Nuclear Age* (1961) that mere possession of a nuclear arsenal, what the French called the *force
de frappe*, was enough to ensure deterrence, and thus concluded that the spread of nuclear weapons could increase
international stability. Some very prominent neo-realist
scholars, such as the late Kenneth Waltz, formerly a Political Science at UC Berkeley and Adjunct Senior Research
Scholar at Columbia University, and John Mearsheimer of
University of Chicago, have also argued along the lines of
Gallois. Specifically, these scholars have advocated some
forms of nuclear proliferation, arguing that it would decrease the likelihood of total war, especially in troubled

regions of the world where there exists a unipolar nuclear weapon state. Aside from the public opinion that opposes proliferation in any form, there are two schools of thought on the matter: those, like Mearsheimer, who favor selective proliferation,*[21] and those of Kenneth Waltz, who was somewhat more non-interventionist.*[22]*[23] Renewed interest in proliferation and the stability-instability paradox that it generates continues as of 2016, with the ongoing debate for a credible indigenous Japanese and South Korean deterrent against North Korea.*[24]

The threat of potentially suicidal terrorists possessing nuclear weapons (a form of nuclear terrorism) complicates the decision process. The prospect of mutually assured destruction may not deter an enemy who expects to die in the confrontation. Further, if the initial act is from a stateless terrorist instead of a sovereign nation, there is no fixed nation or fixed military targets to retaliate against. It has been argued by the New York Times, especially after the September 11, 2001 attacks, that this complication is the sign of the next age of nuclear strategy, distinct from the relative stability of the Cold War.*[25] In 1996, the United States adopted a policy of allowing the targeting of its nuclear weapons at terrorists armed with weapons of mass destruction.*[26]

Robert Gallucci, president of the John D. and Catherine T. MacArthur Foundation, argues that although traditional deterrence is not an effective approach toward terrorist groups bent on causing a nuclear catastrophe, Gallucci believes that "the United States should instead consider a policy of expanded deterrence, which focuses not solely on the would-be nuclear terrorists but on those states that may deliberately transfer or inadvertently lead nuclear weapons and materials to them. By threatening retaliation against those states, the United States may be able to deter that which it cannot physically prevent." .*[27]

Graham Allison makes a similar case, arguing that the key to expanded deterrence is coming up with ways of tracing nuclear material to the country that forged the fissile material. "After a nuclear bomb detonates, nuclear forensics cops would collect debris samples and send them to a laboratory for radiological analysis. By identifying unique attributes of the fissile material, including its impurities and contaminants, one could trace the path back to its origin." *[28] The process is analogous to identifying a criminal by fingerprints. "The goal would be twofold: first, to deter leaders of nuclear states from selling weapons to terrorists by holding them accountable for any use of their own weapons; second, to give leaders every incentive to tightly secure their nuclear weapons and materials." *[28]

4.4 Governance, control, and law

Main articles: Treaty on the Non-Proliferation of Nuclear Weapons, Strategic Arms Limitation Talks, Intermediate-Range Nuclear Forces Treaty, START I, START II, Strategic Offensive Reductions Treaty, Comprehensive Nuclear-Test-Ban Treaty, and New START

Because of the immense military power they can confer,

The International Atomic Energy Agency was created in 1957 to encourage peaceful development of nuclear technology while providing international safeguards against nuclear proliferation.

the political control of nuclear weapons has been a key issue for as long as they have existed; in most countries the use of nuclear force can only be authorized by the head of government or head of state.*[29] Controls and regulations governing nuclear weapons are man-made, and so are imperfect. Therefore, there is an inherent danger of "accidents, mistakes, false alarms, blackmail, theft, and sabotage".*[30]

In the late 1940s, lack of mutual trust was preventing the United States and the Soviet Union from making ground towards international arms control agreements. The Russell–Einstein Manifesto was issued in London on July 9, 1955 by Bertrand Russell in the midst of the Cold War. It highlighted the dangers posed by nuclear weapons and called for world leaders to seek peaceful resolutions to international conflict. The signatories included eleven pre-eminent intellectuals and scientists, including Albert Einstein, who signed it just days before his death on April 18, 1955. A few days after the release, philanthropist Cyrus S. Eaton offered to sponsor a conference—called for in the manifesto—in Pugwash, Nova Scotia, Eaton's birthplace. This conference was to be the first of the Pugwash Conferences on Science and World Affairs, held in July 1957.

By the 1960s steps were being taken to limit both the proliferation of nuclear weapons to other countries and the environmental effects of nuclear testing. The Partial Nuclear Test Ban Treaty (1963) restricted all nuclear testing to underground nuclear testing, to prevent contamination from nuclear fallout, whereas the Treaty on the Non-Proliferation of Nuclear Weapons (1968) attempted to place restrictions on the types of activities signatories could participate in, with the goal of allowing the transference of non-military nuclear technology to member countries without fear of proliferation.

In 1957, the International Atomic Energy Agency (IAEA) was established under the mandate of the United Nations to encourage development of peaceful applications for nuclear technology, provide international safeguards against its misuse, and facilitate the application of safety measures in its use. In 1996, many nations signed the Comprehensive Nuclear-Test-Ban Treaty,[31] which prohibits all testing of nuclear weapons. A testing ban imposes a significant hindrance to nuclear arms development by any complying country.[32] The Treaty requires the ratification by 44 specific states before it can go into force; as of 2012, the ratification of eight of these states is still required.[31]

Additional treaties and agreements have governed nuclear weapons stockpiles between the countries with the two largest stockpiles, the United States and the Soviet Union, and later between the United States and Russia. These include treaties such as SALT II (never ratified), START I (expired), INF, START II (never ratified), SORT, and New START, as well as non-binding agreements such as SALT I and the Presidential Nuclear Initiatives[33] of 1991. Even when they did not enter into force, these agreements helped limit and later reduce the numbers and types of nuclear weapons between the United States and the Soviet Union/Russia.

Nuclear weapons have also been opposed by agreements between countries. Many nations have been declared Nuclear-Weapon-Free Zones, areas where nuclear weapons production and deployment are prohibited, through the use of treaties. The Treaty of Tlatelolco (1967) prohibited any production or deployment of nuclear weapons in Latin America and the Caribbean, and the Treaty of Pelindaba (1964) prohibits nuclear weapons in many African countries. As recently as 2006 a Central Asian Nuclear Weapon Free Zone was established amongst the former Soviet republics of Central Asia prohibiting nuclear weapons.

In the middle of 1996, the International Court of Justice, the highest court of the United Nations, issued an Advisory Opinion concerned with the "Legality of the Threat or Use of Nuclear Weapons". The court ruled that the use or threat of use of nuclear weapons would violate various articles of international law, including the Geneva Conventions, the Hague Conventions, the UN Charter, and the Universal Declaration of Human Rights. In view of the unique, destructive characteristics of nuclear weapons, the International Committee of the Red Cross calls on States

to ensure that these weapons are never used, irrespective of whether they consider them lawful or not.[34]

Additionally, there have been other, specific actions meant to discourage countries from developing nuclear arms. In the wake of the tests by India and Pakistan in 1998, economic sanctions were (temporarily) levied against both countries, though neither were signatories with the Nuclear Non-Proliferation Treaty. One of the stated *casus belli* for the initiation of the 2003 Iraq War was an accusation by the United States that Iraq was actively pursuing nuclear arms (though this was soon discovered not to be the case as the program had been discontinued). In 1981, Israel had bombed a nuclear reactor being constructed in Osirak, Iraq, in what it called an attempt to halt Iraq's previous nuclear arms ambitions; in 2007, Israel bombed another reactor being constructed in Syria.

In 2013, Mark Diesendorf says that governments of France, India, North Korea, Pakistan, UK, and South Africa have used nuclear power and/or research reactors to assist nuclear weapons development or to contribute to their supplies of nuclear explosives from military reactors.[35]

4.4.1 Disarmament

Main article: Nuclear disarmament
See also: Nuclear Tipping Point
For statistics on possession and deployment, see List of states with nuclear weapons.

Nuclear disarmament refers to both the act of reduc-

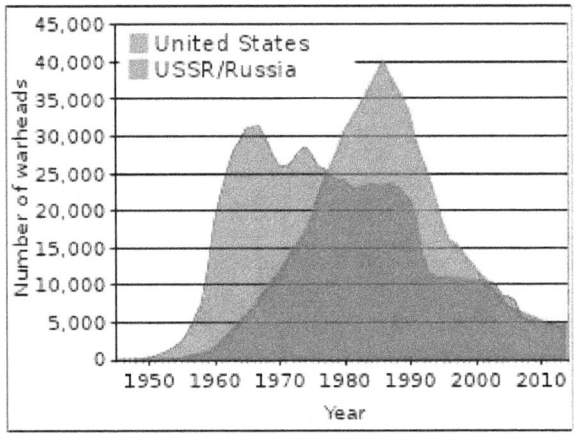

The USSR and United States nuclear weapon stockpiles throughout the Cold War until 2015, with a precipitous drop in total numbers following the end of the Cold War in 1991.

ing or eliminating nuclear weapons and to the end state of a nuclear-free world, in which nuclear weapons are completely eliminated.

Beginning with the 1963 Partial Test Ban Treaty and continuing through the 1996 Comprehensive Test Ban Treaty, there have been many treaties to limit or reduce nuclear weapons testing and stockpiles. The 1968 Nuclear Non-Proliferation Treaty has as one of its explicit conditions that all signatories must "pursue negotiations in good faith" towards the long-term goal of "complete disarmament". The nuclear weapon states have largely treated that aspect of the agreement as "decorative" and without force.[36]

Only one country —South Africa —has ever fully renounced nuclear weapons they had independently developed. The former Soviet republics of Belarus, Kazakhstan, and Ukraine returned Soviet nuclear arms stationed in their countries to Russia after the collapse of the USSR.

Proponents of nuclear disarmament say that it would lessen the probability of nuclear war occurring, especially accidentally. Critics of nuclear disarmament say that it would undermine the present nuclear peace and deterrence and would lead to increased global instability. Various American elder statesmen,[37] who were in office during the Cold War period, have been advocating the elimination of nuclear weapons. These officials include Henry Kissinger, George Shultz, Sam Nunn, and William Perry. In January 2010, Lawrence M. Krauss stated that "no issue carries more importance to the long-term health and security of humanity than the effort to reduce, and perhaps one day, rid the world of nuclear weapons".[38]

Ukrainian workers use equipment provided by the U.S. Defense Threat Reduction Agency to dismantle a Soviet-era missile silo. After the end of the Cold War, Ukraine and the other non-Russian, post-Soviet republics relinquished Soviet nuclear stockpiles to Russia.

In the years after the end of the Cold War, there have been numerous campaigns to urge the abolition of nuclear weapons, such as that organized by the Global Zero movement, and the goal of a "world without nuclear weapons" was advocated by United States President Barack Obama in an April 2009 speech in Prague.[39] A CNN poll from April 2010 indicated that the American public was nearly evenly split on the issue.[40]

Some analysts have argued that nuclear weapons have made the world relatively safer, with peace through deterrence and through the stability–instability paradox, including in south Asia.[41][42] Kenneth Waltz has argued that nuclear weapons have helped keep an uneasy peace, and further nuclear weapon proliferation might even help avoid the large scale conventional wars that were so common prior to their invention at the end of World War II.[23] But former Secretary Henry Kissinger says there is a new danger, which cannot be addressed by deterrence: "The classical notion of deterrence was that there was some consequences before which aggressors and evildoers would recoil. In a world of suicide bombers, that calculation doesn't operate in any comparable way".[43] George Shultz has said, "If you think of the people who are doing suicide attacks, and people like that get a nuclear weapon, they are almost by definition not deterrable".[44]

4.4.2 United Nations

Main article: United Nations Office for Disarmament Affairs

The UN Office for Disarmament Affairs (UNODA) is a department of the United Nations Secretariat established in January 1998 as part of the United Nations Secretary-General Kofi Annan's plan to reform the UN as presented in his report to the General Assembly in July 1997.[45]

Its goal is to promote nuclear disarmament and non-proliferation and the strengthening of the disarmament regimes in respect to other weapons of mass destruction, chemical and biological weapons. It also promotes disarmament efforts in the area of conventional weapons, especially land mines and small arms, which are often the weapons of choice in contemporary conflicts.

4.5 Controversy

See also: Nuclear weapons debate and History of the anti-nuclear movement

4.5.1 Ethics

Main article: Nuclear ethics

Even before the first nuclear weapons had been developed, scientists involved with the Manhattan Project were divided over the use of the weapon. The role of the two atomic bombings of the country in Japan's surrender and the U.S.'s ethical justification for them has been the subject of scholarly and popular debate for decades. The question of whether nations should have nuclear weapons, or test them, has been continually and nearly universally controversial.[46]

4.5.2   Notable nuclear weapons accidents

Main article: Nuclear and radiation accidents and incidents

- February 13, 1950: a Convair B-36B crashed in northern British Columbia after jettisoning a Mark IV atomic bomb. This was the first such nuclear weapon loss in history.

- May 22, 1957: a 42,000-pound Mark-17 hydrogen bomb accidentally fell from a bomber near Albuquerque, New Mexico. The detonation of the device's conventional explosives destroyed it on impact and formed a crater 25-feet in diameter on land owned by the University of New Mexico. According to a researcher at the Natural Resources Defense Council, it was one of the most powerful bombs made to date.[47]

- June 7, 1960: the 1960 Fort Dix IM-99 accident destroyed a Boeing CIM-10 Bomarc nuclear missile and shelter and contaminated the BOMARC Missile Accident Site in New Jersey.

- January 24, 1961: the 1961 Goldsboro B-52 crash occurred near Goldsboro, North Carolina. A Boeing B-52 Stratofortress carrying two Mark 39 nuclear bombs broke up in mid-air, dropping its nuclear payload in the process.[48][49]

- 1965 Philippine Sea A-4 crash, where a Skyhawk attack aircraft with a nuclear weapon fell into the sea.[50] The pilot, the aircraft, and the B43 nuclear bomb were never recovered.[51] It was not until 1989 that the Pentagon revealed the loss of the one-megaton bomb.[52]

- January 17, 1966: the 1966 Palomares B-52 crash occurred when a B-52G bomber of the USAF collided with a KC-135 tanker during mid-air refuelling off the coast of Spain. The KC-135 was completely destroyed when its fuel load ignited, killing all four crew members. The B-52G broke apart, killing three of the seven crew members aboard.[53] Of the four Mk28

type hydrogen bombs the B-52G carried,[54] three were found on land near Almería, Spain. The non-nuclear explosives in two of the weapons detonated upon impact with the ground, resulting in the contamination of a 2-square-kilometer (490-acre) (0.78 square mile) area by radioactive plutonium. The fourth, which fell into the Mediterranean Sea, was recovered intact after a 2½-month-long search.[55]

- January 21, 1968: the 1968 Thule Air Base B-52 crash involved a United States Air Force (USAF) B-52 bomber. The aircraft was carrying four hydrogen bombs when a cabin fire forced the crew to abandon the aircraft. Six crew members ejected safely, but one who did not have an ejection seat was killed while trying to bail out. The bomber crashed onto sea ice in Greenland, causing the nuclear payload to rupture and disperse, which resulted in widespread radioactive contamination.

- September 18–19, 1980: the Damascus Accident, occurred in Damascus, Arkansas, where a Titan missile equipped with a nuclear warhead exploded. The accident was caused by a maintenance man who dropped a socket from a socket wrench down an 80-foot shaft, puncturing a fuel tank on the rocket. Leaking fuel resulted in a hypergolic fuel explosion, jettisoning the W-53 warhead beyond the launch site.[56][57][58]

4.5.3   Nuclear testing and fallout

Main article: Nuclear fallout
See also: Downwinders
Over 500 atmospheric nuclear weapons tests were con-

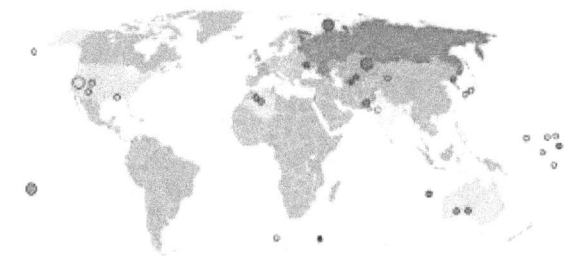

Over 2,000 nuclear tests have been conducted in over a dozen different sites around the world. Red Russia/Soviet Union, blue France, light blue United States, violet Britain, black Israel, orange China, yellow India, brown Pakistan, green North Korea and light green (territories exposed to nuclear bombs)

ducted at various sites around the world from 1945 to 1980. Radioactive fallout from nuclear weapons testing was first drawn to public attention in 1954 when the Castle Bravo hydrogen bomb test at the Pacific Proving Grounds contaminated the crew and catch of the Japanese fishing boat

This view of downtown Las Vegas shows a mushroom cloud in the background. Scenes such as this were typical during the 1950s. From 1951 to 1962 the government conducted 100 atmospheric tests at the nearby Nevada Test Site.

Lucky Dragon.[59] One of the fishermen died in Japan seven months later, and the fear of contaminated tuna led to a temporary boycotting of the popular staple in Japan. The incident caused widespread concern around the world, especially regarding the effects of nuclear fallout and atmospheric nuclear testing, and "provided a decisive impetus for the emergence of the anti-nuclear weapons movement in many countries" .*[59]

As public awareness and concern mounted over the possible health hazards associated with exposure to the nuclear fallout, various studies were done to assess the extent of the hazard. A Centers for Disease Control and Prevention/ National Cancer Institute study claims that fallout from atmospheric nuclear tests would lead to perhaps 11,000 excess deaths amongst people alive during atmospheric testing in the United States from all forms of cancer, including leukemia, from 1951 to well into the 21st century.*[60]*[61] As of March 2009, the U.S. is the only nation that compensates nuclear test victims. Since the Radiation Exposure Compensation Act of 1990, more than $1.38 billion in compensation has been approved. The money is going to people who took part in the tests, notably at the Nevada Test Site, and to others exposed to the radiation.*[62]*[63]

In addition, leakage of byproducts of nuclear weapon production into groundwater has been an ongoing issue, particularly at the Hanford site.*[64]

4.6 Effects of nuclear explosions on human health

Main article: Effects of nuclear explosions on human health

Some scientists estimate that if there were a nuclear war resulting in 100 Hiroshima-size nuclear explosions on cities, it could cause significant loss of life in the tens of millions from long term climatic effects alone. The climatology hypothesis is that *if* each city firestorms, a great deal of soot could be thrown up into the atmosphere which could blanket the earth, cutting out sunlight for years on end, causing the disruption of food chains, in what is termed a Nuclear winter.*[65]*[66]

The medical effects of the atomic bomb on Hiroshima upon humans can be put into the four categories below, with the effects of larger thermonuclear weapons producing blast and thermal effects so large that there would be a negligible number of survivors close enough to the center of the blast who would experience prompt/acute radiation effects, which were observed after the 16 kiloton yield Hiroshima bomb, due to its relatively low yield.*[67]*[68]

- Initial stage—the first 1–9 weeks, in which are the greatest number of deaths, with 90% due to thermal injury and/or blast effects and 10% due to super-lethal radiation exposure.

- Intermediate stage—from 10–12 weeks. The deaths in this period are from ionizing radiation in the median lethal range – LD50

- Late period—lasting from 13–20 weeks. This period has some improvement in survivors' condition.

- Delayed period—from 20+ weeks. Characterized by numerous complications, mostly related to healing of thermal and mechanical injuries, and if the individual was exposed to a few hundred to a thousand Millisieverts of radiation, it is coupled with infertility, sub-fertility and blood disorders. Furthermore, ionizing radiation above a dose of around 50–100 Millisievert exposure has been shown to statistically begin increasing one's chance of dying of cancer sometime in their lifetime over the normal unexposed rate of ~25%, in the long term, a heightened rate of cancer, proportional to the dose received, would begin to be observed after ~5+ years, with lesser problems such as eye cataracts and other more minor effects in other organs and tissue also being observed over the long term.

Fallout exposure – Depending on if further afield individuals Shelter in place or evacuate perpendicular to the direction of the wind, and therefore avoid contact with the fallout plume, and stay there for the days and weeks after the nuclear explosion, their exposure to fallout, and therefore their total dose, will vary. With those who do shelter in place, and or evacuate, experiencing a total dose that would be negligible in comparison to someone who just went about their life as normal.[69][70]

Staying indoors until after the most hazardous fallout isotope, I-131 decays away to 0.1% of its initial quantity after ten half lifes – which is represented by 80 days in I-131s case, would make the difference between likely contracting Thyroid cancer or escaping completely from this substance depending on the actions of the individual.[71]

4.6.1   Public opposition

See also: History of the anti-nuclear movement and International Day against Nuclear Tests
Peace movements emerged in Japan and in 1954 they con-

Women Strike for Peace during the Cuban Missile Crisis

verged to form a unified "Japanese Council Against Atomic and Hydrogen Bombs". Japanese opposition to nuclear weapons tests in the Pacific Ocean was widespread, and "an estimated 35 million signatures were collected on petitions calling for bans on nuclear weapons".[72]

In the United Kingdom, the first Aldermaston March organised by the Campaign for Nuclear Disarmament(CND)

Demonstration against nuclear testing in Lyon, France, in the 1980s.

took place at Easter 1958, when, according to the CND, several thousand people marched for four days from Trafalgar Square, London, to the Atomic Weapons Research Establishment close to Aldermaston in Berkshire, England, to demonstrate their opposition to nuclear weapons.[73][74] The Aldermaston marches continued into the late 1960s when tens of thousands of people took part in the four-day marches.[72]

In 1959, a letter in the *Bulletin of Atomic Scientists* was the start of a successful campaign to stop the Atomic Energy Commission dumping radioactive waste in the sea 19 kilometres from Boston.[75] In 1962, Linus Pauling won the Nobel Peace Prize for his work to stop the atmospheric testing of nuclear weapons, and the "Ban the Bomb" movement spread.[46]

In 1963, many countries ratified the Partial Test Ban Treaty prohibiting atmospheric nuclear testing. Radioactive fallout became less of an issue and the anti-nuclear weapons movement went into decline for some years.[59][76] A resurgence of interest occurred amid European and American fears of nuclear war in the 1980s.[77]

4.7   Costs and technology spin-offs

See also: Global Positioning System, Nuclear weapons delivery, History of computing hardware, ENIAC, and Swords to ploughshares

According to an audit by the Brookings Institution, between 1940 and 1996, the U.S. spent $8.89 trillion in present-day terms[78] on nuclear weapons programs. 57 percent of which was spent on building nuclear weapons delivery systems. 6.3 percent of the total, $557 billion in present-day terms, was spent on environmental remediation and nuclear waste management, for example cleaning up the Hanford site, and 7 percent of the total, $625 billion was spent on making nuclear weapons themselves.[79]

4.8 Non-weapons uses

Main article: Peaceful nuclear explosion

Peaceful nuclear explosions are nuclear explosions conducted for non-military purposes, such as activities related to economic development including the creation of canals. During the 1960s and 70s, both the United States and the Soviet Union conducted a number of PNEs. Six of the explosions by the Soviet Union are considered to have been of an applied nature, not just tests.

Subsequently, the United States and the Soviet Union halted their programs. Definitions and limits are covered in the Peaceful Nuclear Explosions Treaty of 1976.[*][80][*][81] The Comprehensive Nuclear-Test-Ban Treaty of 1996, once it enters into force, will prohibit all nuclear explosions, regardless of whether they are for peaceful purposes or not.[*][82]

4.9 See also

- *The Atomic Age* – Wikipedia book

4.9.1 History

- History of nuclear weapons
  - Manhattan Project
  - Atomic spies
  - German nuclear weapon project
  - Japanese nuclear weapon program
  - Soviet atomic bomb project
  - Nuclear testing at Bikini Atoll
- Timeline of nuclear weapons development
- Los Alamos National Laboratory
- Lawrence Livermore National Laboratory
- Lists of nuclear disasters and radioactive incidents
- Nuclear and radiation accidents, including nuclear weapons accidents
  - Nevada Test Site
  - Project Gnome
- Military strategy
  - Civil Defense
  - Fractional Orbital Bombardment System

- Mutual Assured Destruction
- Weapon of mass destruction
  - Nuclear strategy

4.9.2 More technical details

- Effects of nuclear explosions
- Nuclear winter
- Nuclear Triad
- Intercontinental ballistic missile
- Submarine-launched ballistic missile
- Nuclear torpedo
- Hypersonic glide vehicle
- Nuclear blackout
- Thermonuclear weapon
- Boosted fission weapon
- Cobalt bomb
- Salted bomb
- Neutron bomb
- Pure fusion weapon
- Nuclear bombs and health
- Nuclear weapon yield

4.9.3 In popular culture

- Nuclear weapons in popular culture
- *The Butter Battle Book*

4.9.4 Proliferation and politics

- Agency for the Prohibition of Nuclear Weapons in Latin America and the Caribbean
- Comprehensive Test Ban Treaty
- International Court of Justice advisory opinion on legality of nuclear weapons
- List of states with nuclear weapons
- List of nuclear weapons

- Nth Country Experiment

- Nuclear close calls

- Nuclear Non-Proliferation Treaty

- Nuclear weapons and the United Kingdom

- The Letters of last resort (United Kingdom)

- Nuclear weapons and Russia

- Nuclear weapons and the United States

- Strategic Arms Limitation Talks

- Three Non-Nuclear Principles, of Japan

4.10   Notes and references

[1] Specifically the 1970 to 1980 designed and deployed US B83 nuclear bomb, with a yield of up to 1.2 megatons.

[2] "Frequently Asked Questions #1". Radiation Effects Research Foundation. Retrieved September 18, 2007. total number of deaths is not known precisely ... acute (within two to four months) deaths ... Hiroshima ... 90,000 − 166,000 ... Nagasaki ... 60,000 − 80,000

[3] "Federation of American Scientists: Status of World Nuclear Forces". Fas.org. Retrieved December 29, 2012.

[4] "Nuclear Weapons – Israel". Fas.org. January 8, 2007. Retrieved December 15, 2010.

[5] See also Mordechai Vanunu

[6] Executive release. "South African nuclear bomb". *Nuclear Threat Initiatives*. Nuclear Threat Initiatives, South Africa (NTI South Africa). Retrieved 13 March 2012.

[7] Ian Lowe. "Three minutes to midnight". *Australasian Science*, March 2016, p. 49.

[8] The best overall printed sources on nuclear weapons design are: Hansen, Chuck. *U.S. Nuclear Weapons: The Secret History*. San Antonio, TX: Aerofax, 1988; and the more-updated Hansen, Chuck. "Swords of Armageddon: U.S. Nuclear Weapons Development since 1945" (CD-ROM & download available). PDF. 2,600 pages. Sunnyvale, California. Chuklea Publications, 1995, 2007. ISBN 978-0-9791915-0-3 (2nd Ed.)

[9] David Albright and Kimberly Kramer (August 22, 2005). "Neptunium 237 and Americium: World Inventories and Proliferation Concerns" (PDF). Institute for Science and International Security. Retrieved October 13, 2011.

[10] Carey Sublette. Nuclear Weapons Frequently Asked Questions: 4.5.2 "Dirty" and "Clean" Weapons, accessed May 10, 2011.

[11] On India's alleged hydrogen bomb test, see Carey Sublette, What Are the Real Yields of India's Test?.

[12] McKirdy, Euan. "North Korea announces it conducted nuclear test". *CNN*. Retrieved 7 January 2016.

[13] Sublette, Carey. "The Nuclear Weapon Archive". Retrieved March 7, 2007.

[14] U.S. Department of Energy. Restricted Data Declassification Decisions, 1946 to the Present (RDD-8) (January 1, 2002), accessed November 20, 2011.

[15] "Page discussing the possibility of using antimatter as a trigger for a thermonuclear explosion". Cui.unige.ch. Retrieved May 30, 2013.

[16] Andre Gsponer; Jean-Pierre Hurni (1970). "Paper discussing the number of antiprotons required to ignite a thermonuclear weapon". In G. Velarde and E. Minguez, eds., *Proceedings of the International Conference on Emerging Nuclear Energy Systems, Madrid, June /July*, (World Scientific, Singapore, 1987) 166–169. Arxiv.org. 4 (30): arXiv:physics/0507114. arXiv:physics/0507114. Bibcode:2005physics...7114G.

[17] Keay Davidson: Chronicle Science Writer (October 4, 2004). "Air Force pursuing antimatter weapons: Program was touted publicly, then came official gag order". Sfgate.com. Retrieved May 30, 2013.

[18] "Fourth Generation Nuclear Weapons". Retrieved October 24, 2014.

[19] Stephen I. Schwartz, ed., *Atomic Audit: The Costs and Consequences of U.S. Nuclear Weapons Since 1940*. Washington, D.C.: Brookings Institution Press, 1998. See also Estimated Minimum Incurred Costs of U.S. Nuclear Weapons Programs, 1940–1996, an excerpt from the book. Archived November 21, 2008, at the Wayback Machine.

[20] Creveld, Martin Van (2000). "Technology and War II: Postmodern War?". In Charles Townshend. *The Oxford History of Modern War*. New York, USA: Oxford University Press. p. 349. ISBN 0-19-285373-2.

[21] Mearsheimer, John (2006). "Conversations in International Relations: Interview with John J. Mearsheimer (Part I)" (PDF). *International Relations*. **20** (1): 105–123. doi:10.1177/0047117806060939.See page 116

[22] Kenneth Waltz. "More May Be Better," in Scott Sagan and Kenneth Waltz, eds., *The Spread of Nuclear Weapons* (New York: Norton, 1995).

[23] Kenneth Waltz. "The Spread of Nuclear Weapons: More May Better," *Adelphi Papers*, no. 171 (London: International Institute for Strategic Studies, 1981).

[24] Should We Let the Bomb Spread? Edited by Mr. Henry D. Sokolski. Strategic studies institute. November 2016

[25] See, for example: Feldman, Noah. "Islam, Terror and the Second Nuclear Age," *New York Times Magazine* (October 29, 2006).

[26] Daniel Plesch & Stephen Young, "Senseless policy", *Bulletin of the Atomic Scientists*, November/December 1998, page 4. Fetched from URL on April 18, 2011.

[27] Gallucci, Robert (September 2006). "Averting Nuclear Catastrophe: Contemplating Extreme Responses to U.S. Vulnerability". *Annals of the American Academy of Political and Social Science*. **607**: 51–58. doi:10.1177/0002716206290457. Retrieved January 28, 2013.

[28] Allison, Graham (March 13, 2009). "How to Keep the Bomb From Terrorists". *Newsweek*. Retrieved January 28, 2013.

[29] In the United States, the President and the Secretary of Defense, acting as the National Command Authority, must *jointly* authorize the use of nuclear weapons.

[30] Eric Schlosser, Today's nuclear dilemma, *Bulletin of the Atomic Scientists*, November/December 2015, vol. 71 no. 6, 11–17.

[31] Preparatory Commission for the Comprehensive Nuclear-Test-Ban Treaty Organization (2010). "Status of Signature and Ratification". Accessed May 27, 2010. Of the "Annex 2" states whose ratification of the CTBT is required before it enters into force, China, Egypt, Iran, Israel, and the United States have signed but not ratified the Treaty. India, North Korea, and Pakistan have not signed the Treaty.

[32] Richelson, Jeffrey. *Spying on the bomb: American nuclear intelligence from Nazi Germany to Iran and North Korea.* New York: Norton, 2006.

[33] The Presidential Nuclear Initiatives (PNIs) on Tactical Nuclear Weapons At a Glance, Fact Sheet, Arms Control Association.

[34] Nuclear weapons and international humanitarian law International Committee of the Red Cross

[35] Mark Diesendorf (2013). "Book review: Contesting the future of nuclear power" (PDF). *Energy Policy*.

[36] Gusterson, Hugh, "Finding Article VI" *Bulletin of the Atomic Scientists* (January 8, 2007).

[37] Jim Hoagland (October 6, 2011). "Nuclear energy after Fukushima". *Washington Post.*

[38] Lawrence M. Krauss. The Doomsday Clock Still Ticks, *Scientific American*, January 2010, p. 26.

[39] Graham, Nick (April 5, 2009). "Obama Prague Speech On Nuclear Weapons". Huffingtonpost.com. Retrieved May 30, 2013.

[40] "CNN Poll: Public divided on eliminating all nuclear weapons". Politicalticker.blogs.cnn.com. April 12, 2010. Retrieved May 30, 2013.

[41] Krepon, Michael. "The Stability-Instability Paradox, Misperception, and Escalation Control in South Asia" (PDF). *Stimson*. Retrieved November 20, 2015.

[42] "Michael Krepon • The Stability-Instability Paradox". Retrieved October 24, 2014.

[43] Ben Goddard (January 27, 2010). "Cold Warriors say no nukes". *The Hill.*

[44] Hugh Gusterson (March 30, 2012). "The new abolitionists". *Bulletin of the Atomic Scientists.*

[45] ODS Team. "Renewing the United Nations: A Program for Reform (A/51/950)" (PDF). Daccess-dds-ny.un.org. Retrieved May 30, 2013.

[46] Jerry Brown and Rinaldo Brutoco (1997). *Profiles in Power: The Anti-nuclear Movement and the Dawn of the Solar Age.* Twayne Publishers. pp. 191–192.

[47] "Accident Revealed After 29 Years: H-Bomb Fell Near Albuquerque in 1957". Los Angeles Times. Associated Press. August 27, 1986. Retrieved 31 August 2014.

[48] Barry Schneider (May 1975). "Big Bangs from Little Bombs". *Bulletin of Atomic Scientists*. p. 28. Retrieved July 13, 2009.

[49] James C. Oskins; Michael H. Maggelet (2008). *Broken Arrow —The Declassified History of U.S. Nuclear Weapons Accidents.* lulu.com. ISBN 1-4357-0361-8. Retrieved December 29, 2008.

[50] "Ticonderoga Cruise Reports" (Navy.mil weblist of Aug 2003 compilation from cruise reports). Retrieved April 20, 2012. The National Archives hold[s] deck logs for aircraft carriers for the Vietnam Conflict.

[51] Broken Arrows at www.atomicarchive.com. Accessed August 24, 2007.

[52] "U.S. Confirms '65 Loss of H-Bomb Near Japanese Islands". *The Washington Post*. Reuters. May 9, 1989. p. A-27.

[53] Hayes, Ron (January 17, 2007). "H-bomb incident crippled pilot's career". Palm Beach Post. Archived from the original on June 16, 2011. Retrieved May 24, 2006.

[54] Maydew, Randall C. (1997). *America's Lost H-Bomb: Palomares, Spain, 1966.* Sunflower University Press. ISBN 978-0-89745-214-4.

[55] Long, Tony (January 17, 2008). "Jan. 17, 1966: H-Bombs Rain Down on a Spanish Fishing Village". WIRED. Retrieved February 16, 2008.

[56] Schlosser, Eric (2013). *Command and Control: Nuclear Weapons, the Damascus Accident, and the Illusion of Safety.* Penguin Press. ISBN 978-1-59420-227-8.

[57] Christ, Mark K. "Titan II Missile Explosion". *The Ency-clopedia of Arkansas History & Culture*. Arkansas Historic Preservation Program. Retrieved 31 August 2014.

[58] Stumpf, David K. (2000). Christ, Mark K.; Slater, Cathryn H., eds. *"We Can Neither Confirm Nor Deny" Sentinels of History: Reflections on Arkansas Properties on the National Register of Historic Places*. Fayetteville, Arkansas: University of Arkansas Press.

[59] Rudig, Wolfgang (1990). "Anti-nuclear Movements: A World Survey of Opposition to Nuclear Energy". Longman. pp. 54–55.

[60] "Report on the Health Consequences to the American Population from Nuclear Weapons Tests Conducted by the United States and Other Nations". CDC. Retrieved December 7, 2013.

[61] Committee to Review the CDC-NCI Feasibility Study of the Health Consequences Nuclear Weapons Tests, National Research Council. "Exposure of the American Population to Radioactive Fallout from Nuclear Weapons Tests". Retrieved October 24, 2014.

[62] ABC News. "What governments offer to victims of nuclear tests". *ABC News*. Retrieved October 24, 2014.

[63] Radiation Exposure Compensation System: Claims to Date Summary of Claims Received by 06/11/2009

[64] Coghlan, Andy. "US nuclear dump is leaking toxic waste". *New Scientist*. Retrieved 12 March 2016.

[65] Philip Yam. Nuclear Exchange, *Scientific American*, June 2010, p. 24.

[66] Alan Robock and Owen Brian Toon. Local Nuclear War, Global Suffering. *Scientific American*, January 2010, p. 74-81.

[67] "Remm.nlm.gov".

[68] "Nuclear Warfare" (PDF). *Nd.edu*. p. 3.

[69] 7 hour rule: At 7 hours after detonation the fission product activity will have decreased to about 1/10 (10%) of its amount at 1 hour. At about 2 days (49 hours-7X7) the activity will have decreased to 1% of the 1-hour value. Falloutradiation.com

[70] "Nuclear Warfare" (PDF). p. 22.

[71] Oak Ridge Reservation (USDOE), EPA Facility ID: TN1890090003: Site and Radiological Assessment Branch, Division of Health Assessment and Consultation, Agency for Toxic Substances and Disease Registry. "PUBLIC HEALTH ASSESSMENT Iodine-131 Releases" (PDF). *atsdr.cdc.gov*. U.S. Center for Disease Control. Retrieved 21 May 2016.

[72] Jim Falk (1982). *Global Fission: The Battle Over Nuclear Power*. Oxford University Press. pp. 96–97.

[73] "A brief history of CND". Cnduk.org. Retrieved May 30, 2013.

[74] "Early defections in march to Aldermaston". London: Guardian Unlimited. April 5, 1958.

[75] Jim Falk (1982). *Global Fission: The Battle Over Nuclear Power*. Oxford University Press, p. 93.

[76] Jim Falk (1982). *Global Fission: The Battle Over Nuclear Power*. Oxford University Press, p. 98.

[77] Spencer Weart, *Nuclear Fear: A History of Images* (Cambridge, Mass.: Harvard University Press, 1988), chapters 16 and 19.

[78] Federal Reserve Bank of Minneapolis Community Development Project. "Consumer Price Index (estimate) 1800–". Federal Reserve Bank of Minneapolis. Retrieved January 2, 2017.

[79] "Estimated Minimum Incurred Costs of U.S. Nuclear Weapons Programs, 1940–1996". *Brookings Institution*. Archived from the original on March 5, 2004. Retrieved November 20, 2015.

[80] "Announcement of Treaty on Underground Nuclear Explosions Peaceful Purposes (PNE Treaty)" (PDF). Gerald R. Ford Museum and Library. May 28, 1976.

[81] Peters, Gerhard; Woolley, John T. "Gerald R. Ford: "Message to the Senate Transmitting United States-Soviet Treaty and Protocol on the Limitation of Underground Nuclear Explosions," July 29, 1976". *The American Presidency Project*. University of California – Santa Barbara.

[82] "Status of Signature and Ratification". *ctbto dot org*. CTBT Organization Prepatory Commission. Retrieved December 29, 2016.

4.11   Bibliography

See also: List of books about nuclear issues

- Bethe, Hans Albrecht. *The Road from Los Alamos*. New York: Simon and Schuster, 1991. ISBN 0-671-74012-1

- DeVolpi, Alexander, Minkov, Vladimir E., Simonenko, Vadim A., and Stanford, George S. *Nuclear Shadowboxing: Contemporary Threats from Cold War Weaponry*. Fidlar Doubleday, 2004 (Two volumes, both accessible on Google Book Search) (Content of both volumes is now available in the 2009 trilogy by Alexander DeVolpi: *Nuclear Insights: The Cold War Legacy*)

- Glasstone, Samuel and Dolan, Philip J. *The Effects of Nuclear Weapons (third edition)*. Washington, D.C.: U.S. Government Printing Office, 1977. Available online (PDF).

- *NATO Handbook on the Medical Aspects of NBC Defensive Operations (Part I – Nuclear)*. Departments of the Army, Navy, and Air Force: Washington, D.C., 1996

- Hansen, Chuck. *U.S. Nuclear Weapons: The Secret History*. Arlington, TX: Aerofax, 1988

- Hansen, Chuck. "Swords of Armageddon: U.S. nuclear weapons development since 1945" (CD-ROM & download available). PDF. 2,600 pages. Sunnyvale, California, Chucklea Publications, 1995, 2007. ISBN 978-0-9791915-0-3 (2nd Ed.)

- Holloway, David. *Stalin and the Bomb*. New Haven: Yale University Press, 1994. ISBN 0-300-06056-4

- The Manhattan Engineer District, "The Atomic Bombings of Hiroshima and Nagasaki" (1946)

- (French) Jean-Hugues Oppel, *Réveillez le président*. Éditions Payot et rivages, 2007 (ISBN 978-2-7436-1630-4). The book is a fiction about the nuclear weapons of France; the book also contains about ten chapters on true historical incidents involving nuclear weapons and strategy.

- Smyth, Henry DeWolf. *Atomic Energy for Military Purposes*. Princeton, NJ: Princeton University Press, 1945. (Smyth Report – the first declassified report by the US government on nuclear weapons)

- *The Effects of Nuclear War*. Office of Technology Assessment, May 1979.

- Rhodes, Richard. *Dark Sun: The Making of the Hydrogen Bomb*. New York: Simon and Schuster, 1995. ISBN 0-684-82414-0

- Rhodes, Richard. *The Making of the Atomic Bomb*. New York: Simon and Schuster, 1986 ISBN 0-684-81378-5

- Schultz, George P. and Goodby, James E. *The War that Must Never be Fought*, Hoover Press, 2015, ISBN 978-0-8179-1845-3.

- Weart, Spencer R. *Nuclear Fear: A History of Images*. Cambridge, MA: Harvard University Press, 1988. ISBN 0-674-62836-5

- Weart, Spencer R. *The Rise of Nuclear Fear*. Cambridge, MA: Harvard University Press, 2012. ISBN 0-674-05233-1

4.12 External links

- Nuclear Weapon Archive from Carey Sublette is a reliable source of information and has links to other sources and an informative FAQ.

- The Federation of American Scientists provide solid information on weapons of mass destruction, including nuclear weapons and their effects

- Alsos Digital Library for Nuclear Issues —contains many resources related to nuclear weapons, including a historical and technical overview and searchable bibliography of web and print resources.

- Video archive of US, Soviet, UK, Chinese and French Nuclear Weapon Testing at sonicbomb.com

- The National Museum of Nuclear Science & History (United States)—located in Albuquerque, New Mexico; a Smithsonian Affiliate Museum

- Nuclear Emergency and Radiation Resources

- The Manhattan Project: Making the Atomic Bomb at AtomicArchive.com

- Los Alamos National Laboratory: History (U.S. nuclear history)

- *Race for the Superbomb*, PBS website on the history of the H-bomb

- Recordings of recollections of the victims of Hiroshima and Nagasaki

- The Woodrow Wilson Center's Nuclear Proliferation International History Project or NPIHP is a global network of individuals and institutions engaged in the study of international nuclear history through archival documents, oral history interviews and other empirical sources.

- NUKEMAP3D – a 3D nuclear weapons effects simulator powered by Google Maps.

Chapter 5

Nuclear electromagnetic pulse

This article is about nuclear-generated EMP. For other types, see Electromagnetic pulse

A **nuclear electromagnetic pulse** (commonly abbreviated as nuclear EMP, pronounced /iː.ɛm.piː/, or NEMP) is a characteristic burst of electromagnetic radiation created by nuclear explosions. The resulting rapidly changing electric and magnetic fields may couple with electrical and electronic systems to produce damaging current and voltage surges. The specific characteristics of any particular nuclear EMP event vary according to a number of factors, the most important of which is the altitude of the detonation.

The term "electromagnetic pulse" generally excludes optical (infrared, visible, ultraviolet) and ionizing (such as X-ray and gamma radiation) ranges. In military terminology, a nuclear warhead detonated hundreds of kilometers above the Earth's surface is known as a high-altitude electromagnetic pulse (HEMP) device. Effects of a HEMP device depend on factors including the altitude of the detonation, energy yield, gamma ray output, interactions with the Earth's magnetic field and electromagnetic shielding of targets.

5.1 History

The fact that an electromagnetic pulse is produced by a nuclear explosion was known in the earliest days of nuclear weapons testing. The magnitude of the EMP and the significance of its effects, however, were not immediately realized.[1]

During the first United States nuclear test on 16 July 1945, electronic equipment was shielded because Enrico Fermi expected the electromagnetic pulse. The official technical history for that first nuclear test states, "All signal lines were completely shielded, in many cases doubly shielded. In spite of this many records were lost because of spurious pickup at the time of the explosion that paralyzed the recording equipment." [2] During British nuclear testing in 1952–1953 instrumentation failures were attributed to "radioflash", which was their term for EMP.[3][4]

The first openly reported observation of the unique aspects of high-altitude nuclear EMP occurred during the helium balloon lofted Yucca nuclear test of the Hardtack I series on 28 April 1958. In that test, the electric field measurements from the 1.7 kiloton weapon went off the scale of the test instruments and was estimated to be about 5 times the oscilloscope limits. The Yucca EMP was initially positive-going whereas low-altitude bursts were negative pulses. Also, the polarization of the Yucca EMP signal was horizontal, whereas low-altitude nuclear EMP was vertically polarized. In spite of these many differences, the unique EMP results were dismissed as a possible wave propagation anomaly.[5]

The high-altitude nuclear tests of 1962, as discussed below, confirmed the unique results of the Yucca high-altitude test and increased the awareness of high-altitude nuclear EMP beyond the original group of defense scientists.

The larger scientific community became aware of the significance of the EMP problem after a three-article series on nuclear EMP was published in 1981 by William J. Broad in *Science*.[1][6][7]

5.1.1 Starfish Prime

Main article: Starfish Prime

In July 1962, the US carried out the Starfish Prime test, exploding a 1.44 megaton bomb 400 kilometres (250 mi) above the mid-Pacific Ocean. This demonstrated that the effects of a high-altitude nuclear explosion were much larger than had been previously calculated. Starfish Prime made those effects known to the public by causing electrical damage in Hawaii, about 1,445 kilometres (898 mi) away from the detonation point, knocking out about 300 streetlights, setting off numerous burglar alarms and damaging a microwave link.[8]

Starfish Prime was the first success in the series of United States high-altitude nuclear tests in 1962 known as Operation Fishbowl. Subsequent tests gathered more data on the high-altitude EMP phenomenon.

The Bluegill Triple Prime and Kingfish high-altitude nuclear tests of October and November 1962 in Operation Fishbowl provided data that was clear enough to enable physicists to accurately identify the physical mechanisms behind the electromagnetic pulses.[9]

The EMP damage of the Starfish Prime test was quickly repaired because of the ruggedness (compared to today)[10] of Hawaii's electrical and electronic infrastructure.

The relatively small magnitude of the Starfish Prime EMP in Hawaii (about 5.6 kilovolts/metre) and the relatively small amount of damage (for example, only 1 to 3 percent of streetlights extinguished)[11] led some scientists to believe, in the early days of EMP research, that the problem might not be significant. Newer calculations[10] showed that if the Starfish Prime warhead had been detonated over the northern continental United States, the magnitude of the EMP would have been much larger (22 to 30 kV/m) because of the greater strength of the Earth's magnetic field over the United States, as well as its different orientation at high latitudes. These calculations, combined with the accelerating reliance on EMP-sensitive microelectronics, heightened awareness that EMP could be a significant problem.

5.1.2 Soviet Test 184

Main article: Soviet Project K nuclear tests

In 1962, the Soviet Union also performed three EMP-producing nuclear tests in space over Kazakhstan, the last in the "Soviet Project K nuclear tests".[12] Although these weapons were much smaller (300 kiloton) than the Starfish Prime test, they were over a populated, large land mass and at a location where the Earth's magnetic field was greater: the damage caused by the resulting EMP was reportedly much greater than in Starfish Prime. The geomagnetic storm–like E3 pulse from Test 184 induced a current surge in a long underground power line that caused a fire in the power plant in the city of Karaganda.

After the collapse of the Soviet Union, the level of this damage was communicated informally to U.S. scientists.[13] After the 1991 collapse of the Soviet Union, there was a period of a few years of cooperation between United States and Russian scientists on the HEMP phenomenon. In addition, funding was secured to enable Russian scientists to formally report on some of the Soviet EMP results in international scientific journals.[14] As a result, formal documentation of some of the EMP damage in Kazakhstan exists[15][16] but is still sparse in the open scientific literature, especially in relation to the level of damage that was indicated in the open reports.

For one of the K Project tests, Soviet scientists instrumented a 570-kilometer (350 mi) section of telephone line in the area that they expected to be affected by the pulse. The monitored telephone line was divided into sub-lines of 40 to 80 kilometres (25 to 50 mi) in length, separated by repeaters. Each sub-line was protected by fuses and by gas-filled overvoltage protectors. The EMP from the 22 October (K-3) nuclear test (also known as Test 184) blew all of the fuses and fired all of the overvoltage protectors in all of the sub-lines.[15]

Published reports, including a 1998 IEEE article,[15] have stated that there were significant problems with ceramic insulators on overhead electrical power lines during the tests. A 2010 technical report written for Oak Ridge National Laboratory stated that "Power line insulators were damaged, resulting in a short circuit on the line and some lines detaching from the poles and falling to the ground."[17]

5.2 Characteristics of nuclear EMP

Nuclear EMP is a complex multi-pulse, usually described in terms of three components, as defined by the International Electrotechnical Commission (IEC).[18]

The three components of nuclear EMP, as defined by the IEC, are called "E1", "E2" and "E3".

5.2.1 E1

The E1 pulse is the very fast component of nuclear EMP. E1 is a very brief but intense electromagnetic field that induces very high voltages in electrical conductors. E1 causes most of its damage by causing electrical breakdown voltages to be exceeded. E1 can destroy computers and communications equipment and it changes too quickly (nanoseconds) for ordinary surge protectors to provide effective protection against it, although there are special fast-acting surge protectors that will block the E1 pulse.

E1 is produced when gamma radiation from the nuclear detonation ionizes (strips electrons from) atoms in the upper atmosphere. This is known as the Compton effect and the resulting current is called the "Compton current". The electrons travel in a generally downward direction at relativistic speeds (more than 90 percent of the speed of light). In the absence of a magnetic field, this would produce a large, radial pulse of electric current propagating outward from the burst location confined to the source region (the region over which the gamma photons are attenuated). The Earth's

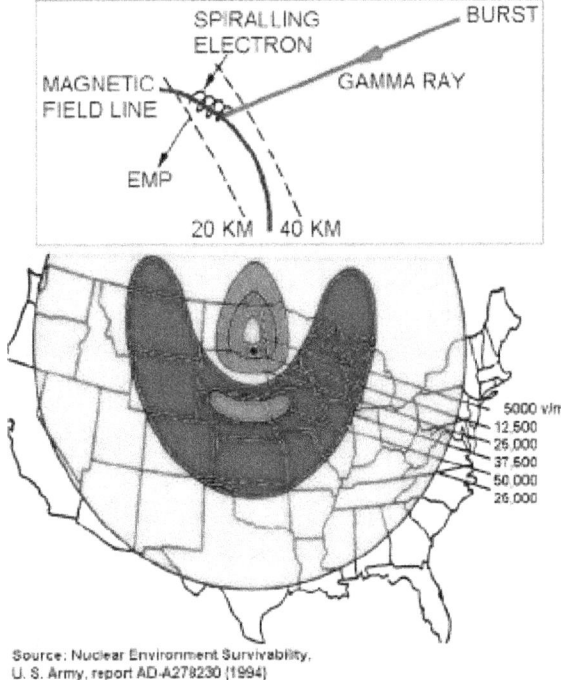

Source: Nuclear Environment Survivability,
U. S. Army, report AD-A278230 (1994)

The mechanism for a 400 km high-altitude burst EMP: gamma rays hit the atmosphere between 20–40 km altitude, ejecting electrons which are then deflected sideways by the Earth's magnetic field. This makes the electrons radiate EMP over a massive area. Because of the curvature and downward tilt of Earth's magnetic field over the USA, the maximum EMP occurs south of the detonation and the minimum occurs to the north.[19]

magnetic field deflects the electron flow at a right angle to the field, leading to synchrotron radiation emitted by the electrons. Because the outward traveling gamma pulse is propagating at the speed of light, the synchrotron radiation of the Compton electrons adds coherently, leading to a radiated electromagnetic signal. This interaction produces a very large, but very brief, electromagnetic pulse over the affected area.[20]

Several physicists worked on the problem of identifying the mechanism of the uniquely large E1 pulse produced by a nuclear weapon detonated at high altitude (HEMP). The correct mechanism was finally identified by Conrad Longmire of Los Alamos National Laboratory in 1963.[9]

Conrad Longmire gives numerical values for a typical case of E1 pulse produced by a second-generation nuclear weapon such as those of Operation Fishbowl in 1962. The typical gamma rays given off by the weapon have an energy of about 2 MeV (mega-electron volts). The gamma rays transfer about half of their energy to the ejected free electrons, giving an energy of about 1 MeV.[20]

In a vacuum and absent a magnetic field, the electrons would travel with a current density of tens of amperes per square metre.[20] Because of the downward tilt of the Earth's magnetic field at high latitudes, the area of peak field strength is a U-shaped region to the equatorial side of the nuclear detonation. As shown in the diagram at the right, for nuclear detonations over the continental United States, this U-shaped region is south of the detonation point. Near the equator, where the Earth's magnetic field is more nearly horizontal, the E1 field strength is more nearly symmetrical around the burst location.

At geomagnetic field strengths typical of the central United States, central Europe or Australia, these initial electrons spiral around the magnetic field lines with a typical radius of about 85 metres (about 280 feet). These initial electrons are stopped by collisions with other air molecules at an average distance of about 170 metres (a little less than 580 feet). This means that most of the electrons are stopped by collisions with air molecules before completing a full spiral around the field lines.[20]

This interaction of the very rapidly moving negatively charged electrons with the magnetic field radiates a pulse of electromagnetic energy. The pulse typically rises to its peak value in some 5 nanoseconds. Its magnitude typically decays to half of its peak value within 200 nanoseconds. (By the IEC definition, this E1 pulse ends 1000 nanoseconds after it begins.) This process occurs simultaneously on about 10^{25} electrons.[20] The simultaneous action of the very large number of electrons causes the resulting electromagnetic pulses from each electron to radiate coherently, thus adding to produce a single very large amplitude, but very narrow, radiated electromagnetic pulse.

Secondary collisions cause subsequent electrons to lose energy before they reach ground level. The electrons generated by these subsequent collisions have such reduced energy that they do not contribute significantly to the E1 pulse.[20]

These 2 MeV gamma rays typically produce an E1 pulse near ground level at moderately high latitudes that peaks at about 50,000 volts per metre. This is a peak power density of 6.6 megawatts per square metre.

The ionization process in the mid-stratosphere causes this region to become an electrical conductor, a process that blocks the production of further electromagnetic signals and causes the field strength to saturate at about 50,000 volts per metre. The strength of the E1 pulse depends upon the number and intensity of the gamma rays and upon the rapidity of the gamma ray burst. Strength is also somewhat dependent upon altitude.

There are reports of "super-EMP" nuclear weapons that are able to exceed the 50,000 volt per metre limit by the nearly instantaneous release of a burst of much higher gamma radiation levels than are known to be produced by second-generation nuclear weapons. The reality and possible construction details of these weapons are classified and unconfirmed in the open scientific literature.[21]

5.2.2   E2

The E2 component is generated by scattered gamma rays and inelastic gammas produced by neutrons. This E2 component is an "intermediate time" pulse that, by the IEC definition, lasts from about 1 microsecond to 1 second after the explosion. E2 has many similarities to lightning, although lightning-induced E2 may be considerably larger than a nuclear E2. Because of the similarities and the widespread use of lightning protection technology, E2 is generally considered to be the easiest to protect against.

According to the United States EMP Commission, the main problem with E2 is the fact that it immediately follows E1, which may have damaged the devices that would normally protect against E2.

The EMP Commission Executive Report of 2004 states, "In general, it would not be an issue for critical infrastructure systems since they have existing protective measures for defense against occasional lightning strikes. The most significant risk is synergistic, because the E2 component follows a small fraction of a second after the first component's insult, which has the ability to impair or destroy many protective and control features. The energy associated with the second component thus may be allowed to pass into and damage systems."[22]

5.2.3   E3

Main article: Geomagnetically induced current

The E3 component is very different from E1 and E2. E3 is a very slow pulse, lasting tens to hundreds of seconds. It is caused by the nuclear detonation's temporary distortion of the Earth's magnetic field. The E3 component has similarities to a geomagnetic storm caused by a solar flare.[23][24] Like a geomagnetic storm, E3 can produce geomagnetically induced currents in long electrical conductors, damaging components such as power line transformers.[25]

Because of the similarity between solar-induced geomagnetic storms and nuclear E3, it has become common to refer to solar-induced geomagnetic storms as "Solar EMP."[26] "Solar EMP," however, does not include an E1 or E2 component.[27]

See also: Coronal mass ejection and Solar flare

5.3   Generation

Factors that control weapon effectiveness include altitude, yield, construction details, target distance, intervening geographical features, and local strength of the Earth's magnetic field.

5.3.1   Weapon altitude

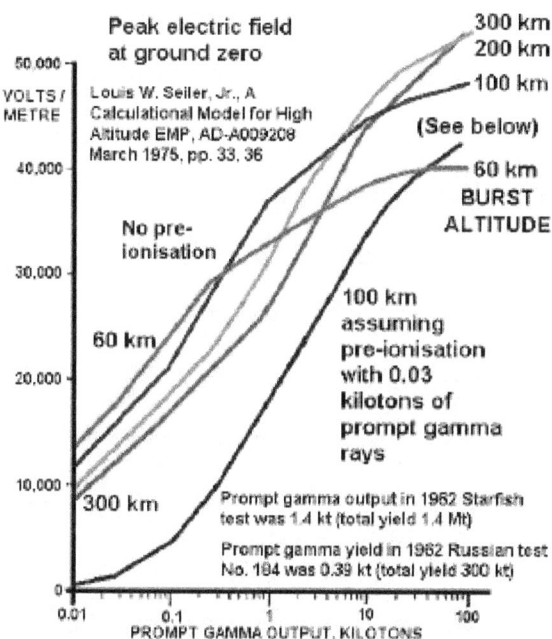

How the peak EMP on the ground varies with the weapon yield and burst altitude. The yield here is the prompt gamma ray output measured in kilotons. This varies from 0.115–0.5% of the total weapon yield, depending on weapon design. The 1.4 Mt total yield 1962 Starfish Prime test had a gamma output of 0.1%, hence 1.4 kt of prompt gamma rays. (The blue 'pre-ionisation' curve applies to certain types of thermonuclear weapons, for which gamma and x-rays from the primary fission stage ionise the atmosphere and make it electrically conductive before the main pulse from the thermonuclear stage. The pre-ionisation in some situations can literally short out part of the final EMP, by allowing a conduction current to immediately oppose the Compton current of electrons.)[28][29]

According to an internet primer published by the Federation of American Scientists[30]

A high-altitude nuclear detonation produces an immediate flux of gamma rays from the nuclear reactions within the device. These photons in turn produce high energy free electrons by Compton scattering at altitudes between (roughly) 20 and 40 km. These electrons are then trapped in the Earth's magnetic field, giving rise to an oscillating *electric current. This current is asymmetric in general and gives rise to a rapidly rising radiated electromagnetic field called an electromagnetic pulse (EMP). Because the electrons are trapped essentially simultaneously, a very large electromagnetic source radiates coherently.*

The pulse can easily span continent-sized areas, and this radiation can affect systems on land, sea, and air. ... A large device detonated at 400–500 km (250 to 312 miles) over Kansas would affect all of the continental U.S. The signal from such an event extends to the visual horizon as seen from the burst point.

Thus, for equipment to be affected, the weapon needs to be above the visual horizon.

The altitude indicated above is greater than that of the International Space Station and many low Earth orbit satellites. Large weapons could have a dramatic impact on satellite operations and communications such as occurred during Operation Fishbowl. The damaging effects on orbiting satellites are usually due to factors other than EMP. In the Starfish Prime nuclear test, most damage was to the satellites' solar panels while passing through radiation belts created by the explosion.[31]

For detonations within the atmosphere, the situation is more complex. Within the range of gamma ray deposition, simple laws no longer hold as the air is ionised and there are other EMP effects, such as a radial electric field due to the separation of Compton electrons from air molecules, together with other complex phenomena. For a surface burst, absorption of gamma rays by air would limit the range of gamma ray deposition to approximately 10 miles, while for a burst in the lower-density air at high altitudes, the range of deposition would be far greater.

5.3.2   Weapon yield

Typical nuclear weapon yields used during Cold War planning for EMP attacks were in the range of 1 to 10 megatons[32] This is roughly 50 to 500 times the size of the Hiroshima and Nagasaki bombs. Physicists have testified at United States Congressional hearings that weapons with yields of 10 kilotons or less can produce a large EMP.[33]

The EMP at a fixed distance from an explosion increases at most as the square root of the yield (see the illustration to the right). This means that although a 10 kiloton weapon has only 0.7% of the energy release of the 1.44-megaton Starfish Prime test, the EMP will be at least 8% as powerful. Since the E1 component of nuclear EMP depends on the

prompt gamma ray output, which was only 0.1% of yield in Starfish Prime but can be 0.5% of yield in low yield pure nuclear fission weapons, a 10 kiloton bomb can easily be 5 x 8% = 40% as powerful as the 1.44 megaton Starfish Prime at producing EMP.[34]

The total prompt gamma ray energy in a fission explosion is 3.5% of the yield, but in a 10 kiloton detonation the triggering explosive around the bomb core absorbs about 85% of the prompt gamma rays, so the output is only about 0.5% of the yield. In the thermonuclear Starfish Prime the fission yield was less than 100% and the thicker outer casing absorbed about 95% of the prompt gamma rays from the pusher around the fusion stage. Thermonuclear weapons are also less efficient at producing EMP because the first stage can pre-ionize the air[34] which becomes conductive and hence rapidly shorts out the Compton currents generated by the fusion stage. Hence, small pure fission weapons with thin cases are far more efficient at causing EMP than most megaton bombs.

This analysis, however, only applies to the fast E1 and E2 components of nuclear EMP. The geomagnetic storm-like E3 component of nuclear EMP is more closely proportional to the total energy yield of the weapon.[35]

5.3.3   Target distance

In nuclear EMP all of the components of the electromagnetic pulse are generated outside of the weapon.[30]

For high-altitude nuclear explosions, much of the EMP is generated far from the detonation (where the gamma radiation from the explosion hits the upper atmosphere). This electric field from the EMP is remarkably uniform over the large area affected.

According to the standard reference text on nuclear weapons effects published by the U.S. Department of Defense, "The peak electric field (and its amplitude) at the Earth's surface from a high-altitude burst will depend upon the explosion yield, the height of the burst, the location of the observer, and the orientation with respect to the geomagnetic field. As a general rule, however, the field strength may be expected to be tens of kilovolts per metre over most of the area receiving the EMP radiation." [36]

The text also states that, "... over most of the area affected by the EMP the electric field strength on the ground would exceed $0.5E_{max}$. For yields of less than a few hundred kilotons, this would not necessarily be true because the field strength at the Earth's tangent could be substantially less than $0.5E_{max}$." [36]

(E_{max} refers to the maximum electric field strength in the affected area.)

In other words, the electric field strength in the entire area that is affected by the EMP will be fairly uniform for weapons with a large gamma ray output. For smaller weapons, the electric field may fall at a faster rate as distance increases.

5.4 Effects

5.4.1 On aircraft

Many nuclear detonations have taken place using aerial bombs. The B-29 aircraft that delivered the nuclear weapons at Hiroshima and Nagasaki did not lose power from electrical damage, because electrons (ejected from the air by gamma rays) are stopped quickly in normal air for bursts below roughly 10 kilometres (6.2 mi), so they are not significantly deflected by the Earth's magnetic field.[37]

If the aircraft carrying the Hiroshima and Nagasaki bombs had been within the intense nuclear radiation zone when the bombs exploded over those cities, then they would have suffered effects from the charge separation (radial) EMP. But this only occurs within the severe blast radius for detonations below about 10 km altitude.

During Operation Fishbowl, EMP disruptions were suffered aboard a KC-135 photographic aircraft flying 300 km (190 mi) from the 410 kt (1,700 TJ) detonations at 48 and 95 km (30 and 59 mi) burst altitudes.[34] The vital electronics were less sophisticated than today's and the aircraft was able to land safely.

5.4.2 Vacuum tube versus solid state electronics

Older, vacuum tube (valve) based equipment is generally much less vulnerable to nuclear EMP than newer solid state equipment. Soviet Cold War–era military aircraft often had avionics based on vacuum tubes because solid-state capabilities were limited and vacuum-tube gear was believed to be more likely to survive.[1]

Other components in vacuum tube circuitry can be damaged by EMP. Vacuum tube equipment was damaged in the 1962 testing.[16] The solid state PRC-77 VHF manpackable 2-way radio survived extensive EMP testing.[38] The earlier PRC-25, nearly identical except for a vacuum tube final amplification stage, was tested in EMP simulators, but was not certified to remain fully functional.

5.5 Post–Cold War attack scenarios

The United States military services developed, and in some cases published, hypothetical EMP attack scenarios.[39]

The United States EMP Commission was created by the United States Congress in 2001. The commission is formally known as the Commission to Assess the Threat to the United States from Electromagnetic Pulse (EMP) Attack.[40]

The Commission brought together notable scientists and technologists to compile several reports. In 2008, the EMP Commission released the "Critical National Infrastructures Report".[35] This report describes the likely consequences of a nuclear EMP on civilian infrastructure. Although this report covered the United States, most of the information can be generalized to other industrialized countries. The 2008 report was a followup to a more generalized report issued by the commission in 2004.[24][41]

In written testimony delivered to the United States Senate in 2005, an EMP Commission staff member reported:

> The EMP Commission sponsored a worldwide survey of foreign scientific and military literature to evaluate the knowledge, and possibly the intentions, of foreign states with respect to electromagnetic pulse (EMP) attack. The survey found that the physics of EMP phenomenon and the military potential of EMP attack are widely understood in the international community, as reflected in official and unofficial writings and statements. The survey of open sources over the past decade finds that knowledge about EMP and EMP attack is evidenced in at least Britain, France, Germany, Israel, Egypt, Taiwan, Sweden, Cuba, India, Pakistan, Iraq under Saddam Hussein, Iran, North Korea, China and Russia.
>
> Many foreign analysts – particularly in Iran, North Korea, China, and Russia – view the United States as a potential aggressor that would be willing to use its entire panoply of weapons, including nuclear weapons, in a first strike. They perceive the United States as having contingency plans to make a nuclear EMP attack, and as being willing to execute those plans under a broad range of circumstances.
>
> Russian and Chinese military scientists in open source writings describe the basic principles of nuclear weapons designed specifically to generate an enhanced-EMP effect, that they term "Super-EMP" weapons. "Super-EMP" weapons, according to these foreign open source writings, can destroy even the best protected U.S. military and civilian electronic systems.[21]

The United States EMP Commission determined that long-known protections are almost completely absent in the civilian infrastructure of the United States and that large parts of US military services were less-protected against EMP than during the Cold War. In public statements, the EMP experts on the EMP Commission recommended making electronic equipment and electrical components resistant to EMP – and maintaining spare parts inventories that would enable prompt repairs.[24][35][42] The United States EMP Commission did not look at the civilian infrastructures of other nations.

In 2011 the Defense Science Board published a report about the ongoing efforts to defend critical military and civilian systems against EMP and other nuclear weapons effects.[43]

5.6   Common misconceptions

A 2010 technical report written for the US government's Oak Ridge National Laboratory included a brief section addressing common EMP myths.[44] The remainder of this section is a direct quotation from that Oak Ridge report regarding common HEMP Myths:

> Much of the literature on HEMP is either classified or not easily accessible. Probably because of this, some of what is openly available tends to vary in accuracy – some, especially from the Internet, has major inaccuracies. Some discussions of HEMP have the right words and concepts, but do not quite have them put together right, or have inaccurate interpretations. Here we will discuss some common misunderstandings. HEMP has also appeared in some movies, and there are on-line discussions about possible errors in their depiction of HEMP. Here we will be concerned with E1 HEMP, and ignore misunderstandings about other types of EMP.
>
> **Extremists:** Some general emphasis of comments fall into either "the world as we know it will come to an end" if there is a high altitude nuclear burst, or the other extreme: "it's not a big deal, nothing much will happen". Since we really have never had a nuclear burst over anything like our current modern infrastructure, no one

really knows for sure what would happen, but both extremes are not very believable.

> **Yield:** There appears to be an assumption that yield is important – it is not for E1. The assumption that E1 is an issue only for cold war type situations, but not for terrorists or rogue nations, is false. Very big bombs might have better area coverage of high fields by going to higher burst heights, but for peak fields the burst yield is only a very minor consideration.
>
> **1962 experience:** Some point to the Starfish event, and the rather minor HEMP effects produced at Hawaii by it. However, there are many problems with extrapolating that experience:
>
> 1. That was about half a century ago. Since then, the use of electronics has increased greatly, and the type of sensitive electronics we currently use did not really exist back then.
> 2. The burst was fairly far away from Hawaii, and the incident E1 HEMP was much less than worst case.
> 3. The island is small – if over the continental U.S., long transmission lines would be exposed (especially an issue for late-time HEMP). In addition, widely separated substations would have been exposed, although with electromechanical relays (not solid state). Also the yield argument has been used – Starfish was a very big weapon, yet it did very little – see the previous item, yield is not really very significant.
>
> **Cars dying:** Some say that all vehicles traveling will come to a halt, with all modern vehicles damaged because of their use of modern electronics (and one movie even had a bulk, non-electronic part dying). Most likely there will be some vehicles affected, but probably just a small fraction of them (although this could create traffic jams in large cities). A car does not

have very long cabling to act as antennas, and there is some protection from metallic construction. As non-metallic materials are used more and more in the future to decrease weight and increase fuel efficiency, this advantage may disappear.

Wristwatch dying: One movie critic pointed out that electronics in a helicopter were affected, but not the star's electronic watch. A watch is much too small for HEMP to affect it.

Electrons present: One critic, with some awareness of the generation process, said that HEMP could not be present unless there were also energetic electrons present. This is true when one is within the source region, which exists for all types of EMP – there are energetic electrons present. However for the HEMP, the radiation and energetic electrons are present at altitudes of 20 to 40 km, not at the ground.

Turn equipment off: There is truth to this recommendation (if there were a way to know that a burst was about to happen). Equipment is more vulnerable if it is operating, because some failure modes involving E1 HEMP trigger the system's energy to damage itself. However, damage can also happen, but not as easily, to systems that are turned off.

Maximum conductor length: There is a suggestion that equipment will be OK if all connected conductors are less than a specific length. Certainly shorter lengths are generally better, but there is no magic length value, with shorter always being better and longer not. Coupling is much too complex for such a blanket statement – instead it should be "the shorter the better, in general". (There can be exceptions, such as resonance effects, which depend on line lengths.)

Stay away from metal: There is

a recommendation to be some distance away from any metal when a HEMP event occurs (assuming there was warning), because very high voltages could be generated. Metal can collect E1 HEMP energy, and easily generate high voltages. However, the "skin effect" (a term not really derived from the skin of humans or any other animal) means that if a human were touching a large "antenna" during an E1 HEMP event, any current flow would not penetrate into the body. Generally E1 HEMP is considered harmless for human bodies.

5.7 Protecting infrastructure

In 2013, the U.S. House of Representatives considered the "Secure High-voltage Infrastructure for Electricity from Lethal Damage Act" that would allow the Federal Energy Regulatory Commission to order emergency measures to provide surge protection for some 300 large transformers around the country.[45] The bill was introduced and referred to committee, but proceeded no further.[46]

The problem of protecting civilian infrastructure from electromagnetic pulse has also been intensively studied throughout the European Union, and in particular by the United Kingdom.[47][48]

5.8 In fiction and popular culture

Main article: Electromagnetic pulse in fiction and popular culture

Especially since the 1980s, Nuclear EMP weapons have gained a significant presence in fiction and popular culture.

The popular media often depict EMP effects incorrectly, causing misunderstandings among the public and even professionals, and official efforts have been made in the United States to set the record straight.[44] See, for example, the Oak Ridge quotation in the above section of this article on "Common Misconceptions." Also, the United States Space Command commissioned science educator Bill Nye to produce a video called "Hollywood vs. EMP" so that Hollywood fiction would not confuse those who must deal with real EMP events.[49] The U.S. Space Command video is not available to the general public.

5.9   See also

- Electromagnetic compatibility (EMC)

- Electromagnetic environment

- Electromagnetic hypersensitivity

- Electromagnetic pulse in fiction and popular culture

- Electromagnetic weapon

- Electromagnetism

- Electronic warfare

- Explosively pumped flux compression generator

- Faraday's law of induction

- Gamma ray burst

- Geomagnetic storm

- High-altitude nuclear explosion

- High-power microwave

- Marx generator

- Operation Fishbowl

- Pulsed power

- Soviet Project K nuclear tests

- Starfish Prime

- Ultrashort pulse

5.10   References

[1] Broad, William J. "Nuclear Pulse (I): Awakening to the Chaos Factor", *Science*. 29 May 1981 212: 1009–1012

[2] Bainbridge, K.T., (Report LA-6300-H), Los Alamos Scientific Laboratory. May 1976. p. 53 *Trinity*

[3] Baum, Carl E., *IEEE Transactions on Electromagnetic Compatibility*. Vol. 49, No. 2, pp. 211–218. May 2007. *Reminiscences of High-Power Electromagnetics*

[4] Baum, Carl E., *Proceedings of the IEEE*, Vol. 80, No. 6, pp. 789–817. June 1992 "From the Electromagnetic Pulse to High-Power Electromagnetics"

[5] Defense Atomic Support Agency. 23 September 1959. "Operation Hardtack Preliminary Report. Technical Summary of Military Effects. Report ADA369152". pp. 346–350.

[6] Broad, William J. "Nuclear Pulse (II): Ensuring Delivery of the Doomsday Signal", *Science*. 5 June 1981 212: 1116–1120

[7] Broad, William J. "Nuclear Pulse (III): Playing a Wild Card", *Science*. 12 June 1981 212: 1248–1251

[8] Vittitoe, Charles N., "Did High-Altitude EMP Cause the Hawaiian Streetlight Incident?" Sandia National Laboratories. June 1989.

[9] Longmire, Conrad L., *NBC Report*, Fall/Winter. 2004. pp. 47–51. U.S. Army Nuclear and Chemical Agency "Fifty Odd Years of EMP"

[10] Theoretical Notes – Note 353. March 1985. "EMP on Honolulu from the Starfish Event" Conrad L. Longmire – Mission Research Corporation

[11] Rabinowitz, Mario (1987) "Effect of the Fast Nuclear Electromagnetic Pulse on the Electric Power Grid Nationwide: A Different View". IEEE Trans. Power Delivery, PWRD-2. 1199–1222 arXiv:physics/0307127

[12] Zak, Anatoly "The K Project: Soviet Nuclear Tests in Space", *The Nonproliferation Review*, Volume 13, Issue. 1 March 2006. pp. 143–150

[13] Subject: US-Russian meeting – HEMP effects on national power grid & telecommunications From: Howard Seguine, 17 Feb. 1995 Memorandum for Record

[14] Pfeffer, Robert and Shaeffer, D. Lynn. Combating WMD Journal. (2009) Issue 3. pp. 33–38. "A Russian Assessment of Several USSR and US HEMP Tests"

[15] Greetsai, Vasily N., et al. IEEE Transactions on Electromagnetic Compatibility. Vol. 40, No. 4, November 1998. "Response of Long Lines to Nuclear High-Altitude Electromagnetic Pulse (HEMP)"

[16] Loborev, Vladimir M. "Up to Date State of the NEMP Problems and Topical Research Directions", Electromagnetic Environments and Consequences: Proceedings of the EUROEM 94 International Symposium, Bordeaux, France, 30 May – 3 June 1994. pp. 15–21

[17] Metatech Corporation (January 2010). *The Early-Time (E1) High-Altitude Electromagnetic Pulse (HEMP) and Its Impact on the U.S. Power Grid." Section 3 - E1 HEMP History* (PDF). Report Meta-R-320. Oak Ridge National Laboratory.

[18] Electromagnetic compatibility (EMC), Part 2: Environment. Section 9: Description of HEMP environment – Radiated disturbance. Basic EMC publication. IEC 61000-2-9

[19] U.S. Army White Sands Missile Range. *Nuclear Environment Survivability*. Report ADA278230. p. D-7. 15 April 1994.

[20] Longmire. Conrad L. LLNL-9323905. Lawrence Livermore National Laboratory. June 1986 "Justification and Verification of High-Altitude EMP Theory. Part 1" (Retrieved 2010-15-12)

[21] March 8, 2005 "Statement. Dr. Peter Vincent Pry. EMP Commission Staff, before the United States Senate Subcommittee on Terrorism, Technology and Homeland Security"

[22] Report of the Commission to Assess the Threat to the United States from Electromagnetic Pulse (EMP) Attack. Volume 1. Executive Report. 2004. p. 6.

[23] High-Altitude Electromagnetic Pulse (HEMP): A Threat to Our Way of Life, 09.07. By William A. Radasky. Ph.D., P.E. - IEEE

[24] Report of the Commission to Assess the Threat to the United States from Electromagnetic Pulse (EMP) Attack

[25] Report Meta-R-321: "The Late-Time (E3) High-Altitude Electromagnetic Pulse (HEMP) and Its Impact on the U.S. Power Grid" January 2010. Written by Metatech Corporation for Oak Ridge National Laboratory.

[26] "EMPACT America, Inc. - Solar EMP". Web.archive.org. 2011-07-26. Archived from the original on July 26, 2011. Retrieved 2013-05-21.

[27] "E3 - ProtecTgrid". *ProtecTgrid*. Retrieved 2017-02-16.

[28] Louis W. Seiler, Jr. *A Calculational Model for High Altitude EMP*. Air Force Institute of Technology. Report ADA009208. pp. 33, 36. March 1975

[29] Glasstone. Samuel and Dolan, Philip J., *The Effects of Nuclear Weapons*. Chapter 11. 1977. United States Department of Defense.

[30] Federation of American Scientists. "Nuclear Weapon EMP Effects"

[31] Hess, Wilmot N. (September 1964). "The Effects of High Altitude Explosions" (PDF). National Aeronautics and Space Administration. NASA TN D-2402. Retrieved 2015-05-13.

[32] U.S. Congressional hearing Transcript H.S.N.C No. 105–18, p. 39

[33] U.S. Congressional hearing Transcript H.A.S.C. No. 106–31, p. 48

[34] Glasstone, Samuel (March 29, 2006). "EMP radiation from nuclear space bursts in 1962". Subsequent tests with lower yield devices [410 kt *Kingfish* at 95 km altitude. 410 kt *Bluegill* at 48 km altitude, and 7 kt *Checkmate* at 147 km] produced electronic upsets on an instrumentation aircraft [presumably the KC-135 that filmed the tests from above the clouds?] that was approximately 300 kilometers away from the detonations.

[35] EMP Commission Critical National Infrastructures Report

[36] Glasstone & Dolan 1977, Chapter 11, section 11.73.

[37] Glasstone & Dolan 1977, Chapter 11, section 11.09.

[38] Seregelyi, J.S. et al. Report ADA266412 "EMP Hardening Investigation of the PRC-77 Radio Set" Retrieved 2009-25-11

[39] Miller, Colin R., Major, USAF "Electromagnetic Pulse Threats in 2010" Air War College, Air University, United States Air Force, November 2005

[40] Commission to Assess the Threat to the United States from Electromagnetic Pulse (EMP) Attack

[41] "Report of the Commission to Assess the Threat to the United States from Electromagnetic Pulse (EMP) Attack" Volume 1: Executive Report 2004

[42] Ross, Lenard H., Jr. and Mihelic, F. Matthew, "Healthcare Vulnerabilities to Electromagnetic Pulse" American Journal of Disaster Medicine. Vol. 3, No. 6, pp. 321–325. November/December 2008.

[43] "Survivability of Systems and Assets to Electromagnetic Pulse (EMP)"

[44] Report Meta-R-320: "The Early-Time (E1) High-Altitude Electromagnetic Pulse (HEMP) and Its Impact on the U.S. Power Grid" January 2010. Written by Metatech Corporation for Oak Ridge National Laboratory. Appendix: E1 HEMP Myths

[45] McCormack, John. "Lights out: House plan would protect nation's electricity from solar flare, nuclear bomb". WashingtonExaminer.com. Retrieved 2013-06-18.

[46] U.S., Congress, All Bill Information (Except Text) for H.R.2417 - Secure High-voltage Infrastructure for Electricity from Lethal Damage Act. Accessed 2016.08.28.

[47] House of Commons Defence Committee, "Developing Threats: Electro-Magnetic Pulses (EMP)" Tenth Report of Session 2010–12.

[48] Extreme Electromagnetics – The Triple Threat to Infrastructure. 14 January 2013 (Proceedings of a seminar)

[49] 2009 Telly Award Winners. (Manitou Motion Picture Company, Ltd.)

- This article incorporates public domain material from the General Services Administration document "Federal Standard 1037C" (in support of MIL-STD-188).

- Vladimir Gurevich "Cyber and Electromagnetic Threaths in Modern Relay Protection" - CRC Press (Taylor & Francis Group), Boca Raton – New York – London, 2014. 222 p.

- Vladimir Gurevich "Protection of Substation Critical Equipment Against Intentional Electromagnetic Threats" - Wiley, London, 2016. 300 p.

5.11   Further reading

- ISBN 978-1-59-248389-1 A 21st Century Complete Guide to Electromagnetic Pulse (EMP) Attack Threats, Report of the Commission to Assess the Threat to the United States from Electromagnetic ... High-Altitude Nuclear Weapon EMP Attacks (CD-ROM)

- ISBN 978-0-16-056127-6 Threat posed by electromagnetic pulse (EMP) to U.S. military systems and civil infrastructure: Hearing before the Military Research and Development Subcommittee - first session, hearing held July 16, 1997 (Unknown Binding)

- ISBN 978-0-471-01403-4 Electromagnetic Pulse Radiation and Protective Techniques

- ISBN 978-0-16-080927-9 Report of the Commission to Assess the Threat to the United States from Electromagnetic Pulse (EMP) Attack

5.12   External links

- Glasstone, Samuel; Dolan, Philip J. (1977). "The Effects of Nuclear Weapons". United States Department of Defense.

- GlobalSecurity.org – Electromagnetic Pulse: From chaos to a manageable solution

- Electromagnetic Pulse (EMP) and Tempest Protection for Facilities – U.S. Army Corps of Engineers

- EMP data from *Starfish* nuclear test measured by Richard Wakefield of LANL, and review of evidence pertaining to the effects 1,300 km away in Hawaii, also review of Russian EMP tests of 1962

- Read Congressional Research Service (CRS) Reports regarding HEMP

- MIL-STD-188-125-1

- Electromagnetic Pulse Risks & Terrorism

- How E-Bombs Work

- Commission to Assess the Threat to the United States from Electromagnetic Pulse (EMP) Attack

- NEMP and Nuclear plant

Chapter 6

Nuclear fallout

"Fallout" redirects here. For the video game series, see Fallout (series). For the first video game in the Fallout series, see Fallout (video game). For other uses, see Fallout (disambiguation).

Nuclear fallout, or simply **fallout**, is the residual radioactive material propelled into the upper atmosphere following a nuclear blast or a nuclear reaction conducted in an unshielded facility, so called because it "falls out" of the sky after the explosion and the shock wave have passed.*[1] It commonly refers to the radioactive dust and ash created when a nuclear weapon explodes, but such dust can also originate from a damaged nuclear plant. Fallout may get entrained with the products of a pyrocumulus cloud and fall as black rain*[2] (rain darkened by soot and other particulates).

This radioactive dust, usually consisting of fission products mixed with bystanding atoms that are neutron activated by exposure, is a highly dangerous kind of radioactive contamination.

6.1 Types of fallout

An air burst (that is, a nuclear detonation far above the surface) can eventually produce worldwide fallout. A ground burst can produce possibly much more severe, local fallout.

6.1.1 Global fallout

After an air burst, fission products, un-fissioned nuclear material, and weapon residues vaporized by the heat of the fireball condense into a fine suspension of small particles 10 nm to 20 μm in diameter. These particles may be quickly drawn up into the stratosphere, particularly if the explosive yield exceeds 10 kt.

Initially little was known about the dispersion of nuclear fallout on a global scale. The AEC assumed that fallout would be dispersed evenly across the globe by atmospheric

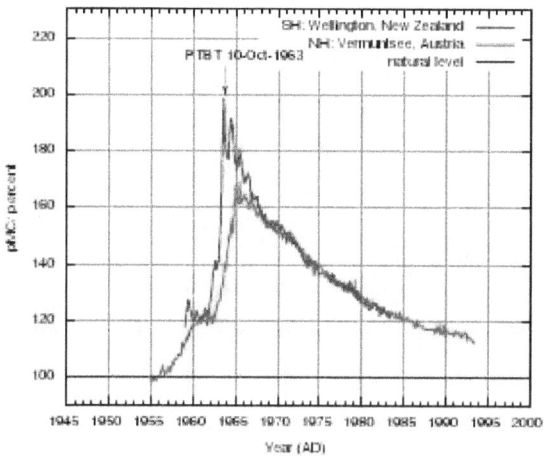

Atmospheric nuclear weapon tests almost doubled the concentration of radioactive ^{14}C in the Northern Hemisphere, before levels slowly declined following the Partial Test Ban Treaty.

winds and gradually settle to the Earth's surface after weeks, months, and even years as worldwide fallout.

The radio-biological hazard of worldwide fallout is a long-term one because of the potential accumulation of long-lived radioisotopes (such as strontium-90 and caesium-137) in the body as a result of ingestion of foods containing the radioactive materials. This hazard is less pertinent than local fallout, which is of much greater immediate operational concern.

6.1.2 Local fallout

In a land or water surface burst, heat vaporizes large amounts of earth or water, which is drawn up into the radioactive cloud. This material becomes radioactive when it condenses with fission products and other radiocontaminants that have become neutron-activated. The table below summarizes the abilities of common isotopes to form fallout. Some radiation taints large amounts of land and drinking water causing formal mutations throughout animal and human life.

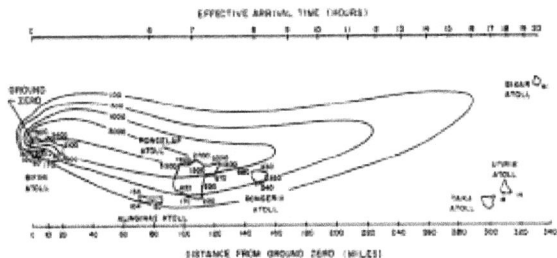

The 450 km (280 mi) fallout plume from 15 Mt shot Castle Bravo, 1954

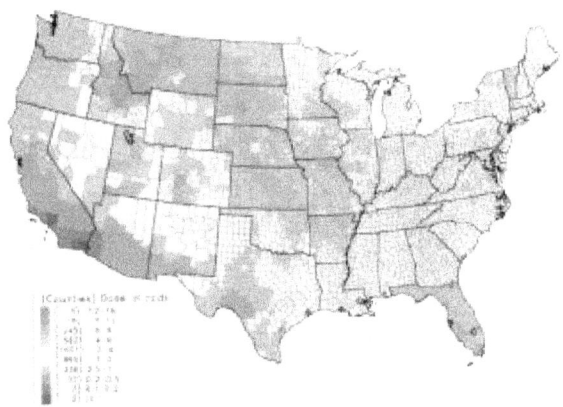

Per capita thyroid doses in the continental United States resulting from all exposure routes from all atmospheric nuclear tests conducted at the Nevada Test Site from 1951-1962.

A surface burst generates large amounts of particulate matter, composed of particles from less than 100 nm to several millimeters in diameter—in addition to very fine particles that contribute to worldwide fallout. The larger particles spill out of the stem and cascade down the outside of the fireball in a downdraft even as the cloud rises, so fallout begins to arrive near ground zero within an hour. More than half the total bomb debris lands on the ground within about 24 hours as local fallout. Chemical properties of the elements in the fallout control the rate at which they are deposited on the ground. Less volatile elements deposit first.

Severe local fallout contamination can extend far beyond the blast and thermal effects, particularly in the case of high yield surface detonations. The ground track of fallout from an explosion depends on the weather from the time of detonation onwards. In stronger winds, fallout travels faster but takes the same time to descend, so although it covers a larger path, it is more spread out or diluted. Thus, the width of the fallout pattern for any given dose rate is reduced where the downwind distance is increased by higher winds. The total amount of activity deposited up to any given time is the same irrespective of the wind pattern, so

overall casualty figures from fallout are generally independent of winds. But thunderstorms can bring down activity as rain more rapidly than dry fallout, particularly if the mushroom cloud is low enough to be below (**"washout"**), or mixed with (**"rainout"**), the thunderstorm.

Whenever individuals remain in a radiologically contaminated area, such contamination leads to an immediate external radiation exposure as well as a possible later internal hazard from inhalation and ingestion of radiocontaminants, such as the rather short-lived iodine-131, which is accumulated in the thyroid.

6.2 Factors affecting fallout

6.2.1 Location

There are two main considerations for the location of an explosion: height and surface composition. A nuclear weapon detonated in the air, called an air burst, produces less fallout than a comparable explosion near the ground.

In case of water surface bursts, the particles tend to be rather lighter and smaller, producing less local fallout but extending over a greater area. The particles contain mostly sea salts with some water; these can have a cloud seeding effect causing local rainout and areas of high local fallout. Fallout from a seawater burst is difficult to remove once it has soaked into porous surfaces because the fission products are present as metallic ions that chemically bond to many surfaces. Water and detergent washing effectively removes less than 50% of this chemically bonded activity from concrete or steel. Complete decontamination requires aggressive treatment like sandblasting, or acidic treatment. After the *Crossroads* underwater test, it was found that wet fallout must be immediately removed from ships by continuous water washdown (such as from the fire sprinkler system on the decks).

Parts of the sea bottom may become fallout. After the Castle Bravo test, white dust—contaminated calcium oxide particles originating from pulverized and calcined corals—fell for several hours, causing beta burns and radiation exposure to the inhabitants of the nearby atolls and the crew of the Daigo Fukuryū Maru fishing boat. The scientists called the fallout **Bikini snow**.

For subsurface bursts, there is an additional phenomenon present called "base surge". The base surge is a cloud that rolls outward from the bottom of the subsiding column, which is caused by an excessive density of dust or water droplets in the air. For underwater bursts, the visible surge is, in effect, a cloud of liquid (usually water) droplets with the property of flowing almost as if it were a homogeneous fluid. After the water evaporates, an invisible base surge of small radioactive particles may persist.

For subsurface land bursts, the surge is made up of small solid particles, but it still behaves like a fluid. A soil earth medium favors base surge formation in an underground burst. Although the base surge typically contains only about 10% of the total bomb debris in a subsurface burst, it can create larger radiation doses than fallout near the detonation, because it arrives sooner than fallout, before much radioactive decay has occurred.

6.2.2 Meteorological

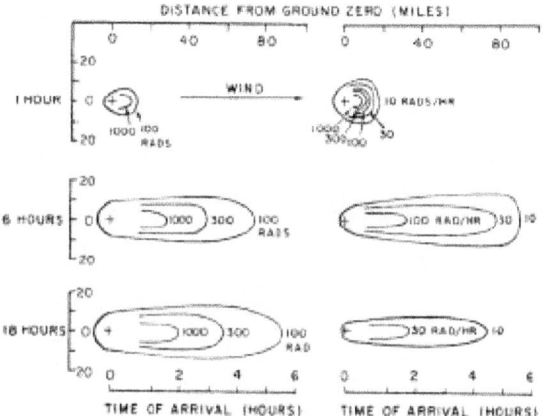

Comparison of fallout gamma dose and dose rate contours for a 1 Mt fission land surface burst, based on DELFIC calculations. Because of radioactive decay, the dose rate contours contract after fallout has arrived, but dose contours continue to grow

Meteorological conditions greatly influence fallout, particularly local fallout. Atmospheric winds are able to bring fallout over large areas. For example, as a result of a *Castle Bravo* surface burst of a 15 Mt thermonuclear device at Bikini Atoll on March 1, 1954, a roughly cigar-shaped area of the Pacific extending over 500 km downwind and varying in width to a maximum of 100 km was severely contaminated. There are three very different versions of the fallout pattern from this test, because the fallout was only measured on a small number of widely spaced Pacific Atolls. The two alternative versions both ascribe the high radiation levels at north Rongelap to a downwind hotspot caused by the large amount of radioactivity carried on fallout particles of about 50-100 micrometres size.[3]

After *Bravo*, it was discovered that fallout landing on the ocean disperses in the top water layer (above the thermocline at 100 m depth), and the land equivalent dose rate can be calculated by multiplying the ocean dose rate at two days after burst by a factor of about 530. In other 1954 tests, including *Yankee* and *Nectar*, hotspots were mapped out by ships with submersible probes, and similar hotspots

occurred in 1956 tests such as *Zuni* and *Tewa*. [4] However, the major U.S. 'DELFIC' (Defence Land Fallout Interpretive Code) computer calculations use the natural size distributions of particles in soil instead of the afterwind sweep-up spectrum, and this results in more straightforward fallout patterns lacking the downwind hotspot.

Snow and rain, especially if they come from considerable heights, accelerate local fallout. Under special meteorological conditions, such as a local rain shower that originates above the radioactive cloud, limited areas of heavy contamination just downwind of a nuclear blast may be formed.

6.3 Effects

A wide range of biological changes may follow the irradiation of animals. These vary from rapid death following high doses of penetrating whole-body radiation, to essentially normal lives for a variable period of time until the development of delayed radiation effects, in a portion of the exposed population, following low dose exposures.

The unit of actual *exposure* is the röntgen, defined in ionisations per unit volume of air. All ionisation based instruments (including geiger counters and ionisation chambers) measure exposure. However, effects depend on the energy per unit mass, not the exposure measured in air. A deposit of 1 joule per kilogram has the unit of 1 gray (Gy). For 1 MeV energy gamma rays, an exposure of 1 röntgen in air produces a dose of about 0.01 gray (1 centigray, cGy) in water or surface tissue. Because of shielding by the tissue surrounding the bones, the bone marrow only receives about 0.67 cGy when the air exposure is 1 röntgen and the surface skin dose is 1 cGy. Some lower values reported for the amount of radiation that would kill 50% of personnel (the LD_{50}) refer to bone marrow dose, which is only 67% of the air dose.

6.3.1 Short term

Further information: LD_{50}

The dose that would be lethal to 50% of a population is a common parameter used to compare the effects of various fallout types or circumstances. Usually, the term is defined for a specific time, and limited to studies of acute lethality. The common time periods used are 30 days or less for most small laboratory animals and to 60 days for large animals and humans. The LD_{50} figure assumes that the individuals did not receive other injuries or medical treatment.

In the 1950s, the LD_{50} for gamma rays was set at 3.5 Gy, while under more dire conditions of war (a bad diet, little medical care, poor nursing) the LD_{50} was 2.5 Gy (250

Fallout shelter sign on a building in New York City.

rad). There have been few documented cases of survival beyond 6 Gy. One person at Chernobyl survived a dose of more than 10 Gy, but many of the persons exposed there were not uniformly exposed over their entire body. If a person is exposed in a non-homogeneous manner then a given dose (averaged over the entire body) is less likely to be lethal. For instance, if a person gets a hand/low arm dose of 100 Gy, which gives them an overall dose of 4 Gy, they are more likely to survive than a person who gets a 4 Gy dose over their entire body. A hand dose of 10 Gy or more would likely result in loss of the hand. A British industrial radiographer who was estimated to have received a hand dose of 100 Gy over the course of his lifetime lost his hand because of radiation dermatitis.[5] Most people become ill after an exposure to 1 Gy or more. The fetuses of pregnant women are often more vulnerable to radiation and may miscarry, especially in the first trimester.

One hour after a surface burst, the radiation from fallout in the crater region is 30 grays per hour (Gy/h). Civilian dose rates in peacetime range from 30 to 100 µGy per year.

Fallout radiation decays exponentially relatively quickly with time. Most areas become fairly safe for travel and decontamination after three to five weeks.

For yields of up to 10 kt, prompt radiation is the dominant producer of casualties on the battlefield. Humans receiving an acute incapacitating dose (30 Gy) have their performance degraded almost immediately and become ineffective within several hours. However, they do not die until five to six days after exposure, assuming they do not receive any other injuries. Individuals receiving less than a total of 1.5 Gy are not incapacitated. People receiving doses greater than 1.5 Gy become disabled, and some eventually die.

A dose of 5.3 Gy to 8.3 Gy is considered lethal but not immediately incapacitating. Personnel exposed to this amount of radiation have their performance degraded in two to three hours, depending on how physically demanding the tasks they must perform are, and remain in this disabled state at least two days. However, at that point they experience a recovery period and can perform non-demanding tasks for about six days, after which they relapse for about four weeks. At this time they begin exhibiting symptoms of radiation poisoning of sufficient severity to render them totally ineffective. Death follows at approximately six weeks after exposure, although outcomes may vary.

6.3.2   Long term

See also: Project GABRIEL, Bellesrad, Enewetak Atoll § History, Bikini Atoll § Current habitable state, and Project 4.1

Late or delayed effects of radiation occur following a wide

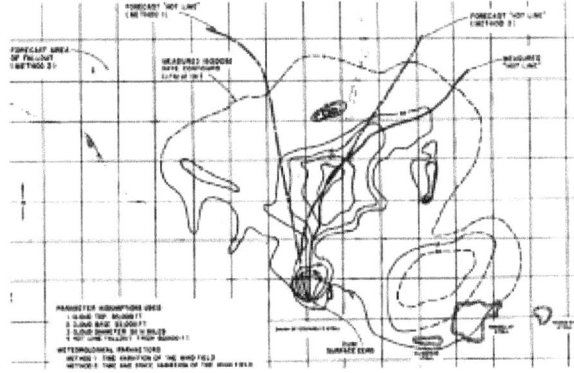

Comparison of predicted fallout "hotline" with test results in the 3.53 Mt 15% fission Zuni test at Bikini in 1956. The predictions were made under simulated tactical nuclear war conditions aboard ship by Edward A. Schuert.

range of doses and dose rates. Delayed effects may appear months to years after irradiation and include a wide variety of effects involving almost all tissues or organs. Some of the *possible* delayed consequences of radiation injury are life shortening, carcinogenesis, cataract formation, chronic radiodermatitis, decreased fertility, and genetic mutations.[6]

Presently, the only teratological effect observed in humans following nuclear attacks on highly populated areas is microcephaly which is the only proven malformation, or congenital abnormality, found in the in utero developing human fetuses present during the Hiroshima and Nagasaki bombings. Of all the pregnant women who were close enough to be exposed to the *prompt* burst of intense neutron and gamma doses in the two cities, the total number of children born with microcephaly was below 50.[7] No statistically demonstrable increase of congenital malformations was found among the *later conceived children* born to survivors of the nuclear detonations at Hiroshima and Na-

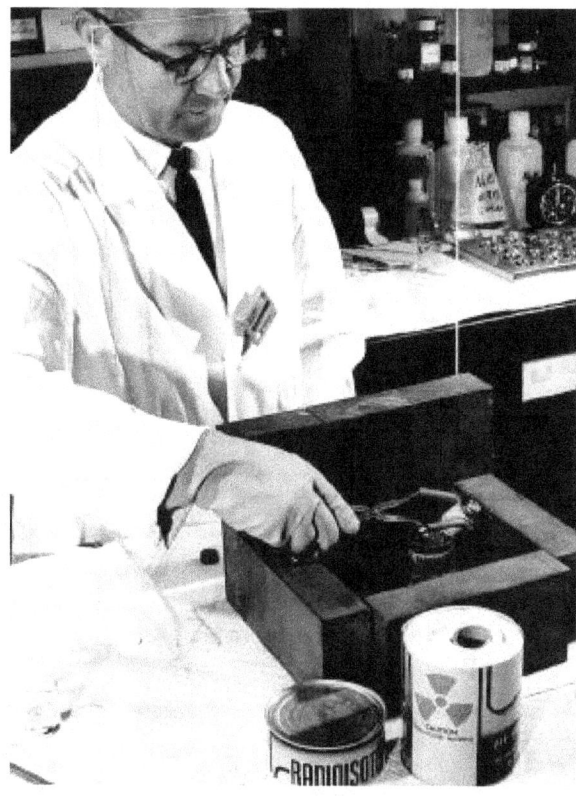

Following the detonation of the first atomic bomb, pre-war steel and post-war steel which is manufactured without atmospheric air, became a valuable commodity for scientists wishing to make instruments that detect radioactive emissions, since these two types of steel are the only steels that do not contain trace amounts of fallout.

gasaki.[7][8][9] The surviving women of Hiroshima and Nagasaki who could conceive and were exposed to substantial amounts of radiation went on and had children with no higher incidence of abnormalities than the Japanese average.[10][11]

The Baby Tooth Survey founded by the husband and wife team of physicians Eric Reiss and Louise Reiss, was a research effort focused on detecting the presence of strontium-90, a cancer-causing radioactive isotope created by the more than 400 atomic tests conducted above ground that is absorbed from water and dairy products into the bones and teeth given its chemical similarity to calcium. The team sent collection forms to schools in the St. Louis, Missouri area, hoping to gather 50,000 teeth each year. Ultimately, the project collected over 300,000 teeth from children of various ages before the project was ended in 1970.[12]

Preliminary results of the Baby Tooth Survey were published in the November 24, 1961, edition of the journal *Science*, and showed that levels of strontium 90 had risen steadily in children born in the 1950s, with those born later showing the most pronounced increases.[13] The results of

a more comprehensive study of the elements found in the teeth collected showed that children born after 1963 had levels of strontium 90 in their baby teeth that was 50 times higher than that found in children born before large-scale atomic testing began. The findings helped convince U.S. President John F. Kennedy to sign the Partial Nuclear Test Ban Treaty with the United Kingdom and Soviet Union, which ended the above-ground nuclear weapons testing that created the greatest amounts of atmospheric nuclear fallout.[14]

The baby tooth survey was a "campaign [that] effectively employed a variety of media advocacy strategies" to alarm the public and "galvanized" support against atmospheric nuclear testing,[15] with putting an end to such testing being commonly viewed as a positive outcome for a myriad of other reasons. The survey could not show then at the time, nor in the decades that have elapsed, that the levels of global strontium-90 or fallout in general, were in any way life threatening, primarily because "50 times the strontium-90 from *before* nuclear testing" is a minuscule number, and multiplication of minuscule numbers results in only a slightly larger minuscule number. Moreover, the Radiation and Public Health Project which currently retains the teeth has had their stance and publications heavily criticized: A 2003 article in *The New York Times* states that the group's work has been controversial and has little credibility with the scientific establishment.[16] Similarly, in an April 2014 article in *Popular Science*, Sarah Fecht explains that the group's work, specifically the widely discussed case of cherry-picking data to suggest that fallout from the 2011 Fukushima accident caused infant deaths in America, is "junk science", as despite their papers being peer-reviewed, all attempts to independently corroborate their results return findings that are not in agreement with what the organization suggests.[17] The organization had earlier also tried to suggest the same thing occurred after the 1979 Three Mile Island accident but this was likewise exposed to be without merit.[18] The tooth survey, and the expansion of the organization into attempting the same test-ban approach with US nuclear electric power stations as the new target, is likewise detailed and critically labelled as the "Tooth Fairy issue" by the Nuclear Regulatory Commission.[19]

6.4 Fallout protection

Main article: Fallout shelter
See also: Fallout Protection, Nuclear warfare § Survival, and nuclear famine

During the Cold War, the governments of the U.S., the USSR, Great Britain, and China attempted to educate their citizens about surviving a nuclear attack by providing procedures on minimizing short-term exposure to fallout. This effort commonly became known as Civil Defense.

Fallout protection is almost exclusively concerned with protection from radiation. Radiation from fallout is encountered in the forms of alpha, beta, and gamma radiation, and as ordinary clothing affords protection from alpha and beta radiation.[20] most fallout protection measures deal with reducing exposure to gamma radiation.[21] For the purposes of radiation shielding, many materials have a characteristic *halving thickness*: the thickness of a layer of a material sufficient to reduce gamma radiation exposure by 50%. Halving thicknesses of common materials include: 1 cm (0.4 inch) of lead, 6 cm (2.4 inches) of concrete, 9 cm (3.6 inches) of packed earth or 150 m (500 ft) of air. When multiple thicknesses are built, the shielding multiplies. A practical fallout shield is ten halving-thicknesses of a given material, such as 90 cm (36 inches) of packed earth, which reduces gamma ray exposure by approximately 1024 times (2^{10}).[22][23] A shelter built with these materials for the purposes of fallout protection is known as a fallout shelter.

6.4.1 The Seven Ten Rule

The danger of radiation from fallout also decreases with time, as radioactivity decays exponentially with time, such that for each factor of seven increase in time, the radiation is reduced by a factor of ten. For example, after 7 hours, the average dose rate is reduced by a factor of ten; after 49 hours, it is reduced by a further factor of ten (to 1/100th); after two weeks the radiation from the fallout will have reduced by a factor of 1000 compared the initial level; and after 14 weeks the average dose rate will have reduced to 1/10,000th of the initial level.[23]

6.5 Nuclear reactor accident

See also: Comparison of Chernobyl and other radioactivity releases

Fallout can also refer to nuclear accidents, although a nuclear reactor does not explode like a nuclear weapon. The isotopic signature of bomb fallout is very different from the fallout from a serious power reactor accident (such as Chernobyl or Fukushima). The Fukushima plants have tons of nuclear fuel, thousands of Fuel Assemblies, more than 6,000 fuel rods[24] in spent fuel pools.

The key differences are in volatility and half-life.

6.5.1 Volatility

The boiling point of an element (or its compounds) is able to control the percentage of that element a power reactor accident releases. The ability of an element to form a solid, controls the rate it is deposited on the ground after having been injected into the atmosphere by a nuclear detonation or accident.

6.5.2 Half-life

A half life is the time it takes half of the radiation of a specific substance to decay. A large amount of short-lived isotopes such as ^{97}Zr are present in bomb fallout. This isotope and other short-lived isotopes are constantly generated in a power reactor, but because the criticality occurs over a long length of time, the majority of these short lived isotopes decay before they can be released.

6.6 See also

- *Atomic Cafe*—Documentary film about nuclear civil defense films.

- Castle Bravo —largest nuclear fallout accident by United States.

- Dirty bomb

- *Fallout: An American Nuclear Tragedy*

- *Fallout Protection*—U.S. government booklet

- Effects of nuclear explosions

- *Fallout* (RTÉ drama)—Irish drama exploring some inane and improbable scenarios following a nuclear accident at Sellafield.

- Fallout (series)

- Fallout shelter

- Fission product

- Hot Particle

- Human radiation experiments

- List of nuclear accidents

- Lists of nuclear disasters and radioactive incidents

- Neutron bomb

- Nuclear terrorism

- *Nuclear war survival skills* by Cresson Kearny

- Nuclear weapon design
- Potassium iodide
- Project GABRIEL
- *Protect and Survive*, a series of booklets and a public information film series produced for the British government in the 1970s and 1980s.
- Radioactive contamination
- Radiation poisoning
- Radiation biology
- Radioactive waste
- Radiological weapon
- Joseph Rotblat
- *Survival Under Atomic Attack*, an official U.S. government booklet regarding the effects of a nuclear attack.

6.7 References

[1] "Radioactive Fallout | Effects of Nuclear Weapons | atomicarchive.com". *www.atomicarchive.com*. Retrieved 2016-12-31.

[2] "AtomicBombMuseum.org - Destructive Effects". *atomicbombmuseum.org*. Retrieved 2016-12-31.

[3] Howard A. Hawthorne, Editor (May 1979). "COMPILATION OF LOCAL FALLOUT DATA FROM TEST DETONATIONS 1945-1962 - EXTRACTED FROM DASA 1251 - Volume II - Oceanic U.S. Tests" (PDF). General Electric Company. Archived from the original (PDF) on 2008-04-10.

[4] Project Officer T. Triffet, P. D. LaRiviere (March 1961). "OPERATION REDWING - Project 2.63, Characterization of Fallout - Pacific Proving Grounds, May-July 1956" (PDF). US Naval Radiological Defense Laboratory. Archived from the original (PDF) on 2008-04-10.

[5] "Death of a classified worker probably caused by overexposure to gamma radiation" (PDF). *British Medical Journal*. **54**: 713–718. 1994.

[6] Simon, Steven L.; Bouville, André; Land, Charles E. (2006). *Fallout from Nuclear Weapons Tests and Cancer Risks*, **94** (1). American Scientist. pp. 48–57

[7] Teratology in the Twentieth Century Plus Ten

[8] JAMA Network | JAMA | The Children of Atomic Bomb Survivors: A Genetic Study

[9] British Journal of Cancer - Sex ratio among offspring of childhood cancer survivors treated with radiotherapy

[10] Birth defects among the children of atomic-bomb survivors (1948-1954) - Radiation Effects Research Foundation

[11] NUCLEAR CRISIS: Hiroshima and Nagasaki cast long shadows over radiation science - Monday, April 11, 2011 - www.eenews.net

[12] Staff. "Teeth to Measure Fall-Out". *The New York Times*. March 18, 1969.

[13] Sullivan, Walter. "Babies Surveyed for Strontium 90; Ratio to Calcium in Bones Is Discovered to Be Low A survey has shown that pregnant mothers and their unborn children absorb radioactive strontium, as a substitute for calcium, only about 10 per cent of the time", *The New York Times*, November 25, 1961.

[14] Hevesi, Dennis. "Dr. Louise Reiss, Who Helped Ban Atomic Testing, Dies at 90", *The New York Times*, January 10, 2011.

[15] Gerl, E. (2014). Scientist-citizen advocacy in the atomic age: A case study of the Baby Tooth Survey, 1958-1963.

[16] Andy Newman (2003-11-11). "In Baby Teeth, a Test of Fallout; A Long-Shot Search for Nuclear Peril in Molars and Cuspids". The New York Times. Retrieved 2008-12-31.

[17] Sarah Fecht (2014-04-08). "What Can We Do About Junk Science". Popular Science. Retrieved 2014-05-21.

[18] "Scientists challenge baby deaths at Three Mile Island". *New Scientist*. London. **86** (1204): 180. 24 April 1980.

[19] "Backgrounder on Radiation Protection and the "Tooth Fairy" Issue". U.S. Nuclear Regulatory Commission. 2010-02-17. Retrieved 2010-11-07.

[20] Kearny, Cresson H (1986). *Nuclear War Survival Skills*. Oak Ridge, TN: Oak Ridge National Laboratory. p. 44. ISBN 0-942487-01-X.

[21] Kearny, Cresson H (1986). *Nuclear War Survival Skills*. Oak Ridge, TN: Oak Ridge National Laboratory. p. 131. ISBN 0-942487-01-X.

[22] "Halving-thickness for various materials". "The Compass DeRose Guide to Emergency Preparedness - Hardened Shelters".

[23] Kearny, Cresson H (1986). *Nuclear War Survival Skills*. Oak Ridge, TN: Oak Ridge National Laboratory. pp. 11–20. ISBN 0-942487-01-X.

[24] "No. 1 fuel pool power to be restored: Tepco". Retrieved 20 March 2013.

6.8 Further reading

- Glasstone, Samuel and Dolan, Philip J., *The Effects of Nuclear Weapons (third edition)*, U.S. Government Printing Office, 1977. (Available Online)

- *NATO Handbook on the Medical Aspects of NBC Defensive Operations (Part I - Nuclear)*, Departments of the Army, Navy, and Air Force, Washington, D.C., 1996, (Available Online)

- Smyth, H. DeW., *Atomic Energy for Military Purposes*, Princeton University Press, 1945. (Smyth Report)

- *The Effects of Nuclear War*, Office of Technology Assessment (May 1979), (Available Online)

- T. Imanaka, S. Fukutani, M. Yamamoto, A. Sakaguchi and M. Hoshi, *J. Radiation Research*, 2006, **47**, Suppl A121-A127.

- Sheldon Novick, *The Careless Atom* (Boston MA: Houghton Mifflin Co., 1969), p. 98

6.9 External links

- NUKEMAP3D – a 3D nuclear weapons effects simulator powered by Google Maps. It simulates the effects of nuclear weapons upon geographic areas.

Chapter 7

Weapon of mass destruction

For the Hip-hop album, see Weapons of Mass Destruction (album).

A **weapon of mass destruction** (**WMD**) is a nuclear, radiological, chemical, biological or other weapon that can kill and bring significant harm to a large number of humans or cause great damage to human-made structures (e.g. buildings), natural structures (e.g. mountains), or the biosphere. The scope and usage of the term has evolved and been disputed, often signifying more politically than technically. Originally coined in reference to aerial bombing with chemical explosives, since World War II it has come to refer to large-scale weaponry of other technologies, such as chemical, biological, radiological, or nuclear.

7.1 Early uses of term

The first use of the term "weapon of mass destruction" on record is by Cosmo Gordon Lang, Archbishop of Canterbury, in 1937 in reference to the aerial bombardment of Guernica, Spain:

> Who can think at this present time without a sickening of the heart of the appalling slaughter, the suffering, the manifold misery brought by war to Spain and to China? Who can think without horror of what another widespread war would mean, waged as it would be with all the new weapons of mass destruction?[1]

At the time, nuclear weapons had not been developed. Japan conducted research on biological weapons (see Unit 731),[2] and chemical weapons had seen wide battlefield use in World War I. They were outlawed by the Geneva Protocol of 1925.[3] Italy used mustard gas against civilians and soldiers in Ethiopia in 1935–36.[4]

Following the atomic bombings of Hiroshima and Nagasaki that ended World War II and during the Cold War, the term came to refer more to non-conventional weapons. The application of the term to specifically nuclear and radiological weapons is traced by William Safire to the Russian phrase "Оружие массового поражения" – *oruzhiye massovogo porazheniya* (weapons of mass destruction).

He credits James Goodby (of the Brookings Institution) with tracing what he considers the earliest known English-language use soon after the nuclear bombing of Hiroshima and Nagasaki (although it is not quite verbatim): a communique from a 15 November 1945, meeting of Harry Truman, Clement Attlee and Mackenzie King (probably drafted by Vannevar Bush, as Bush claimed in 1970) referred to "weapons adaptable to mass destruction."

That exact phrase, says Safire, was also used by Bernard Baruch in 1946 (in a speech at the United Nations probably written by Herbert Bayard Swope).[5] The phrase found its way into the very first resolution adopted by the United Nations General assembly in January 1946 in London, which used the wording "the elimination from national armaments of atomic weapons and of all other weapons adaptable to mass destruction."[6] The resolution also created the Atomic Energy Commission (predecessor of the International Atomic Energy Agency (IAEA)).

An exact use of this term was given in a lecture "Atomic Energy as an Atomic Problem" by J. Robert Oppenheimer. The lecture was delivered to the Foreign Service and the State Department, on 17 September 1947. The lecture is reprinted in *The Open Mind* (New York: Simon & Schuster, 1955).

> "It is a very far reaching control which would eliminate the rivalry between nations in this field, which would prevent the surreptitious arming of one nation against another, which would provide some cushion of time before atomic attack, and presumably therefore before any attack with weapons of mass destruction, and which would go a long way toward removing atomic energy at least as a source of conflict between the powers."

The term was also used in the introduction to the hugely influential U.S. government document known as NSC-68 written in April 1950.*[7]

During a speech at Rice University on 12 September 1962, President John F. Kennedy spoke of not filling space "with weapons of mass destruction, but with instruments of knowledge and understanding." *[8] The following month, during a televised presentation about the Cuban Missile Crisis on 22 October 1962, Kennedy made reference to "offensive weapons of sudden mass destruction." *[9]

An early use of the exact phrase in an international treaty was in the Outer Space Treaty of 1967, but no definition was provided.

7.1.1 Evolution of its use

During the Cold War, the term "weapons of mass destruction" was primarily a reference to nuclear weapons. At the time, in the West the euphemism "strategic weapons" was used to refer to the American nuclear arsenal, which was presented as a necessary deterrent against nuclear or conventional attack from the Soviet Union under Mutual Assured Destruction.

Subsequent to Operation Opera, the destruction of a pre-operational nuclear reactor inside Iraq by the Israeli Air Force, the Israeli prime minister, Menachem Begin, countered criticism by saying that "on no account shall we permit an enemy to develop weapons of mass destruction against the people of Israel." This policy of pre-emptive action against real or perceived weapons of mass destruction became known as the Begin Doctrine.

The term "weapons of mass destruction" continued to see periodic use, usually in the context of nuclear arms control; Ronald Reagan used it during the 1986 Reykjavík Summit, when referring to the 1967 Outer Space Treaty.*[10] Reagan's successor, George H.W. Bush, used the term in an 1989 speech to the United Nations, primarily in reference to chemical arms.*[11]

The end of the Cold War reduced U.S. reliance on nuclear weapons as a deterrent, causing it to shift its focus to disarmament. With the 1990 invasion of Kuwait and 1991 Gulf War, Iraq's nuclear, biological, and chemical weapons programs became a particular concern of the first Bush Administration.*[12] Following the war, Bill Clinton and other western politicians and media continued to use the term, usually in reference to ongoing attempts to dismantle Iraq's weapons programs.

After the 11 September 2001 attacks and the 2001 anthrax attacks in the United States, an increased fear of nonconventional weapons and asymmetrical warfare took hold in many countries. The fear reached a crescendo with the 2002 Iraq disarmament crisis and the alleged existence of weapons of mass destruction in Iraq that became the primary justification for the 2003 invasion of Iraq. However, American forces found none in Iraq. (Old stockpiles of chemical munitions including sarin and mustard agents were found, but all were considered to be unusable because of corrosion.)*[13] Iraq, however, declared a chemical weapons stockpile in 2009. The stockpile contained mainly chemical precursors, but some warheads remained usable.*[14]

Because of its prolific use and (worldwide) public profile during this period, the American Dialect Society voted "weapons of mass destruction" (and its abbreviation, "WMD") the word of the year in 2002.*[15] and in 2003 Lake Superior State University added WMD to its list of terms banished for "*Mis-use, Over-use and General Uselessness*".*[16]

In its criminal complaint against the main suspect of the Boston Marathon bombing of 15 April 2013, the FBI refers to a pressure-cooker improvised bomb as a "weapon of mass destruction." *[17]

7.2 Definitions of the term

7.2.1 United States

Strategic

The most widely used definition of "weapons of mass destruction" is that of nuclear, biological, or chemical weapons (NBC) although there is no treaty or customary international law that contains an authoritative definition. Instead, international law has been used with respect to the specific categories of weapons within WMD, and not to WMD as a whole. While nuclear, chemical and biological weapons are regarded as the three major types of WMDs.*[18] some analysts have argued that radiological materials as well as missile technology and delivery systems such as aircraft and ballistic missiles could be labeled as WMDs as well.*[18]

The abbreviations NBC (for nuclear, biological and chemical) or CBR (chemical, biological, radiological) are used with regards to battlefield protection systems for armored vehicles, because all three involve insidious toxins that can be carried through the air and can be protected against with vehicle air filtration systems.

However, there is an argument that nuclear and biological weapons do not belong in the same category as chemical and "dirty bomb" radiological weapons, which have limited destructive potential (and close to none, as far as property is concerned), whereas nuclear and biological weapons have the unique ability to kill large numbers of people with very small amounts of material, and thus could be said to belong in a class by themselves.

The NBC definition has also been used in official U.S. documents, by the U.S. President,[19][20] the U.S. Central Intelligence Agency,[21] the U.S. Department of Defense,[22][23] and the U.S. Government Accountability Office.[24]

Other documents expand the definition of WMD to also include radiological or conventional weapons. The U.S. military refers to WMD as:

Chemical, biological, radiological, or nuclear weapons capable of a high order of destruction or causing mass casualties and exclude the means of transporting or propelling the weapon where such means is a separable and divisible part from the weapon. Also called WMD.[25]

This may also refer to nuclear ICBMs (intercontinental ballistic missiles).

The significance of the words *separable and divisible part of the weapon* is that missiles such as the Pershing II and the SCUD are considered weapons of mass destruction, while aircraft capable of carrying bombloads are not.

In 2004, the United Kingdom's Butler Review recognized the "considerable and long-standing academic debate about the proper interpretation of the phrase 'weapons of mass destruction' ". The committee set out to avoid the general term but when using it, employed the definition of United Nations Security Council Resolution 687, which defined the systems which Iraq was required to abandon:

- "Nuclear weapons or nuclear-weapons-usable material or any sub-systems or components or any research, development, support or manufacturing facilities relating to [nuclear weapons].

- Chemical and biological weapons and all stocks of agents and all related subsystems and components and all research,development,support and manufacturing facilities.

- Ballistic missiles with a range greater than 150 kilometres and related major parts, and repair and production facilities." [26]

Chemical weapons expert Gert G. Harigel considers only nuclear weapons true weapons of mass destruction, because "only nuclear weapons are completely indiscriminate by their explosive power, heat radiation and radioactivity, and only they should therefore be called a weapon of mass destruction". He prefers to call chemical and biological

weapons "weapons of terror" when aimed against civilians and "weapons of intimidation" for soldiers.

Testimony of one such soldier expresses the same viewpoint.[27] For a period of several months in the winter of 2002–2003, U.S. Deputy Secretary of Defense Paul Wolfowitz frequently used the term "weapons of mass terror," apparently also recognizing the distinction between the psychological and the physical effects of many things currently falling into the WMD category.

Gustavo Bell Lemus, the Vice President of Colombia, at 9 July 2001 United Nations Conference on the Illicit Trade in Small Arms and Light Weapons in All Its Aspects, quoted the Millennium Report of the UN Secretary-General to the General Assembly, in which Kofi Annan said that small arms could be described as WMD because the fatalities they cause *"dwarf that of all other weapons systems – and in most years greatly exceed the toll of the atomic bombs that devastated Hiroshima and Nagasaki"*.[28]

An additional condition often implicitly applied to WMD is that the use of the weapons must be strategic. In other words, they would be designed to *"have consequences far outweighing the size and effectiveness of the weapons themselves"*.[29] The strategic nature of WMD also defines their function in the military doctrine of total war as targeting the means a country would use to support and supply its war effort, specifically its population, industry, and natural resources.

Within U.S. civil defense organizations, the category is now **Chemical, Biological, Radiological, Nuclear, and Explosive (CBRNE)**, which defines WMD as:

(1) Any explosive, incendiary, poison gas, bomb, grenade, or rocket having a propellant charge of more than four ounces [113 g], missile having an explosive or incendiary charge of more than one-quarter ounce [7 g], or mine or device similar to the above. (2) Poison gas. (3) Any weapon involving a disease organism. (4) Any weapon that is designed to release radiation at a level dangerous to human life.[30]

Military

For the general purposes of national defense,[31] the U.S. Code[32] defines a weapon of mass destruction as:

- any weapon or device that is intended, or has the capability, to cause death or serious bodily injury to a significant number of people through the release, dissemination, or impact of:

  - toxic or poisonous chemicals or their precursors

- a disease organism

- radiation or radioactivity[33]

For the purposes of the prevention of weapons proliferation,[34] the U.S. Code defines weapons of mass destruction as "chemical, biological, and nuclear weapons, and chemical, biological, and nuclear materials used in the manufacture of such weapons."[35]

Criminal (civilian)

For the purposes of U.S. criminal law concerning terrorism,[36] weapons of mass destruction are defined as:

- any "destructive device" defined as any explosive, incendiary, or poison gas – bomb, grenade, rocket having a propellant charge of more than four ounces, missile having an explosive or incendiary charge of more than one-quarter ounce, mine, or device similar to any of the devices described in the preceding clauses[37]

- any weapon that is designed or intended to cause death or serious bodily injury through the release, dissemination, or impact of toxic or poisonous chemicals, or their precursors

- any weapon involving a biological agent, toxin, or vector

- any weapon that is designed to release radiation or radioactivity at a level dangerous to human life[38]

The Federal Bureau of Investigation's definition is similar to that presented above from the terrorism statute:[39]

- any "destructive device" as defined in Title 18 USC Section 921: any explosive, incendiary, or poison gas – bomb, grenade, rocket having a propellant charge of more than four ounces, missile having an explosive or incendiary charge of more than one-quarter ounce, mine, or device similar to any of the devices described in the preceding clauses

- any weapon designed or intended to cause death or serious bodily injury through the release, dissemination, or impact of toxic or poisonous chemicals or their precursors

- any weapon involving a disease organism

- any weapon designed to release radiation or radioactivity at a level dangerous to human life

- any device or weapon designed or intended to cause death or serious bodily injury by causing a malfunction of or destruction of an aircraft or other vehicle that carries humans or of an aircraft or other vehicle whose malfunction or destruction may cause said aircraft or other vehicle to cause death or serious bodily injury to humans who may be within range of the vector in its course of travel or the travel of its debris.

Indictments and convictions for possession and use of WMD such as truck bombs,[40] pipe bombs,[41] shoe bombs,[42] and cactus needles coated with a biological toxin[43] have been obtained under 18 USC 2332a.

As defined by 18 USC §2332 (a), a Weapon of Mass Destruction is:

- (a) any destructive device as defined in section 921 of the title;

- (B) any weapon that is designed or intended to cause death or serious bodily injury through the release, dissemination, or impact of toxic or poisonous chemicals, or their precursors;

- (C) any weapon involving a biological agent, toxin, or vector (as those terms are defined in section 178 of this title); or

- (D) any weapon that is designed to release radiation or radioactivity at a level dangerous to human life;

Under the same statute, conspiring, attempting, threatening, or using a Weapon of Mass Destruction may be imprisoned for any term of years or for life, and if resulting in death, be punishable by death or by imprisonment for any terms of years or for life. They can also be asked to pay a maximum fine of $250,000.[44]

The Washington Post reported on 30 March 2006: "Jurors asked the judge in the death penalty trial of Zacarias Moussaoui today to define the term 'weapons of mass destruction' and were told it includes airplanes used as missiles". Moussaoui was indicted and tried for the use of airplanes as WMD.

The surviving Boston Marathon bombing perpetrator, Dzhokhar Tsarnaev, was charged in June 2013 with the federal offense of "use of a weapon of mass destruction" after he and his brother Tamerlan Tsarnaev allegedly placed crude shrapnel bombs, made from pressure cookers packed with ball bearings and nails, near the finish line of the Boston Marathon. He was convicted in April 2015. The bombing resulted in three deaths and at least 264 injuries.

7.3 Treaties

See also: Arms control and List of weapons of mass destruction treaties

The development and use of WMD is governed by several international conventions and treaties, although not all countries have signed and ratified them:

- Partial Test Ban Treaty

- Outer Space Treaty

- Nuclear Non-Proliferation Treaty (NPT)

- Seabed Arms Control Treaty

- Comprehensive Test Ban Treaty (CTBT, has not entered into force as of 2015)

- Biological and Toxin Weapons Convention (BWC)

- Chemical Weapons Convention (CWC)

7.4 Use, possession and access

7.4.1 Nuclear weapons

Main article: List of countries with nuclear weapons
The only country to have used a nuclear weapon in war is

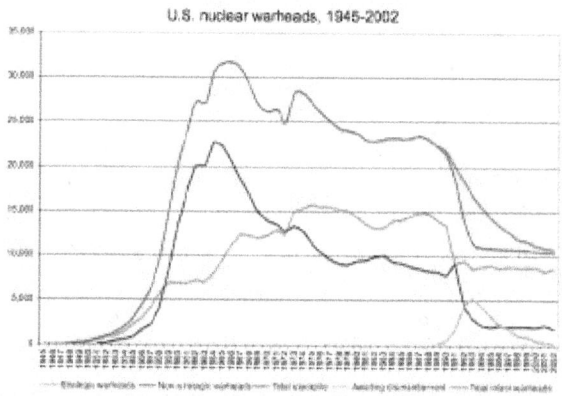

U.S. nuclear warheads, 1945–2002

the United States, which dropped two atomic bombs on the Japanese cities of Hiroshima and Nagasaki during World War II. There are eight countries that have declared they possess nuclear weapons and are known to have tested a nuclear weapon, only five of which are members of the NPT. The eight are China, France, India, North Korea, Pakistan, Russia, the United Kingdom, and the United States.

Israel is considered by most analysts to have nuclear weapons numbering in the low hundreds as well, but maintains an official policy of nuclear ambiguity, neither denying nor confirming its nuclear status.

South Africa developed a small nuclear arsenal in the 1980s but disassembled them in the early 1990s, making it the only country to have fully given up an independently developed nuclear weapons arsenal. Belarus, Kazakhstan, and Ukraine inherited stockpiles of nuclear arms following the break-up of the Soviet Union, but relinquished them to the Russian Federation.

Countries with access to nuclear weapons through nuclear sharing agreements include Belgium, Germany, Italy, the Netherlands, and Turkey.

7.5 United States politics

Due to the indiscriminate impact of WMD, the fear of a WMD attack has shaped political policies and campaigns, fostered social movements, and has been the central theme of many films. Support for different levels of WMD development and control varies nationally and internationally. Yet understanding of the nature of the threats is not high, in part because of imprecise usage of the term by politicians and the media.

FIGURE 8-11.—An atomic bomb.

An atomic-bomb blueprint

Fear of WMD, or of threats diminished by the possession of WMD, has long been used to catalyze public support for various WMD policies. They include mobilization of pro- and anti-WMD campaigners alike, and generation of popular political support. The term WMD may be used as a powerful buzzword*[45] or to generate a culture of fear.*[46] It is also used ambiguously, particularly by not distinguishing among the different types of WMD.*[47]

A television commercial called *Daisy*, promoting Democrat Lyndon Johnson's 1964 presidential candidacy, invoked the fear of a nuclear war and was an element in Johnson's subsequent election.

More recently, the threat of potential WMD in Iraq was used by President George W. Bush as justification for the 2003 invasion of Iraq.[48] Broad reference to Iraqi WMD in general was seen as an element of President Bush's arguments.[47] The claim that Iraq possessed Weapons of Mass Destruction (WMD) led to the invasion of Iraq in 2003 by Coalition forces.

Over 500 munitions were discovered throughout Iraq since 2003 containing chemical agents mustard and Sarin gas, produced in the 1980s and no longer usable as originally intended.[49]

In 2004, Polish troops found nineteen 1980s-era rocket warheads, thwarting an attempt by militants to buy them at $5000 each. Some of the rockets contained extremely deteriorated nerve agent.[50]

The American Heritage Dictionary defines a weapon of mass destruction as: "a weapon that can cause widespread destruction or kill large numbers of people, especially a nuclear, chemical, or biological weapon." [51] In other words, it does not have to be nuclear, biological or chemical (NBC). For example, the terrorist for the Boston Marathon bombings was charged under United States law *18 U.S.C. 2332A*[52] for using a weapon of mass destruction[53] and that was a pressure cooker bomb. In other words, it was a weapon that caused large-scale death and destruction, without being an NBC weapon.

7.6 Media coverage

In 2004, the Center for International and Security Studies at Maryland (CISSM) released a report[54] examining the media's coverage of WMD issues during three separate periods: nuclear weapons tests by India and Pakistan in May 1998; the U.S. announcement of evidence of a North Korean nuclear weapons program in October 2002; and revelations about Iran's nuclear program in May 2003. The CISSM report notes that poor coverage resulted less from political bias among the media than from tired journalistic conventions. The report's major findings were that:

1. *Most media outlets represented WMD as a monolithic menace, failing to adequately distinguish between weapons programs and actual weapons or to address the real differences among chemical, biological, nuclear, and radiological weapons.*

2. *Most journalists accepted the Bush administration's formulation of the "War on Terror" as a campaign against WMD, in contrast to coverage during the Clinton era, when many journalists made careful distinctions between acts of terrorism and the acquisition and use of WMD.*

3. *Many stories stenographically reported the incumbent administration's perspective on WMD, giving too little critical examination of the way officials framed the events, issues, threats, and policy options.*

4. *Too few stories proffered alternative perspectives to official line, a problem exacerbated by the journalistic prioritizing of breaking-news stories and the "inverted pyramid" style of storytelling.*

In a separate study published in 2005,[55] a group of researchers assessed the effects reports and retractions in the media had on people's memory regarding the search for WMD in Iraq during the 2003 Iraq War. The study focused on populations in two coalition countries (Australia and the United States) and one opposed to the war (Germany). Results showed that U.S. citizens generally did not correct initial misconceptions regarding WMD, even following disconfirmation; Australian and German citizens were more responsive to retractions. Dependence on the initial source of information led to a substantial minority of Americans exhibiting false memory that WMD were indeed discovered, while they were not. This led to three conclusions:

1. *The repetition of tentative news stories, even if they are subsequently disconfirmed, can assist in the creation of false memories in a substantial proportion of people.*

2. *Once information is published, its subsequent correction does not alter people's beliefs unless they are suspicious about the motives underlying the events the news stories are about.*

3. *When people ignore corrections, they do so irrespective of how certain they are that the corrections occurred.*

A poll conducted between June and September 2003 asked people whether they thought evidence of WMD had been discovered in Iraq since the war ended. They were also asked which media sources they relied upon. Those who obtained their news primarily from Fox News were three times as likely to believe that evidence of WMD had been discovered in Iraq than those who relied on PBS and NPR for their news, and one third more likely than those who primarily watched CBS.

Based on a series of polls taken from June–September 2003.[56]

In 2006 Fox News reported the claims of two Republican lawmakers that WMDs had been found in Iraq.[57] based upon unclassified portions of a report by the National

Ground Intelligence Center. Quoting from the report, Senator Rick Santorum said "Since 2003, coalition forces have recovered approximately 500 weapons munitions which contain degraded mustard or sarin nerve agent". According to David Kay, who appeared before the U.S. House Armed Services Committee to discuss these badly corroded munitions, they were leftovers, many years old, improperly stored or destroyed by the Iraqis.[58] Charles Duelfer agreed, stating on NPR's *Talk of the Nation*: "When I was running the ISG – the Iraq Survey Group – we had a couple of them that had been turned in to these IEDs, the improvised explosive devices. But they are local hazards. They are not a major, you know, weapon of mass destruction."[59]

Later, wikileaks would show that WMDs of these kinds continued to be found as the Iraqi occupation continued.[60]

Many news agencies, including Fox News, reported the conclusions of the CIA that, based upon the investigation of the Iraq Survey Group, WMDs are yet to be found in Iraq.[61][62]

7.7 Public perceptions

Awareness and opinions of WMD have varied during the course of their history. Their threat is a source of unease, security, and pride to different people. The anti-WMD movement is embodied most in nuclear disarmament, and led to the formation of the British Campaign for Nuclear Disarmament in 1957.

In order to increase awareness of all kinds of WMD, in 2004 the nuclear physicist and Nobel Peace Prize winner Joseph Rotblat inspired the creation of The WMD Awareness Programme[63] to provide trustworthy and up to date information on WMD worldwide.

In 1998 University of New Mexico's Institute for Public Policy released their third report[64] on U.S. perceptions – including the general public, politicians and scientists – of nuclear weapons since the breakup of the Soviet Union. Risks of nuclear conflict, proliferation, and terrorism were seen as substantial.

While maintenance of the U.S. nuclear arsenal was considered above average in importance, there was widespread support for a reduction in the stockpile, and very little support for developing and testing new nuclear weapons.

Also in 1998, but after the UNM survey was conducted, nuclear weapons became an issue in India's election of March,[65] in relation to political tensions with neighboring Pakistan. Prior to the election the Bharatiya Janata Party (BJP) announced it would "declare India a nuclear weapon state" after coming to power.

BJP won the elections, and on 14 May, three days after India tested nuclear weapons for the second time, a public opinion poll reported that a majority of Indians favored the country's nuclear build-up.

On 15 April 2004, the Program on International Policy Attitudes (PIPA) reported[66] that U.S. citizens showed high levels of concern regarding WMD, and that preventing the spread of nuclear weapons should be "a very important U.S. foreign policy goal", accomplished through multilateral arms control rather than the use of military threats.

A majority also believed the United States should be more forthcoming with its biological research and its Nuclear Non-Proliferation Treaty commitment of nuclear arms reduction.

A Russian opinion poll conducted on 5 August 2005 indicated half the population believes new nuclear powers have the right to possess nuclear weapons.[67] 39% believes the Russian stockpile should be reduced, though not fully eliminated.

7.8 In popular culture

Main article: Weapons of mass destruction in popular culture

Weapons of mass destruction and their related impacts have been a mainstay of popular culture since the beginning of the Cold War, as both political commentary and humorous outlet. The actual phrase "weapons of mass destruction" has been used similarly and as a way to characterise any powerful force or product since the Iraqi weapons crisis in the lead up to the Coalition invasion of Iraq in 2003.

7.9 Common hazard symbols

Main article: Hazard symbol

7.9.1 Radioactive weaponry/hazard symbol

The international radioactivity symbol (also known as trefoil) first appeared in 1946, at the University of California, Berkeley Radiation Laboratory. At the time, it was rendered as magenta, and was set on a blue background.[68]

It is drawn with a central circle of radius R, the blades having an internal radius of $1.5R$ and an external radius of $5R$,

Radioactivity

2007 ISO radioactivity danger symbol.

and separated from each other by 60°.[69] It is meant to represent a radiating atom.

The International Atomic Energy Agency found that the trefoil radiation symbol is unintuitive and can be variously interpreted by those uneducated in its meaning; therefore, its role as a hazard warning was compromised as it did not clearly indicate "danger" to many non-Westerners and children who encountered it. As a result of research, a new radiation hazard symbol was developed in 2007 to be placed near the most dangerous parts of radiation sources featuring a skull, someone running away, and using a red rather than yellow background.[70]

The red background is intended to convey urgent danger.

and the sign is intended to be used on equipment where very strong radiation fields can be encountered if the device is dismantled or otherwise tampered with. The intended use of the sign is not in a place where the normal user will see it, but in a place where it will be seen by someone who has started to dismantle a radiation-emitting device or equipment. The aim of the sign is to warn people such as scrap metal workers to stop work and leave the area.[71]

7.9.2   Biological weaponry/hazard symbol

Biohazard

Developed by Dow Chemical company in the 1960s for their containment products.[72]

According to Charles Dullin, an environmental-health engineer who contributed to its development:[69]

> We wanted something that was memorable but meaningless, so we could educate people as to what it means.

7.10   See also

- Commission on the Intelligence Capabilities of the United States Regarding Weapons of Mass Destruction

- Commission on the Prevention of WMD proliferation and terrorism
- Ethnic bioweapon
- Fallout shelter
- NBC suit
- New physical principles weapons
- Nuclear terrorism
- Orbital bombardment
- Russia and weapons of mass destruction
- *The Bomb* (film)
- United States and weapons of mass destruction
- Weapons of Mass Destruction Commission

7.11 References

[1] "Archbishop's Appeal." *Times* (London). 28 December 1937. p. 9.

[2] "Biological Weapons Program – Japan". Fas.org. Retrieved 5 August 2010.

[3] Eric Croddy (1997). *Chemical and Biological Warfare: An Annotated Bibliography*. Scarecrow Press. p. 30.

[4] William R. Cullen (2008). *Is Arsenic an Aphrodisiac?: The Sociochemistry of an Element*. Royal Society of Chemistry. p. 241.

[5] "Weapons of Mass Destruction". *New York Times Magazine*. 19 April 1998. p.22. Retrieved 24 February 2007.

[6] "UNODA – Nuclear Weapons Home". Un.org. Retrieved 14 May 2012.

[7] "NSC-68 United States Objectives and Programs for National Security". Fas.org. Retrieved 5 August 2010.

[8] "John F. Kennedy Moon Speech—Rice Stadium". nasa.gov. Retrieved 30 June 2015.

[9] Kennedy JF (1962-10-22). Televised remarks to the American people re "the Soviet military buildup on the island of Cuba"

[10] "CNN Cold War – Historical Documents: Reagan-Gorbachev transcripts". Web.archive.org. 18 May 2008. Archived from the original on 18 May 2008. Retrieved 14 May 2012.

[11] "Excerpts From Bush's Speech at the Opening of the U.N. General Assembly –". *New York Times*. Union Of Soviet Socialist Republics (Ussr). 26 September 1989. Retrieved 5 August 2010.

[12] MICHAEL WINES. Special to The New York Times (30 September 1990). "Confrontation in the Gulf; U.S. Explores New Strategies to Limit Weapons of Mass Destruction –". *New York Times*. IRAQ. Retrieved 5 August 2010.

[13] *Munitions Found in Iraq Meet WMD Criteria*, Military.com, report filed by American Forces Press Service. 29 June 2006

[14] http://www.nti.org/gsn/article/india-completes-chemical-weapons-disposal-iraq-declares

[15] "American Dialect Society". Americandialect.org. 13 January 2003. Retrieved 5 August 2010.

[16] "Lake Superior State University:: Banished Words List:: 2003". Lssu.edu. Retrieved 5 August 2010.

[17] "Criminal Complaint United States vs Dzhokhar Tsarnaev". *The Washington Post*. Retrieved 23 April 2013.

[18] Reed, Laura (2014). "Weapons of Mass Destruction". *Hampshire College*. Hampshire College. Retrieved 21 October 2014.

[19] "Archived copy". Archived from the original on 2 April 2010. Retrieved 6 February 2016.

[20] "Weekly Compilation of Presidential Documents Volume 37, Issue 19 (May 14, 2001)" (PDF). Frwebgate.access.gpo.gov. Retrieved 14 May 2012.

[21] CIA Site Redirect – Central Intelligence Agency

[22] "Archived copy". Archived from the original on 1 October 2004. Retrieved 6 February 2016.

[23] "Archived copy" (PDF). Archived from the original (PDF) on 1 June 2006. Retrieved 6 February 2016.

[24] "Weapons of Mass Destruction: State Department Oversight of Science Centers Program" (PDF). Retrieved 5 August 2010.

[25] "Department of Defense Dictionary of Military and Associated Terms". Dtic.mil. 12 April 2001. Retrieved 5 August 2010.

[26] *Review of Intelligence on Weapons of Mass Destruction: Report of a Committee of Privy Counsellors* (HC 898), London: The Stationery Office, 2004, §14.

[27] "A Soldier's Viewpoint on Surviving Nuclear, Chemical and Biological Attacks". Sightm1911.com. Retrieved 5 August 2010.

[28] "Colombia". Web.archive.org. 2 September 2007. Archived from the original on 2 September 2007. Retrieved 14 May 2012.Template:Dead link 2013.06.08

[29] What makes a weapon one of mass destruction? – Times Onlinehttp://www.timesonline.co.uk/tol/news/uk/article1013136.ece at the Wayback Machine (archived 11 March 2007)

[30] Capt. G. Shane Hendricks, Dr. Margot J. Hall (2007). "The History and Science of CBRNE Agents, Part I" (PDF). American Institute of Chemists. p. 1. Retrieved 4 July 2014.

[31] "US CODE: Title 50 —War and National Defense". .law.cornell.edu. 23 March 2010. Retrieved 5 August 2010.

[32] "US CODE: 50, ch. 40—Defense Against Weapons of Mass Destruction". .law.cornell.edu. 23 March 2010. Retrieved 5 August 2010.

[33] "US CODE: 50, ch. 40, § 2302. Definitions". .law.cornell.edu. 23 March 2010. Retrieved 5 August 2010.

[34] "US CODE: 50, ch. 43—Preventing Weapons of Mass Destruction Proliferation and Terrorism". .law.cornell.edu. 23 March 2010. Retrieved 5 August 2010.

[35] "US CODE: 50, ch. 43; § 2902. Definitions". .law.cornell.edu. 23 March 2010. Retrieved 5 August 2010.

[36] "US CODE: Chapter 113B—Terrorism". .law.cornell.edu. 28 June 2010. Retrieved 5 August 2010.

[37] "US CODE: Title 18, § 921. Definitions". .law.cornell.edu. 13 September 1994. Retrieved 5 August 2010.

[38] "US CODE: Title 18, § 2332a. Use of weapons of mass destruction". .law.cornell.edu. 28 June 2010. Retrieved 5 August 2010.

[39] "What is A Weapon of Mass Destruction". Fbi.gov. 30 March 2007. Retrieved 5 August 2010.

[40] "8/95 Grand Jury Indictment Of McVeigh & Nichols". Lectlaw.com. Retrieved 5 August 2010.

[41] "FindLaw for Legal Professionals – Case Law, Federal and State Resources, Forms, and Code". Caselaw.lp.findlaw.com. Retrieved 5 August 2010.

[42] "U.S. v. Richard C. Reid" (PDF). Retrieved 5 August 2010.

[43] The Free Lance-Star – 14 Jul 1998

[44]

[45] "David T. Wright – Weapons of mass distraction". Thornwalker.com. 13 April 1998. Retrieved 5 August 2010.

[46] "Weapons of Mass Destruction Are Overrated as a Threat to America: Newsroom: The Independent Institute". Independent.org. 28 January 2004. Retrieved 5 August 2010.

[47] Archived 11 October 2007 at the Wayback Machine.

[48] "War Pimps, by Jeffrey St. Clair [Weapons of Mass Deception: The Uses of Propaganda in President Bush's War on Iraq, by John Stauber and Sheldon Rampton]". Theava.com. 13 August 2003. Retrieved 5 August 2010.

[49] "Munitions Found in Iraq Meet WMD Criteria, Official Says". *US Department of Defense*. Retrieved 1 April 2014.

[50] "Troops 'foil Iraq nerve gas bid'". BBC. 2 July 2004. Retrieved 7 December 2007.

[51] American Heritage Dictionary: Weapon of mass destruction

[52] 18 U.S.C. 2332A

[53] Court case]

[54] Archived 22 October 2004 at the Wayback Machine. by Prof. Susan Moeller

[55] "Psychological Science – Journal Information". Blackwellpublishing.com. Retrieved 5 August 2010.

[56] *Misperceptions, the Media and the Iraq War* at the Wayback Machine (archived 10 February 2006). PIPA. 2 October 2003

[57] "Report: Hundreds of WMDs Found in Iraq". Fox News. 22 June 2006.

[58] Kay, David. "House Armed Services Committee Hearing", 29 June 2006

[59] Duelfer, Charles. Expert: Iraq WMD Find Did Not Point to Ongoing Program NPR. 22 June 2006

[60] Shachtman, Noah (23 October 2010). "WikiLeaks Show WMD Hunt Continued in Iraq – With Surprising Results". Wired.com.

[61] "CIA's Final Report: No WMD Found in Iraq". MSNBC. 25 April 2005.

[62] "Iraq WMD Inspectors End Search, Find Nothing". Fox News. 26 April 2005.

[63] wmdawareness.org.uk

[64] John Pike. "Sandia National Laboratories – News Releases". Globalsecurity.org. Retrieved 5 August 2010.

[65] John Pike. "17 Days in May – India Nuclear Forces". Globalsecurity.org. Retrieved 5 August 2010.

[66] "The Pipa/Knowledge Networks Poll" (PDF). Web.archive.org. 29 September 2005. Archived from the original (PDF) on 29 September 2005. Retrieved 14 May 2012.

[67] Russian public opinion on nuclear weapons (5 August 2005). "Russian public opinion on nuclear weapons – Blog – Russian strategic nuclear forces". Russianforces.org. Retrieved 5 August 2010.

[68] "Origin of the Radiation Warning Symbol (Trefoil)".

[69] "Biohazard and radioactive Symbol, design and proportions" (PDF). Archived from the original (PDF) on 31 December 2013.

[70] Linda Lodding, "Drop it and Run! New Symbol Warns of Radiation Dangers and Aims to Save Lives," *IAEA Bulletin* 482 (March 2007): 70–72.

[71] IAEA news release Feb 2007

[72] "Biohazard Symbol History".

7.12 Bibliography

- Bentley, Michelle. *Weapons of Mass Destruction: The Strategic Use of a Concept* (Routledge, 2014.) On the usage of the term in American policy

- Cirincione, Joseph, ed. *Repairing the Regime: Preventing the Spread of Weapons of Mass Destruction* (Routledge, 2014)

- Croddy, Eric A. ed. *Weapons of Mass Destruction: An Encyclopedia of Worldwide Policy, Technology, and History* (2 vol 2004); 1024pp excerpt

- Curley, Robert, ed. *Weapons of Mass Destruction* (Britannica Educational Publishing, 2011)

- Graham Jr, Thomas, and Thomas Graham. *Common sense on weapons of mass destruction* (University of Washington Press, 2011)

- Horowitz, Michael C., and Neil Narang. "Poor Man's atomic bomb? exploring the relationship between "weapons of mass destruction"." *Journal of Conflict Resolution* (2013) online

- Hutchinson, Robert. *Weapons of Mass Destruction: The no-nonsense guide to nuclear, chemical and biological weapons today* (Hachette UK, 2011)

7.13 Further reading

7.13.1 Definition and origin

- "WMD: Words of mass dissemination" (12 February 2003), BBC News.

- Bentley, Michelle, "War and/of Worlds: Constructing WMD in U.S. Foreign Policy," *Security Studies* 22 (Jan. 2013), 68–97.

- Michael Evans, "What makes a weapon one of mass destruction?" (6 February 2004), *The Times*.

- Bruce Schneier, "Definition of 'Weapon of Mass Destruction'" (6 April 2009), *Schneier on Security*.

- Stefano Felician, Le armi di distruzione di massa, CEMISS, Roma, 2010.

- George Moraetes, "Nuclear Power Plant Cybersecurity'" (30 December 2014), *Pulse on LinkedIn – Featured in Oil & Energy*.

7.13.2 International law

- United Nations Security Council Resolution 1540

- David P. Fidler, "Weapons of Mass Destruction and International Law" (February 2003), American Society of International Law.

- Joanne Mariner, "FindLaw Forum: Weapons of mass destruction and international law's principle that civilians cannot be targeted" (20 November 2001), CNN.

7.13.3 Media

- Media Coverage of Weapons of Mass Destruction at the Wayback Machine (archived 17 February 2006), by Susan D. Moeller, Center for International and Security Studies at Maryland, 2004.

- Memory for fact, fiction, and misinformation, by Stephan Lewandowsky, Werner G.K. Stritzke, Klaus Oberauer, and Michael Morales, *Psychological Science*, 16(3): 190–195, 2005.

7.13.4 Ethics

- Jacob M. Appel, "Is All Fair in Biological Warfare?," *Journal of Medical Ethics*, June 2009.

7.13.5 Public perceptions

- Steven Kull et al., Americans on WMD Proliferation (15 April 2004), Program on International Policy Attitudes/Knowledge Networks survey.

7.14 External links

- Journal dedicated to CBRNE issues

- United Nations: Disarmament at the Wayback Machine (archived 24 June 2005)

- US Department of State at the Wayback Machine (archived 13 March 2007)

- Nuclear Threat Initiative (NTI)

- Nuclear Threat Initiative (NTI)

- Federation of American Scientists (FAS)

- Carnegie Endowment for International Peace

- GlobalSecurity.org

- Avoiding Armageddon, PBS

- FAS assessment of countries that own weapons of mass destruction

- National Counterproliferation Center – Office of the Director of National Intelligence

- HLSWatch.com: Homeland Security Watch policy and current events resource

- Office of the Special Assistant for Chemical Biological Defense and Chemical Demilitarization Programs. Official Department of Defense web site that provides information about the DoD Chemical Biological Defense Program

- Terrorism and the Threat From Weapons of Mass Destruction in the Middle East at the Wayback Machine (archived 29 April 2001)

- Iranian Chemical Attacks Victims (*Payvand News Agency*)

- Iran: 'Forgotten Victims' Of Saddam Hussein Era Await Justice

- Comparison of Chinese, Japanese and Vietnamese translations

- Nuclear Age Peace Foundation

- The WMD Awareness Programme. Inspired by the 1995 Nobel Peace Prize winner Professor Sir Joseph Rotblat. The WMD Awareness Programme is dedicated to providing trustworthy and up to date information on Weapons of Mass Destruction worldwide.

- Radius Engineering International Inc. Radius Engineering International Inc. ed. "Nuclear Weapons Effects" (PDF). Archived from the original (PDF) on 14 December 2010. Retrieved 20 December 2010. These tables describe the effects of various nuclear blast sizes. All figures are for 15 mph (13 kn; 24 km/h) winds. Thermal burns represent injuries to an unprotected person. The legend describes the data.

- Gareth Porter, *Documents linking Iran to nuclear weapons push may have been fabricated*, TheRawStory, 10 November 2008

- Gareth Porter. *The Iranian Nuke Forgeries: CIA Determines Documents were Fabricated*, CounterPunch, 29 December 2009

Chapter 8

Radioactive contamination

The Hanford site represents two-thirds of the United States' high-level radioactive waste by volume. Nuclear reactors line the river-bank at the Hanford Site along the Columbia River in January 1960.

As of 2013, the Fukushima nuclear disaster site remains highly radioactive, with some 160,000 evacuees still living in temporary housing, and some land will be unfarmable for centuries. The difficult cleanup job will take 40 or more years, and cost tens of billions of dollars. [1] [2]

Radioactive contamination, also called **radiological contamination**, is the deposition of, or presence of radioactive substances on surfaces or within solids, liquids or gases (including the human body), where their presence is unin-

tended or undesirable (from the International Atomic Energy Agency - IAEA - definition). [3]

Such contamination presents a hazard because of the radioactive decay of the contaminants, which emit harmful ionising radiation such as alpha particles or beta particles, gamma rays or neutrons. The degree of hazard is determined by the concentration of the contaminants, the energy of the radiation being emitted, the type of radiation, and the proximity of the contamination to organs of the body. It is important to be clear that the contamination gives rise to the radiation hazard, and the terms "radiation" and "contamination" are not interchangeable.

Contamination may affect a person, a place, an animal, or an object such as clothing. Following an atmospheric nuclear weapon discharge or a nuclear reactor containment breach, the air, soil, people, plants, and animals in the vicinity will become contaminated by nuclear fuel and fission products. A spilled vial of radioactive material like uranyl nitrate may contaminate the floor and any rags used to wipe up the spill. Cases of widespread radioactive contamination include the Bikini Atoll, the Rocky Flats Plant in Colorado, the Fukushima Daiichi nuclear disaster, the Chernobyl disaster, and the area around the Mayak facility in Russia.

8.1 Sources of contamination

Radioactive contamination is typically the result of a spill or accident during the production, or use of, radionuclides (radioisotopes); these have unstable nuclei which are subject to radioactive decay.

Less typically, nuclear fallout is the distribution of radioactive contamination by a nuclear explosion. The amount of radioactive material released in an accident is called the *source term*.

Contamination may occur from radioactive gases, liquids or particles. For example, if a radionuclide used in nuclear medicine is spilled (accidentally or, as in the case of the Goiânia accident, through ignorance), the material could

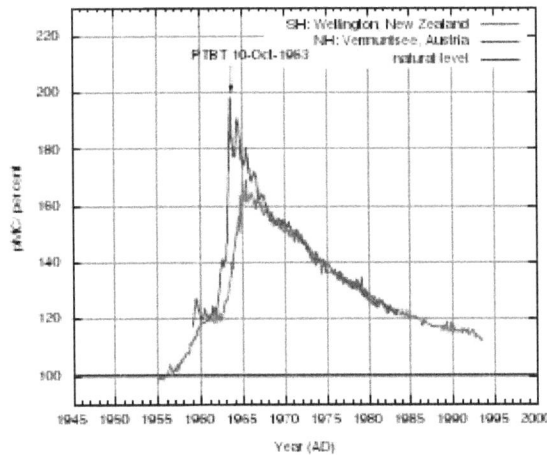

Global airborne contamination Atmospheric nuclear weapon tests almost doubled the concentration of ^{14}C in the Northern Hemisphere. Plot of atmospheric ^{14}C, New Zealand [4] and Austria. [5] *The New Zealand curve is representative for the Southern Hemisphere, the Austrian curve is representative for the Northern Hemisphere.* [6]

be spread by people as they walk around. Radioactive contamination may also be an inevitable result of certain processes, such as the release of radioactive xenon in nuclear fuel reprocessing. In cases that radioactive material cannot be contained, it may be diluted to safe concentrations. For a discussion of environmental contamination by alpha emitters please see actinides in the environment.

Contamination does not include residual radioactive material remaining at a site after the completion of decommissioning. Therefore, radioactive material in sealed and designated containers is not properly referred to as contamination, although the units of measurement might be the same.

8.1.1   Containment

Containment is the primary way of preventing contamination being released into the environment or coming into contact or being ingested by humans.

Being within the intended Containment differentiates radioactive *material* from radioactive *contamination*. When radioactive materials are concentrated to a detectable level outside a containment, the area affected is generally referred to as "contaminated".

There are a large number of techniques for containing radioactive material so that it does not spread beyond the containment and become contamination. In the case of liquids this is by the use of high integrity tanks or containers, usually with a sump system so that leakage can be detected by

Large industrial glovebox in the nuclear industry

radiometric or conventional instrumentation.

Where material is likely to become airborne, then extensive use is made of the glovebox, which is a common technique in hazardous laboratory and process operations in many industries. The gloveboxes are kept under a slight negative pressure and the vent gas is filtered in high efficiency filters, which are monitored by radiological instrumentation to ensure they are functioning correctly.

8.1.2   Naturally occurring radioactivity

Main article: Environmental radioactivity

A variety of radionuclides occur naturally in the environment. Elements like uranium and thorium, and their decay products, are present in rock and soil. Potassium-40, a primordial nuclide, makes up a small percentage of all potassium and is present in the human body. Other nuclides, like carbon-14, which is present in all living organisms, are continuously created by cosmic rays.

These levels of radioactivity pose little danger but can confuse measurement. A particular problem is encountered with naturally generated radon gas which can affect instruments which are set to detect contamination close to normal background levels and can cause false alarms. Because of this skill is required by the operator of radiological survey equipment to differentiate between background radiation and the radiation which emanates from contamination.

Naturally occurring radioactive materials (NORM) can be brought to the surface or concentrated by human activities like mining and oil and gas extraction.

8.2 Control and monitoring of contamination

G-M counters being used as gamma survey monitors, seeking radioactive satellite debris

Radioactive contamination may exist on surfaces or in volumes of material or air, and specialist techniques are used to measure the levels of contamination by detection of the emitted radiation.

8.2.1 Contamination monitoring

Contamination monitoring depends entirely upon the correct and appropriate deployment and utilisation of radiation monitoring instruments.

Surface contamination

Surface contamination may either be fixed or "free". In the case of fixed contamination, the radioactive material cannot by definition be spread, but its radiation is still measurable. In the case of free contamination there is the hazard of contamination spread to other surfaces such as skin or clothing, or entrainment in the air. A concrete surface contaminated by radioactivity can be shaved to a specific depth, removing the contaminated material for disposal.

For occupational workers controlled areas are established where there may be a contamination hazard. Access to such areas is controlled by a variety of barrier techniques, sometimes involving changes of clothing and foot wear as required. The contamination within a controlled area is normally regularly monitored. Radiological protection instrumentation (RPI) plays a key role in monitoring and detecting any potential contamination spread, and combinations of hand held survey instruments and permanently installed area monitors such as Airborne particulate monitors and area gamma monitors are often installed. Detection and measurement of surface contamination of personnel and plant is normally by Geiger counter, scintillation counter or proportional counter. Proportional counters and dual phosphor scintillation counters can discriminate between alpha and beta contamination, but the Geiger counter cannot. Scintillation detectors are generally preferred for hand held monitoring instruments, and are designed with a large detection window to make monitoring of large areas faster. Geiger detectors tend to have small windows, which are more suited to small areas of contamination.

Exit monitoring

The spread of contamination by personnel exiting controlled areas in which nuclear material is used or processed is monitored by specialised installed exit control instruments such as frisk probes, hand contamination monitors and whole body exit monitors. These are used to check that persons exiting controlled areas do not carry contamination on their body or clothes.

In the United Kingdom the HSE has issued a user guidance note on selecting the correct portable radiation measurement instrument for the application concerned.[7] This covers all radiation instrument technologies, and is a useful comparative guide for selecting the correct technology for the contamination type.

The UK NPL publishes a guide on the alarm levels to be used with instruments for checking personnel exiting controlled areas in which contamination may be encountered.[8] Surface contamination is usually expressed in units of radioactivity per unit of area for alpha or beta emitters. For SI, this is becquerels per square meter (or Bq/m^2). Other units such as picoCuries per 100 cm^2 or disintegrations per minute per square centimeter (1 $dpm/cm^2 = 167 \ Bq/m^2$) may be used.

8.2.2 Airborne contamination

Main article: Airborne particulate radioactivity monitoring

The air can be contaminated with radioactive isotopes in particulate form, which poses a particular inhalation hazard. Respirators with suitable air filters, or completely self-contained suits with their own air supply can mitigate these dangers.

Airborne contamination is measured by specialist radiological instruments that continuously pump the sampled air through a filter. Airborne particles accumulate on the filter and can be measured in a number of ways:

1. The filter paper is periodically manually removed to an instrument such as a "scaler" which measures any accumulated radioactivity.

2. The filter paper is static and is measured in situ by a radiation detector.

3. The filter is a slowly moving strip and is measured by a radiation detector. These are commonly called "moving filter" devices and automatically advance the filter to present a clean area for accumulation, and thereby allow a plot of airborne concentration over time.

Commonly a semiconductor radiation detection sensor is used that can also provide spectrographic information on the contamination being collected.

A particular problem with airborne contamination monitors designed to detect alpha particles is that naturally occurring radon can be quite prevalent and may appear as contamination when low contamination levels are being sought. Modern instruments consequently have "radon compensation" to overcome this effect.

See the article on Airborne particulate radioactivity monitoring for more information.

8.2.3 Internal human contamination

Main article: committed dose

Radioactive contamination can enter the body through ingestion, inhalation, absorption, or injection. This will result in a committed dose of radiation.

For this reason, it is important to use personal protective equipment when working with radioactive materials. Radioactive contamination may also be ingested as the result of eating contaminated plants and animals or drinking contaminated water or milk from exposed animals. Following a major contamination incident, all potential pathways of internal exposure should be considered.

8.3 Decontamination

Cleaning up contamination results in radioactive waste unless the radioactive material can be returned to commercial use by reprocessing. In some cases of large areas of contamination, the contamination may be mitigated by burying and covering the contaminated substances with concrete, soil, or rock to prevent further spread of the contamination to the environment. If a person's body is contaminated by ingestion or by injury and standard cleaning cannot reduce the

contamination further, then the person may be permanently contaminated.

Contamination control products have been used by the U.S. Department of Energy (DOE) and the commercial nuclear industry for decades to minimize contamination on radioactive equipment and surfaces and fix contamination in place. "Contamination control products" is a broad term that includes fixatives, strippable coatings, and decontamination gels. A *fixative* product functions as a permanent coating to stabilize residual loose/transferable radioactive contamination by fixing it in place; this aids in preventing the spread of contamination and reduces the possibility of the contamination becoming airborne, reducing workforce exposure and facilitating future deactivation and decommissioning (D&D) activities. *Strippable coating* products are loosely adhered paint-like films and are used for their decontamination abilities. They are applied to surfaces with loose/transferable radioactive contamination and then, once dried, are peeled off, which removes the loose/transferable contamination along with the product. The residual radioactive contamination on the surface is significantly reduced once the strippable coating is removed. Modern strippable coatings show high decontamination efficiency and can rival traditional mechanical and chemical decontamination methods. *Decontamination gels* work in much the same way as other strippable coatings. The results obtained through the use of contamination control products is variable and depends on the type of substrate, the selected contamination control product, the contaminants, and the environmental conditions (e.g., temperature, humidity, etc.).

Some of the largest areas committed to be decontaminated are in the Fukushima Prefecture, Japan. The national government is under pressure to clean up radioactivity due to the Fukushima nuclear accident of March 2011 from as much land as possible so that some of the 110,000 displaced people can return. Stripping out the key radioisotope threatening health (caesium-137) from low level waste could also dramatically decrease the volume of waste requiring special disposal. A goal is to find techniques that might be able to strip out 80 to 95% of the caesium from contaminated soil and other materials, efficiently and without destroying the organic content in the soil. One being investigated is termed hydrothermal blasting. The caesium is broken away from soil particles and then precipitated with ferric ferricyanide (Prussian blue). It would be the only component of the waste requiring special burial sites."[9] The aim is to get annual exposure from the contaminated environment down to one millisievert (mSv) above background. The most contaminated area where radiation doses are greater than 50 mSv/year must remain off limits, but some areas that are currently less than 5 mSv/year may be decontaminated allowing 22,000 residents to return.

To help with protection of people living in geographical areas which have been radioactively contaminated the International Commission on Radiological Protection has published a guide: "Publication 111 - Application of the Commission's Recommendations to the Protection of People Living in Long-term Contaminated Areas after a Nuclear Accident or a Radiation Emergency" .[10]

8.4 Contamination hazards

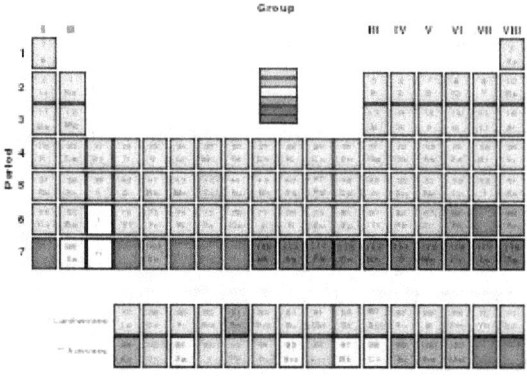

Periodic table with elements colored according to the half-life of their most stable isotope.
Elements which contain at least one stable isotope.
Radioactive elements: the most stable isotope is very long-lived, with half-life of over four million years.
Radioactive elements: the most stable isotope has half-life between 800 and 34,000 years.
Radioactive elements: the most stable isotope has half-life between one day and 103 years.
Highly radioactive elements: the most stable isotope has half-life between several minutes and one day.
Extremely radioactive elements: the most stable isotope has half-life less than several minutes.

8.4.1 Low-level contamination

The hazards to people and the environment from radioactive contamination depend on the nature of the radioactive contaminant, the level of contamination, and the extent of the spread of contamination. Low levels of radioactive contamination pose little risk, but can still be detected by radiation instrumentation. If a survey or map is made of a contaminated area, random sampling locations may be labeled with their activity in becquerels or curies on contact. Low levels may be reported in counts per minute using a scintillation counter.

In the case of low-level contamination by isotopes with a short half-life, the best course of action may be to simply allow the material to naturally decay. Longer-lived isotopes should be cleaned up and properly disposed of, because even a very low level of radiation can be life-threatening when in long exposure to it.

Facilities and physical locations that are deemed to be contaminated may be cordoned off by a health physicist and labeled "Contaminated area." Persons coming near such an area would typically require anti-contamination clothing ("anti-Cs").

8.4.2 High-level contamination

High levels of contamination may pose major risks to people and the environment. People can be exposed to potentially lethal radiation levels, both externally and internally, from the spread of contamination following an accident (or a deliberate initiation) involving large quantities of radioactive material. The biological effects of external exposure to radioactive contamination are generally the same as those from an external radiation source not involving radioactive materials, such as x-ray machines, and are dependent on the absorbed dose.

When radioactive contamination is being measured or mapped *in situ*, any location that appears to be a point source of radiation is likely to be heavily contaminated. A highly contaminated location is colloquially referred to as a "hot spot." On a map of a contaminated place, hot spots may be labeled with their "on contact" dose rate in mSv/h. In a contaminated facility, hot spots may be marked with a sign, shielded with bags of lead shot, or cordoned off with warning tape containing the radioactive trefoil symbol.

The hazard from contamination is the emission of ionising radiation. The principal radiations which will be encountered are alpha, beta and gamma, but these have quite different characteristics. They have widely differing penetrating powers and radiation effect, and the accompanying diagram shows the penetration of these radiations in simple terms. For an understanding of the different ionising effects of these radiations and the weighting factors applied, see the article on absorbed dose.

Radiation monitoring involves the measurement of radiation dose or radionuclide contamination for reasons related to the assessment or control of exposure to radiation or radioactive substances, and the interpretation of the results. The methodological and technical details of the design and operation of environmental radiation monitoring programmes and systems for different radionuclides, environmental media and types of facility are given in IAEA Safety Standards Series No. RS-G-1.8[11] and in IAEA

8.5   Health effects of contamination

8.5.1   Biological effects

See also: Sievert, Radiation poisoning, and committed dose

Radioactive contamination by definition emits ionizing radiation, which can irradiate the human body from an external or internal origin.

External irradiation

This is due to radiation from contamination located outside the human body. The source can be in the vicinity of the body or can be on the skin surface. The level of health risk is dependent on duration and the type and strength of irradiation. Penetrating radiation such as gamma rays, X-rays, neutrons or beta particles pose the greatest risk from an external source. Low penetrating radiation such as alpha particles have a low external risk due to the shielding effect of the top layers of skin. See the article on sievert for more information on how this is calculated.

Internal irradiation

Radioactive contamination can be ingested into the human body if it is airborne or is taken in as contamination of food or drink, and will irradiate the body internally. The art and science of assessing internally generated radiation dose is Internal dosimetry.

The biological effects of ingested radionuclides depend greatly on the activity, the biodistribution, and the removal rates of the radionuclide, which in turn depends on its chemical form, the particle size, and route of entry. Effects may also depend on the chemical toxicity of the deposited material, independent of its radioactivity. Some radionuclides may be generally distributed throughout the body and rapidly removed, as is the case with tritiated water.

Some organs concentrate certain elements and hence radionuclide variants of those elements. This action may lead to much lower removal rates. For instance, the thyroid gland takes up a large percentage of any iodine that enters the body. Large quantities of inhaled or ingested radioactive iodine may impair or destroy the thyroid, while other tissues are affected to a lesser extent. Radioactive iodine-131 is a common fission product; it was a major component of the radioactivity released from the Chernobyl disaster, leading to nine fatal cases of pediatric thyroid cancer and hypothyroidism. On the other hand, radioactive iodine is used in the diagnosis and treatment of many diseases of the thyroid precisely because of the thyroid's selective uptake

The radiation warning symbol (trefoil)

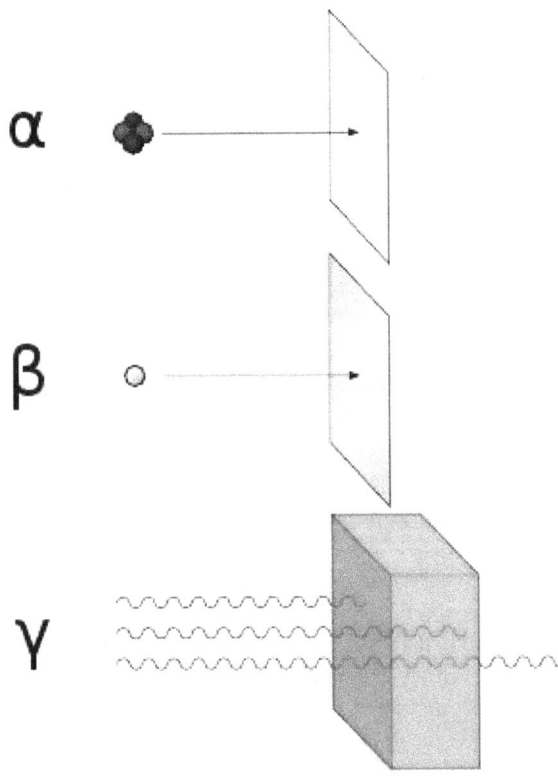

Alpha radiation consists of helium-4 nucleus and is readily stopped by a sheet of paper. Beta radiation, consisting of electrons, is halted by an aluminium plate. Gamma radiation is eventually absorbed as it penetrates a dense material. Lead is good at absorbing gamma radiation, due to its density.

Safety Reports Series No. 64.*[12]

of iodine.

The radiation risk proposed by the International Commission on Radiological Protection (ICRP) predicts that an effective dose of one sievert (100 rem) carries a 5.5% chance of developing cancer. Such a risk is the sum of both internal and external radiation dose.[13]

The ICRP states "Radionuclides incorporated in the human body irradiate the tissues over time periods determined by their physical half-life and their biological retention within the body. Thus they may give rise to doses to body tissues for many months or years after the intake. The need to regulate exposures to radionuclides and the accumulation of radiation dose over extended periods of time has led to the definition of committed dose quantities".[14] The ICRP further states "For internal exposure, committed effective doses are generally determined from an assessment of the intakes of radionuclides from bioassay measurements or other quantities (e.g., activity retained in the body or in daily excreta). The radiation dose is determined from the intake using recommended dose coefficients".[15]

The ICRP defines two dose quantities for individual committed dose:

Committed equivalent dose. $H_T(t)$ is the time integral of the equivalent dose rate in a particular tissue or organ that will be received by an individual following intake of radioactive material into the body by a Reference Person, where t is the integration time in years.[16] This refers specifically to the dose in a specific tissue or organ, in a similar way to external equivalent dose.

Committed effective dose, $E(t)$ is the sum of the products of the committed organ or tissue equivalent doses and the appropriate tissue weighting factors W_T, where t is the integration time in years following the intake. The commitment period is taken to be 50 years for adults, and to age 70 years for children.[16] This refers specifically to the dose to the whole body, in a similar way to external effective dose.

8.5.2 Social and psychological effects

A 2015 report in *Lancet* explained that serious impacts of nuclear accidents were often not directly attributable to radiation exposure, but rather social and psychological effects.[17] The consequences of low-level radiation are often more psychological than radiological. Because damage from very-low-level radiation cannot be detected, people exposed to it are left in anguished uncertainty about what will happen to them. Many believe they have been fundamentally contaminated for life and may refuse to have children for fear of birth defects. They may be shunned by others in their community who fear a sort of mysterious contagion.[18]

Forced evacuation from a radiological or nuclear accident may lead to social isolation, anxiety, depression, psychosomatic medical problems, reckless behavior, even suicide. Such was the outcome of the 1986 Chernobyl nuclear disaster in the Ukraine. A comprehensive 2005 study concluded that "the mental health impact of Chernobyl is the largest public health problem unleashed by the accident to date".[18] Frank N. von Hippel, a U.S. scientist, commented on the 2011 Fukushima nuclear disaster, saying that "fear of ionizing radiation could have long-term psychological effects on a large portion of the population in the contaminated areas".[19] Evacuation and long-term displacement of affected populations create problems for many people, especially the elderly and hospital patients.[17]

Such great psychological danger does not accompany other materials that put people at risk of cancer and other deadly illness. Visceral fear is not widely aroused by, for example, the daily emissions from coal burning, although, as a National Academy of Sciences study found, this causes 10,000 premature deaths a year in the US population of 317,413,000. Medical errors leading to death in U.S. hospitals are estimated to be between 44,000 and 98,000. It is "only nuclear radiation that bears a huge psychological burden —for it carries a unique historical legacy".[18]

8.6   See also

- Chemical hazard
- Criticality accident
- Human decontamination
- List of Milestone nuclear explosions
- Lists of nuclear disasters and radioactive incidents
- Low-background steel
- Nuclear and radiation accidents
- Nuclear debate (disambiguation)
- Nuclear power
- Radiation biology
- Radiation exposure (disambiguation)
- Radiophobia
- Relative biological effectiveness
- Rongelap Atoll
- Soviet submarine K-19
- Category:Victims of radiological poisoning

8.7   References

[1] Richard Schiffman (12 March 2013). "Two years on, America hasn't learned lessons of Fukushima nuclear disaster". *The Guardian.*

[2] Martin Fackler (June 1, 2011). "Report Finds Japan Underestimated Tsunami Danger". *New York Times.*

[3] International Atomic Energy Agency (2007). *IAEA Safety Glossary: Terminology Used in Nuclear Safety and Radiation Protection* (PDF). Vienna: IAEA. ISBN 92-0-100707-8.

[4] "Atmospheric δ^{14}C record from Wellington". *Trends: A Compendium of Data on Global Change. Carbon Dioxide Information Analysis Center.* Oak Ridge National Laboratory. 1994. Retrieved 2007-06-11.

[5] Levin, I.; et al. (1994). "δ^{14}C record from Vermunt". *Trends: A Compendium of Data on Global Change. Carbon Dioxide Information Analysis Center.*

[6] "Radiocarbon dating". University of Utrecht. Retrieved 2008-02-19.

[7] http://www.hse.gov.uk/pubns/irp7.pdf

[8] Operational Monitoring Good Practice Guide "The Selection of Alarm Levels for Personnel Exit Monitors" Dec 2009 - National Physical Laboratory. Teddington UK

[9] Dennis Normile, "Cooling a Hot Zone." Science, 339 (1 March 2013) pp. 1028-1029.

[10] ICRP Protection of people living in long term contaminated areas

[11] International Atomic Energy Agency (2005). *Environmental and Source Monitoring for Purposes of Radiation Protection, IAEA Safety Standards Series No. RS–G-1.8* (PDF). Vienna: IAEA.

[12] International Atomic Energy Agency (2010). *Programmes and Systems for Source and Environmental Radiation Monitoring. Safety Reports Series No. 64.* Vienna: IAEA. p. 234. ISBN 978-92-0-112409-8.

[13] ICRP publication 103 - Paragraph 83.

[14] ICRP Publication 103 paragraph 140

[15] ICRP publication 103 - Paragraph 144.

[16] ICRP publication 103 - Glossary.

[17] Arifumi Hasegawa, Koichi Tanigawa, Akira Ohtsuru, Hirooki Yabe, Masaharu Maeda, Jun Shigemura, et al. Health effects of radiation and other health problems in the aftermath of nuclear accidents, with an emphasis on Fukushima. *The Lancet,* 1 August 2015.

[18] Andrew C. Revkin (March 10, 2012). "Nuclear Risk and Fear, from Hiroshima to Fukushima". *New York Times.*

[19] Frank N. von Hippel (September–October 2011). *The radiological and psychological consequences of the Fukushima Daiichi accident. Bulletin of the Atomic Scientists.* vol. 67 no. 5. pp. 27–36.

• Measurement Good Practice Guide No. 30 "Practical Radiation Monitoring" Oct 2002 - National Physical Laboratory. Teddington UK

8.8   External links

- Q&A: Health effects of radiation exposure, *BBC News,* 21 July 2011.

- Alliance for Nuclear Responsibility

- training guide Brookhaven National Laboratory Training Guide.

- International Fund for Animal Welfare report on impact of radiation on animals

Chapter 9

Half-life

This article is about radioactive decay of isotopes. For other uses, see Half-Life (disambiguation).

Half-life (abbreviated $t_{1/2}$) is the time required for a quantity to reduce to half its initial value. The term is commonly used in nuclear physics to describe how quickly unstable atoms undergo, or how long stable atoms survive, radioactive decay. The term is also used more generally to characterize any type of exponential or non-exponential decay. For example, the medical sciences refer to the biological half-life of drugs and other chemicals in the body. The converse of half-life is doubling time.

The original term, *half-life period*, dating to Ernest Rutherford's discovery of the principle in 1907, was shortened to *half-life* in the early 1950s.[1] Rutherford applied the principle of a radioactive element's half-life to studies of age determination of rocks by measuring the decay period of radium to lead-206.

Half-life is constant over the lifetime of an exponentially decaying quantity, and it is a characteristic unit for the exponential decay equation. The accompanying table shows the reduction of a quantity as a function of the number of half-lives elapsed.

9.1 Probabilistic nature

A half-life usually describes the decay of discrete entities, such as radioactive atoms. In that case, it does not work to use the definition that states "half-life is the time required for exactly half of the entities to decay". For example, if there are 3 radioactive atoms with a half-life of one second, there will not be "1.5 atoms" left after one second.

Instead, the half-life is defined in terms of probability: "Half-life is the time required for exactly half of the entities to decay *on average*". In other words, the *probability* of a radioactive atom decaying within its half-life is 50%.

For example, the image on the right is a simulation of many

identical atoms undergoing radioactive decay. Note that after one half-life there are not *exactly* one-half of the atoms remaining, only *approximately*, because of the random variation in the process. Nevertheless, when there are many identical atoms decaying (right boxes), the law of large numbers suggests that it is a *very good approximation* to say that half of the atoms remain after one half-life.

There are various simple exercises that demonstrate probabilistic decay, for example involving flipping coins or running a statistical computer program.[2][3][4]

9.2 Formulas for half-life in exponential decay

Main article: Exponential decay

An exponential decay can be described by any of the following three equivalent formulas:

$$N(t) = N_0 \left(\frac{1}{2}\right)^{\frac{t}{t_{1/2}}}$$
$$N(t) = N_0 e^{-\frac{t}{\tau}}$$
$$N(t) = N_0 e^{-\lambda t}$$

where

- N_0 is the initial quantity of the substance that will decay (this quantity may be measured in grams, moles, number of atoms, etc.),

- $N(t)$ is the quantity that still remains and has not yet decayed after a time t,

- $t_{1/2}$ is the half-life of the decaying quantity,

- τ is a positive number called the mean lifetime of the decaying quantity,

71

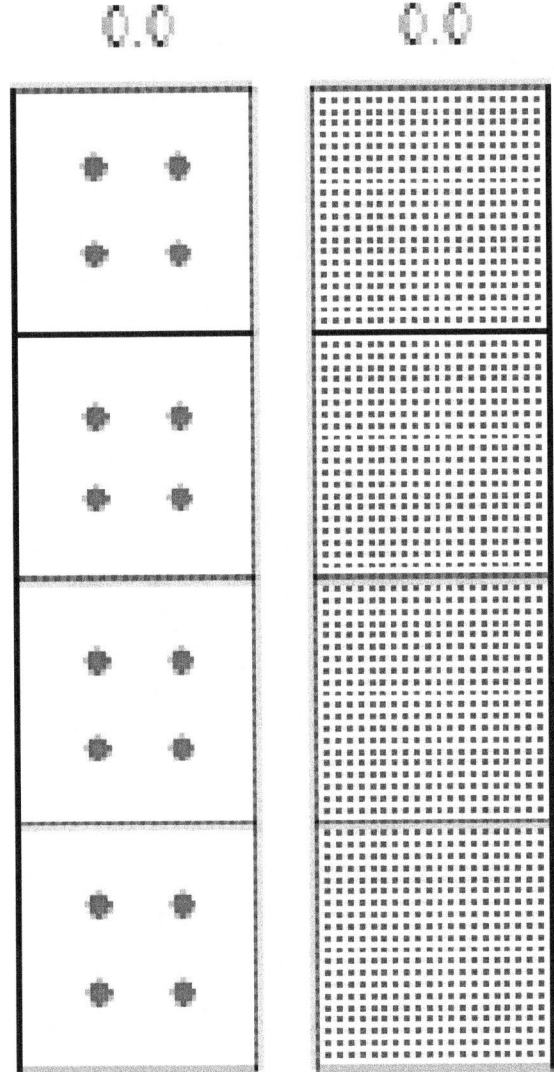

By plugging in and manipulating these relationships, we get all of the following equivalent descriptions of exponential decay, in terms of the half-life:

$$N(t) = N_0 \left(\frac{1}{2}\right)^{\frac{t}{t_{1/2}}} = N_0 2^{-t/t_{1/2}}$$
$$= N_0 e^{-t \ln(2)/t_{1/2}}$$

$$t_{1/2} = \frac{t}{\log_2(N_0/N(t))} = \frac{t}{\log_2(N_0) - \log_2(N(t))}$$
$$= \frac{1}{\log_{2^t}(N_0) - \log_{2^t}(N(t))} = \frac{t \ln(2)}{\ln(N_0) - \ln(N(t))}$$

Regardless of how it's written, we can plug into the formula to get

- $N(0) = N_0$ as expected (this is the definition of "initial quantity")

- $N(t_{1/2}) = \frac{1}{2}N_0$ as expected (this is the definition of half-life)

- $\lim_{t \to \infty} N(t) = 0$; i.e., amount approaches zero as t approaches infinity as expected (the longer we wait, the less remains).

9.2.1   Decay by two or more processes

Some quantities decay by two exponential-decay processes simultaneously. In this case, the actual half-life $T_{1/2}$ can be related to the half-lives t_1 and t_2 that the quantity would have if each of the decay processes acted in isolation:

$$\frac{1}{T_{1/2}} = \frac{1}{t_1} + \frac{1}{t_2}$$

For three or more processes, the analogous formula is:

$$\frac{1}{T_{1/2}} = \frac{1}{t_1} + \frac{1}{t_2} + \frac{1}{t_3} + \cdots$$

For a proof of these formulas, see Exponential decay § Decay by two or more processes.

9.2.2   Examples

Further information: Exponential decay § Applications and examples

There is a half-life describing any exponential-decay process. For example:

Simulation of many identical atoms undergoing radioactive decay, starting with either 4 atoms per box (left) or 400 (right). The number at the top is how many half-lives have elapsed. Note the consequence of the law of large numbers: with more atoms, the overall decay is more regular and more predictable.

- λ is a positive number called the decay constant of the decaying quantity.

The three parameters $t_{1/2}$, τ, and λ are all directly related in the following way:

$$t_{1/2} = \frac{\ln(2)}{\lambda} = \tau \ln(2)$$

where $\ln(2)$ is the natural logarithm of 2 (approximately 0.693).

Half life demonstrated using dice in a classroom experiment

- The current flowing through an RC circuit or RL circuit decays with a half-life of $RC\ln(2)$ or $\ln(2)L/R$, respectively. For this example, the term half time might be used instead of "half life", but they mean the same thing.

- In a first-order chemical reaction, the half-life of the reactant is $\ln(2)/\lambda$, where λ is the reaction rate constant.

- In radioactive decay, the half-life is the length of time after which there is a 50% chance that an atom will have undergone nuclear decay. It varies depending on the atom type and isotope, and is usually determined experimentally. See List of nuclides.

The half life of a species is the time it takes for the concentration of the substance to fall to half of its initial value.

9.3 In non-exponential decay

Main article: Rate equation

The decay of many physical quantities is not exponential —for example, the evaporation of water from a puddle, or (often) the chemical reaction of a molecule. In such cases, the half-life is defined the same way as before: as the time elapsed before half of the original quantity has decayed. However, unlike in an exponential decay, the half-life depends on the initial quantity, and the prospective half-life will change over time as the quantity decays.

As an example, the radioactive decay of carbon-14 is exponential with a half-life of 5,730 years. A quantity of carbon-14 will decay to half of its original amount (on average) after 5,730 years, regardless of how big or small the original

quantity was. After another 5,730 years, one-quarter of the original will remain. On the other hand, the time it will take a puddle to half-evaporate depends on how deep the puddle is. Perhaps a puddle of a certain size will evaporate down to half its original volume in one day. But on the second day, there is no reason to expect that one-quarter of the puddle will remain; in fact, it will probably be much less than that. This is an example where the half-life reduces as time goes on. (In other non-exponential decays, it can increase instead.)

The decay of a mixture of two or more materials which each decay exponentially, but with different half-lives, is not exponential. Mathematically, the sum of two exponential functions is not a single exponential function. A common example of such a situation is the waste of nuclear power stations, which is a mix of substances with vastly different half-lives. Consider a mixture of a rapidly decaying element A, with a half-life of 1 second, and a slowly decaying element B, with a half-life of 1 year. In a couple of minutes, almost all atoms of element A will have decayed after repeated halving of the initial number of atoms, but very few of the atoms of element B will have done so as only a tiny fraction of its half-life has elapsed. Thus, the mixture taken as a whole will not decay by halves.

9.4 In biology and pharmacology

Main article: Biological half-life

A biological half-life or elimination half-life is the time it takes for a substance (drug, radioactive nuclide, or other) to lose one-half of its pharmacologic, physiologic, or radiological activity. In a medical context, the half-life may also describe the time that it takes for the concentration of a substance in blood plasma to reach one-half of its steady-state value (the "plasma half-life").

The relationship between the biological and plasma half-lives of a substance can be complex, due to factors including accumulation in tissues, active metabolites, and receptor interactions.[5]

While a radioactive isotope decays almost perfectly according to so-called "first order kinetics" where the rate constant is a fixed number, the elimination of a substance from a living organism usually follows more complex chemical kinetics.

For example, the biological half-life of water in a human being is about 9 to 10 days, though this can be altered by behavior and various other conditions. The biological half-life of cesium in human beings is between one and four months.

9.5   See also

- Half time (physics)

- List of isotopes by half-life

- Mean lifetime

9.6   References

[1] John Ayto. " 20th Century Words" (1989). Cambridge University Press.

[2] Chivers, Sidney (March 16, 2003). " Re: What happens durring half lifes [sic] when there is only one atom left?". MADSCI.org.

[3] "Radioactive-Decay Model". Exploratorium.edu. Retrieved 2012-04-25.

[4] Wallin, John (September 1996). "Assignment #2: Data, Simulations, and Analytic Science in Decay". Astro.GLU.edu.

[5] Lin VW; Cardenas DD (2003). *Spinal cord medicine*. Demos Medical Publishing, LLC. p. 251. ISBN 1-888799-61-7.

9.7   External links

- Nucleonica.net, Nuclear Science Portal

- Nucleonica.net, wiki: Decay Engine

- Bucknell.edu, System Dynamics – Time Constants

- Subotex.com, Half-Life elimination of drugs in blood plasma – Simple Charting Tool

Chapter 10

Dirty bomb

For other uses, see Dirty bomb (disambiguation).
Not to be confused with Salted bomb.

A **dirty bomb** or **radiological dispersal device (RDD)** is a speculative radiological weapon that combines radioactive material with conventional explosives. The purpose of the weapon is to contaminate the area around the dispersal agent/conventional explosion with radioactive material, serving primarily as an area denial device against civilians. It is however not to be confused with a nuclear explosion, such as a fission bomb, which by releasing nuclear energy produces blast effects far in excess of what is achievable by the use of conventional explosives.

Though an RDD would be designed to disperse radioactive material over a large area, a bomb that uses conventional explosives and produces a blast wave would be far more lethal to people than the hazard posed by radioactive material that may be mixed with the explosive.[1] At levels created from probable sources, not enough radiation would be present to cause severe illness or death. A test explosion and subsequent calculations done by the United States Department of Energy found that assuming nothing is done to clean up the affected area and everyone stays in the affected area for one year, the radiation exposure would be "fairly high", but not fatal.[2][3] Recent analysis of the nuclear fallout from the Chernobyl disaster confirms this, showing that the effect on many people in the surrounding area, although not those in close proximity, was almost negligible.[4]

Since a dirty bomb is unlikely to cause many deaths by radiation exposure, many do not consider this to be a weapon of mass destruction.[2] Its purpose would presumably be to create psychological, not physical, harm through ignorance, mass panic, and terror. For this reason dirty bombs are sometimes called "weapons of mass disruption". Additionally, containment and decontamination of thousands of victims, as well as decontamination of the affected area might require considerable time and expense, rendering areas partly unusable and causing economic damage.

10.1 Dirty bombs and terrorism

Further information: Nuclear terrorism

Since the 9/11 attacks the fear of terrorist groups using dirty bombs has increased immensely, which has been frequently reported in the media.[5] The meaning of terrorism used here, is described by the U.S. Department of Defense's definition, which is "the calculated use of unlawful violence or threat of unlawful violence to inculcate fear; intended to coerce or to intimidate governments or societies in the pursuit of goals that are generally political, religious, or ideological objectives".[6] There have only ever been two cases of caesium-containing bombs, and neither was detonated. Both involved Chechnya. The first attempt of radiological terror was carried out in November 1995 by a group of Chechen separatists, who buried a caesium-137 source wrapped in explosives at the Izmaylovsky Park in Moscow. A Chechen rebel leader alerted the media, the bomb was never activated, and the incident amounted to a mere publicity stunt.[7]

In December 1998, a second attempt was announced by the Chechen Security Service, who discovered a container filled with radioactive materials attached to an explosive mine. The bomb was hidden near a railway line in the suburban area Argun, ten miles east of the Chechen capital of Grozny. The same Chechen separatist group was suspected to be involved.[8] Despite the increased fear of a dirty bombing attack, it is hard to assess whether the actual risk of such an event has increased significantly.[9] The following discussions on implications, effects and probability of an attack, as well as indications of terror groups planning such, are based mainly on statistics, qualified guessing and a few comparable scenarios.

10.1.1 Effect of a dirty bomb explosion

When dealing with the implications of a dirty bomb attack, there are two main areas to be addressed: (i) the civilian impact, not only dealing with immediate casualties and long term health issues, but also the psychological effect and then (ii) the economic impact. With no prior event of a dirty bomb detonation, it is considered difficult to predict the impact. Several analyses have predicted that RDDs will neither sicken nor kill many people.[10]

10.1.2 Accidents with radioactives

The effects of uncontrolled radioactive contamination have been reported several times.

See also: Lists of nuclear disasters and radioactive incidents

One example is the radiological accident occurring in Goiânia, Brazil, between September 1987 and March 1988: Two metal scavengers broke into an abandoned radiotherapy clinic and removed a teletherapy source capsule containing powdered caesium-137 with an activity of 50 TBq. They brought it back to the home of one of the men to take it apart and sell as scrap metal. Later that day both men were showing acute signs of radiation illness with vomiting and one of the men had a swollen hand and diarrhea. A few days later one of the men punctured the 1 mm thick window of the capsule, allowing the caesium chloride powder to leak out and when realizing the powder glowed blue in the dark, brought it back home to his family and friends to show it off. After 2 weeks of spread by contact contamination causing an increasing number of adverse health effects, the correct diagnosis of acute radiation sickness was made at a hospital and proper precautions could be put into procedure. By this time 249 people were contaminated, 151 exhibited both external and internal contamination of which 20 people were seriously ill and 5 people died.[11]

The Goiânia incident to some extent predicts the contamination pattern if it is not immediately realized that the explosion spread radioactive material, but also how fatal even very small amounts of ingested radioactive powder can be.[12] This raises worries of terrorists using powdered alpha emitting material, that if ingested can pose a serious health risk,[13] as in the case of deceased former K.G.B. spy Alexander Litvinenko, who either ate, drank or inhaled polonium-210. "Smoky bombs" based on alpha emitters might easily be just as dangerous as beta or gamma emitting dirty bombs.[14]

Main article: Goiânia accident

10.1.3 Public perception of risks

For the majority involved in an RDD incident, the radiation health risks (i.e. increased probability of developing cancer later in life due to radiation exposure) are comparatively small, comparable to the health risk from smoking five packages of cigarettes on a daily basis.[15] The fear of radiation is not always logical. Although the exposure might be minimal, many people find radiation exposure especially frightening because it is something they cannot see or feel, and it therefore becomes an unknown source of danger. Dealing with public fear may prove the greatest challenge in case of an RDD event.[16] Policy, science and media may inform the public about the real danger and thus reduce the possible psychological and economic effects.

Statements from the U.S. government after 9/11 may have contributed unnecessarily to the public fear of a dirty bomb. When United States Attorney General John Ashcroft on June 10, 2002, announced the arrest of José Padilla, allegedly plotting to detonate such a weapon, he said:

> [A] radioactive "dirty bomb" (...) spreads radioactive material that is highly toxic to humans and can cause mass death and injury.
> —Attorney General John Ashcroft[12]

This public fear of radiation also plays a big role in why the costs of an RDD impact on a major metropolitan area (such as lower Manhattan) might be equal to or even larger than that of the 9/11 attacks.[12] Assuming the radiation levels are not too high and the area does not need to be abandoned such as the town of Pripyat near the Chernobyl reactor,[17] an expensive and time consuming cleanup procedure will begin. This will mainly consist of tearing down highly contaminated buildings, digging up contaminated soil and quickly applying sticky substances to remaining surfaces so that radioactive particles adhere before radioactivity penetrates the building materials.[18] These procedures are the current state of the art for radioactive contamination cleanup, but some experts say that a complete cleanup of external surfaces in an urban area to current decontamination limits may not be technically feasible.[12] Loss of working hours will be vast during cleanup, but even after the radiation levels reduce to an acceptable level, there might be residual public fear of the site including possible unwillingness to conduct business as usual in the area. Tourist traffic is likely never to resume.[12]

There is also a psychological warfare element to radioactive substances. Visceral fear is not widely aroused by the daily emissions from coal burning, for example, even though a National Academy of Sciences study found this causes 10,000 premature deaths a year in the US population of 317,413,000. Medical errors leading to death in U.S. hospitals are estimated to be between 44,000 and 98,000. It is "only nuclear radiation that bears a huge psychological burden —for it carries a unique historical legacy" ."[19]

10.1.4 Constructing and obtaining material for a dirty bomb

In order for a terrorist organization to construct and detonate a dirty bomb, it must acquire radioactive material. Possible RDD material could come from the millions of radioactive sources used worldwide in the industry, for medical purposes and in academic applications mainly for research."[20] Of these sources, only nine reactor produced isotopes stand out as being suitable for radiological terror: americium-241, californium-252, caesium-137, cobalt-60, iridium-192, plutonium-238, polonium-210, radium-226 and strontium-90,"[9] and even from these it is possible that radium-226 and polonium-210 do not pose a significant threat."[21] Of these sources the U.S. Nuclear Regulatory Commission has estimated that within the U.S., approximately one source is lost, abandoned or stolen every day of the year. Within the European Union the annual estimate is 70."[22] There exist thousands of such "orphan" sources scattered throughout the world, but of those reported lost, no more than an estimated 20 percent can be classified as a potential high security concern if used in a RDD."[21] Especially Russia is believed to house thousands of orphan sources, which were lost following the collapse of the Soviet Union. A large but unknown number of these sources probably belong to the high security risk category. Noteworthy are the beta emitting strontium-90 sources used as radioisotope thermoelectric generators for beacons in lighthouses in remote areas of Russia."[23] In December 2001, three Georgian woodcutters stumbled over such a power generator and dragged it back to their camp site to use it as a heat source. Within hours they suffered from acute radiation sickness and sought hospital treatment. The International Atomic Energy Agency (IAEA) later stated that it contained approximately 40 kilocuries (1.5 PBq) of strontium,"[24] equivalent to the amount of radioactivity released immediately after the Chernobyl accident (though the total radioactivity release from Chernobyl was 2500 times greater at around 100 MCi (3,700 PBq)"[25]).

Although a terrorist organization might obtain radioactive material through the "black market","[26] and there has been a steady increase in illicit trafficking of radioactive sources from 1996 to 2004, these recorded trafficking incidents mainly refer to rediscovered orphan sources without any sign of criminal activity."[9] and it has been argued that there is no conclusive evidence for such a market."[27] In addition to the hurdles of obtaining usable radioactive material, there are several conflicting requirements regarding the properties of the material the terrorists need to take into consideration: First, the source should be "sufficiently" radioactive to create direct radiological damage at the explosion or at least to perform societal damage or disruption. Second, the source should be transportable with enough shielding to protect the carrier, but not so much that it will be too heavy to maneuver. Third, the source should be sufficiently dispersible to effectively contaminate the area around the explosion."[28]

An example of a worst-case scenario is a terror organization possessing a source of very highly radioactive material, e.g. a strontium-90 thermal generator, with the ability to create an incident comparable to the Chernobyl accident. Although the detonation of a dirty bomb using such a source might seem terrifying, it would be hard to assemble the bomb and transport it without severe radiation damage and possible death of the perpetrators involved. Shielding the source effectively would make it almost impossible to transport and a lot less effective if detonated.

Due to the three constraints of making a dirty bomb, RDDs might still be defined as "high-tech" weapons and this is probably why they have not been used up to now."[28]

10.1.5 Possibility of terrorist groups using dirty bombs

The present assessment of the possibility of terrorists using a dirty bomb is based on cases involving Al-Qaeda. This is because the attempts by this group to acquire a dirty bomb are the most well-described in the literature, in part due to the attention this group received for their involvement in the 9/11 attacks.

On 8 May 2002, José Padilla (a.k.a. Abdulla al-Muhajir) was arrested on suspicion that he was an Al-Qaeda terrorist planning to detonate a dirty bomb in the U.S. This suspicion was raised by information obtained from an arrested top Al-Qaeda official in U.S. custody, Abu Zubaydah, who under interrogation revealed that the organization was close to constructing a dirty bomb. Although Padilla had not obtained radioactive material or explosives at the time of arrest, law enforcement authorities uncovered evidence that he was on reconnaissance for usable radioactive material and possible locations for detonation."[29] It has been doubted whether José Padilla was preparing such an attack, and it has been claimed that the arrest was highly politically motivated, given the pre-9/11 security lapses by the CIA and FBI."[30]

Later, these charges against José Padilla were dropped. Although there was no hard evidence for Al-Qaeda possessing a dirty bomb, there is a broad agreement that Al-Qaeda poses a potential dirty bomb attack threat"[31] because they need to overcome the alleged image that the U.S. and its allies are winning the war against terror.[5] A further concern is the argument, that "if suicide bombers are prepared to die flying airplanes into building, it is also conceivable that they are prepared to forfeit their lives building dirty bombs".[32] If this would be the case, both the cost and complexity of any protective systems needed to allow the perpetrator to survive long enough to both build the bomb and carry out the attack, would be significantly reduced.[12]

Several other captives were alleged to have played a role in this plot.[33] Guantanamo captive Binyam Mohammed has alleged he was subjected to extraordinary rendition, and that his confession of a role in the plot was coerced through torture.[34][35] He sought access through the American and United Kingdom legal systems to provide evidence he was tortured.[36][37] Guantanamo military commission prosecutors continue to maintain the plot was real, and charged Binyam for his alleged role in 2008. However they dropped this charge in October 2008, but maintain they could prove the charge and were only dropping the charge to expedite proceedings. US District Court Judge Emmet G. Sullivan insisted that the administration still had to hand over the evidence that justified the dirty bomb charge, and admonished United States Department of Justice lawyers that dropping the charge "raises serious questions in this court's mind about whether those allegations were ever true."

In 2006, Dhiren Barot from North London pleaded guilty of conspiring to murder innocent people within the United Kingdom and United States using a radioactive dirty bomb. He planned to target underground car parks within the UK and buildings in the U.S. such as the International Monetary Fund, World Bank buildings in Washington D.C., the New York Stock Exchange, Citigroup buildings and the Prudential Financial buildings in Newark, New Jersey. He also faces 12 other charges including, conspiracy to commit public nuisance, seven charges of making a record of information for terrorist purposes and four charges of possessing a record of information for terrorist purposes. Experts say if the plot to use the dirty bomb was carried out "it would have been unlikely to cause deaths, but was designed to affect about 500 people."[38]

In January 2009, a leaked FBI report described the results of a search of the Maine home of James G. Cummings, a white supremacist who had been shot and killed by his wife. Investigators found four one-gallon containers of 35 percent hydrogen peroxide, uranium, thorium, lithium metal, aluminum powder, beryllium, boron, black iron oxide and magnesium as well as literature on how to build dirty bombs and information about cesium-137, strontium-90 and cobalt-60, radioactive materials.[39] Officials confirmed the veracity of the report but stated that the public

was never at risk.[40]

In April 2009, the Security Service of Ukraine announced the arrest of a legislator and two businessmen from the Ternopil Oblast. Seized in the undercover sting operation was 3.7 kilograms of what was claimed by the suspects during the sale as plutonium-239, used mostly in nuclear reactors and nuclear weapons, but was determined by experts to be probably americium, a "widely used" radioactive material which is commonly used in amounts of less than 1 milligram in smoke detectors, but can also be used in a dirty bomb. The suspects reportedly wanted US$ 10 million for the material, which the Security Service determined was produced in Russia during the era of the Soviet Union and smuggled into Ukraine through a neighboring country.[41][42]

In July 2014, ISIS militants seized 88 pounds of uranium compounds from Mosul University. The material was unenriched and so could not be used to build a conventional fission bomb, but a dirty bomb is a theoretical possibility. However, uranium's relatively low radioactivity makes it a poor candidate for use in a dirty bomb.[43][44]

Little is known about civil preparedness to respond to a dirty bomb attack. The Boston Marathon appeared to many to be a situation with high potential for use of a dirty bomb as a terrorist weapon.[45] However, the bombing attack that occurred on April 15, 2013 did not involve use of dirty bombs. Any radiological testing or inspections that may have occurred following the attack were either conducted sub rosa or not at all. Also, there was no official dirty bomb "all clear" issued by the Obama administration. Massachusetts General Hospital had, apparently under their own disaster plan, issued instructions to their emergency room to be prepared for incoming radiation poisoning cases.[46]

10.2 Dirty bomb tests

Israel carried out a four-year series of tests on nuclear explosives to measure the effects were "hostile forces" ever to use them against Israel, Israel's Haaretz daily newspaper reported June 8, 2015.[47]

10.3 Other uses of the term

The term has also been used historically to refer to certain types of nuclear weapons. Due to the inefficiency of early nuclear weapons, only a small amount of the nuclear material would be consumed during the explosion. Little Boy had an efficiency of only 1.4%. Fat Man, which used a different design and a different fissile material, had an efficiency of 14%. Thus, they tended to disperse large amounts of unused fissile material, and the fission products, which are on average much more dangerous, in the form of nuclear fallout. During the 1950s, there was considerable debate over whether "clean" bombs could be produced and these were often contrasted with "dirty" bombs. "Clean" bombs were often a stated goal and scientists and administrators said that high-efficiency nuclear weapon design could create explosions which generated almost all of their energy in the form of nuclear fusion, which does not create harmful fission products.

But the *Castle Bravo* accident of 1954, in which a thermonuclear weapon produced a large amount of fallout which was dispersed among human populations, suggested that this was not what was actually being used in modern thermonuclear weapons, which derive around half of their yield from a final fission stage of the fast fissioning of the uranium tamper of the secondary. While some proposed producing "clean" weapons, other theorists noted that one could make a nuclear weapon intentionally "dirty" by "salting" it with a material, which would generate large amounts of long-lasting fallout when irradiated by the weapon core. These are known as salted bombs; a specific subtype often noted is a cobalt bomb.

10.4 In popular culture

- In the 1964 British movie *Goldfinger*, both Auric Goldfinger and James Bond refer to the nuclear device being smuggled into Fort Knox as "dirty."

- The crime drama television series *Numb3rs* has an episode that revolves around a dirty bomb (season 1, episode 10).

- In a two-part 2011 episode of *Castle*, former US soldiers plot to detonate a dirty bomb in New York City and frame a Syrian immigrant for the crime.

- In the 2013 Indian movie *Vishwaroopam*, the plot revolves around a dirty bomb developed by scraping caesium from oncological equipment to trigger a blast in New York City.

- In the 2014 movie, Batman: Assault on Arkham, Joker has a dirty bomb which he plans on detonating in Gotham.

- In the January 14, 2016 Republican presidential debates, Ben Carson referenced dirty bombs twice when speaking on US foreign policy.

- In the June 1, 2015 game by Splash Damage Dirty Bomb, the game is played in a dirty bomb fallout area in London.

- In the *Madam Secretary* episode "Right of the Boom", a dirty bomb is detonated at a women's education conference in Washington, D.C.

10.5 See also

- Lists of nuclear disasters and radioactive incidents
- Nuclear warfare
- Radiation hormesis
- Radiation poisoning
- Nuclear weapon design
- Depleted uranium
- Gammator
- 1968 Thule Air Base B-52 crash
- German nuclear energy project

10.6 References

10.6.1 Notes

[1] http://www.bioterrorism.slu.edu/bt/products/ahec_rad/ppt/Dirty%20Bomb.ppt

[2] "NRC: Fact Sheet on Dirty Bombs" .

[3] "Yahoo Screen - Watch videos online" . *Yahoo Screen*. 23 March 2015. Retrieved 30 March 2015.

[4] " BBC NEWS - Science/Nature - Chernobyl's 'nuclear nightmares'". Retrieved 30 March 2015.

[5] Petroff (2007)

[6] "404w DTIC Maintenanc" . Retrieved 30 March 2015.

[7] King (2004); NOVA, Chronology of events

[8] Edwards (2004); NOVA, Chronology of events

[9] Frost (2005)

[10] Reshetin (2005); Dingle (2005)

[11] King (2004); Zimmerman and Loeb (2004); Sohier and Hardeman (2006)

[12] Zimmerman and Loeb (2004)

[13] Mullen et al. (2002); Reshetin (2005)

[14] Zimmerman (2006)

[15] Ring (2004)

[16] Johnson (2003)

[17] "The Lifeless Silence of Pripyat". *Time Magazine*, June 23, 1986 Online article from Time Magazine

[18] Vantine and Crites (2002); Zimmerman and Loeb (2004); Weiss (2005)

[19] Andrew C. Revkin (March 10, 2012). "Nuclear Risk and Fear, from Hiroshima to Fukushima". *New York Times*.

[20] Ferguson et al. (2003); Frost (2005)

[21] Ferguson et al. (2003)

[22] Ferguson et al. (2003); Zimmerman and Loeb (2004)

[23] Burgess (2003); Van Tuyle and Mullen (2003); Sohier and Hardeman (2006)

[24] "NOVA - Dirty Bomb - Chronology of Events - PBS". Retrieved 30 March 2015.

[25] "Chernobyl". Retrieved 30 March 2015.

[26] King (2004); Hoffman (2006)

[27] Belyaninov (1994); Frost (2005)

[28] Sohier and Hardeman (2006)

[29] Ferguson et al. (2003); Hosenball et al. (2002)

[30] Burgess (2003); King (2004)

[31] King (2004); Ferguson et al. (2003)

[32] Burgess (2003)

[33] "Judge in Guantanamo case questions dirty bomb allegations". The Statesman. 2008-10-31. Retrieved 2008-11-01.

[34] Peter Finn, Del Quentin Wilbur (2008-10-31). "Motives of Justice Lawyers Questioned in Detainee's Case". Washington Post. Retrieved 2008-11-01.

[35] William Glaberson (2008-10-31). "Questioning 'dirty bomb' plot, judge orders U.S. to yield papers on detainee". International Herald Tribune. Retrieved 2008-11-01.

[36] Debra Cassens Weiss (2008-10-23). "UK Court: US Should Release Documents Relating to Detainee's Torture Claim". American Bar Association Journal. Retrieved 2008-11-01.

[37] Robert Verkaik (2008-10-31). "CIA officers could face trial in Britain over torture allegations". London: The Independent. Retrieved 2008-11-01.

[38] "Man admits UK-US terror bomb plot". *BBC News*. 2006-10-12. Retrieved 2010-04-01.

[39] Report: 'Dirty bomb' parts found in slain man's home. *Bangor Daily News*. 10 February 2009

[40] Officials verify dirty bomb probe results. *Bangor Daily News*. 11 February 2009

[41] "Three arrested in Ukraine for trying to sell radioactive material". Xinhua. 2009-04-25. Retrieved 2009-04-17.

[42] "Ukraine arrests 3 in radioactive material sale". AP. 2009-04-14. Retrieved 2009-04-17.

[43] "Iraqi 'Terrorist Groups' Have Seized Nuclear Materials".

[44] "ISIS seizes uranium from lab; experts downplay 'dirty bomb' threat".

[45] "Terror response study spurs concern". *Boston.com*. Retrieved 30 March 2015.

[46] "Determined Response: MGH and the Boston Marathon Bombing". *Massachusetts General Hospital Giving*. Retrieved 30 March 2015.

[47] "Israel tested 'dirty bombs' in the Negev Desert". Retrieved 9 June 2015.

10.6.2   Works cited

- Belyaninov, K. (1994). "Nuclear nonsense, black-market bombs, and fissile flim-flam". *Bulletin of the Atomic Scientists*. **50** (2), pp. 44–50.

- Burgess, M. (2003) "Pascal's New Wager: The Dirty Bomb Threat Heightens". Center for Defense Information.

- Dingle, J. (2005). "DIRTY BOMBS: real threat?". *Security*. **42** (4), p. 48.

- Edwards, R. (2004). "Only a matter of time?". *New Scientist*. **182** (2450), pp. 8–9.

- Adam Curtis's The Power of Nightmares. Part III – Video/Transcript at Information Clearing House

- Ferguson, C.D., Kazi, T. and Perera J. (2003) *Commercial Radioactive Sources: Surveying the Security Risks*, Monterey Institute of International Studies, Center for Nonproliferation Studies, Occasional Paper #11, ISBN 1-885350-06-6, Webpage with PDF file of paper.

- Frost, R. M. (2005), *Nuclear Terrorism After 9/11*, Routledge for The International Institute for Strategic Studies, ISBN 0-415-39992-0.

- Hoffman, B. (2006), *Inside Terrorism*, Columbia University Press, N.Y., ISBN 0-231-12698-0.

- Hosenball, M., Hirsch, M. and Moreau, R. (2002) "War on Terror: Nabbing a "Dirty Bomb" Suspect", *Newsweek (Int. ed.)*, **ID: X7835733**: 28-33.

- Johnson, Jr., R.H. (2003), "Facing the Terror of Nuclear Terrorism", *Occupational Health & Safety*, **72** (5), pp. 44–50.

- King, G. (2004), *Dirty Bomb: Weapon of Mass Disruption*, Chamberlain Bros., Penguin Group, ISBN 1-59609-000-6.

- Liolios, T.E. (2008) *The effects of using Cesium-137 teletherapy sources as a radiological weapon (dirty bomb)*, Hellenic Arms Control Center, Occasional Paper May 2008, .

- Mullen, E., Van Tuyle, G. and York, R. (2002) "Potential radiological dispersal device (RDD) threats and related technology", *Transactions of the American Nuclear Society*, **87**: 309.

- Petroff, D.M. (2003), "Responding to 'dirty bombs'", *Occupational Health and Safety*, **72** (9), pp. 82–87.

- Reshetin, V.P. (2005), "Estimation of radioactivity levels associated with a ^{90}Sr dirty bomb event", *Atmospheric Environment*, **39** (25), pp. 4471–4477, doi:10.1016/j.atmosenv.2005.03.047.

- Ring, J.P. (2004), "Radiation Risks and Dirty Bombs", *The Radiation Safety Journal, Health Physics*, **86** (suppl. 1), pp. S42–S47.

- Sohier, A. and Hardeman, F. (2006) "Radiological Dispersion Devices: are we prepared?", *Journal of Environmental Radioactivity*, **85**: 171-181.

- Van Tuylen, G.J. and Mullen, E. (2003) "Large radiological source applications: RDD implications and proposed alternative technologies", *Global 2003: Atoms for Prosperity: Updating Eisenhouwer's Global Vision for Nuclear Energy*, **LA-UR-03-6281**: 622-631, ISBN 0-89448-677-2.

- Vantine, H.C. and Crites, T.R. (2002) "Relevance of nuclear weapons cleanup experience to dirty bomb response", *Transactions of the American Nuclear Society*, **87**: 322-323.

- Weiss, P. (2005), "Ghost town busters", *Science news*, Science News, Vol. 168, No. 18, **168** (18), pp. 282–284, doi:10.2307/4016859, JSTOR 4016859.

- Zimmerman, P.D. and Loeb, C. (2004) "Dirty Bombs: The Threat Revisited", *Defense Horizons*, **38**: 1-11.

- Zimmerman, P.D. (2006), "The Smoky Bomb Threat", *New York Times*, **156** (53798), p. 33.

10.7 External links

- U.S. Nuclear Regulatory Commission, Factsheet on Dirty Bombs

- Al Qaeda's Nuclear Options - Crusade Media News - http://www.crusade-media.com/news1.html

- Council on Foreign Relations, Terrorism Q&A: Dirty Bombs

- U.S. Dep't of Labor Occupational Safety & Health Administration, Radiological Dispersal Devices / Dirty Bombs

- Federation of American Scientists, Dirty bomb threat analysis

- Health Physics Society, Factsheet

- Health Physics Society, January 2004 study, Dirty Bombs Could Cause Devastating Economic Damage

- CNN, Explosion, not radiation, "dirty bomb's" worst fallout

- PBS, NOVA, Dirty Bomb This Web site was produced for PBS Online by WGBH. Web site © 1996–2003 WGBH Educational Foundation

- Lost and stolen nuclear materials in the US Three Mile Island Alert describes the problem

- The making of the terror myth

- Annotated bibliography for dirty bombs from the Alsos Digital Library for Nuclear Issues

- "Dirty Bombs": Background in Brief Congressional Research Service

- "Dirty Bombs": Technical Background, Attack Prevention and Response, Issues for Congress Congressional Research Service

Chapter 11

Salted bomb

Not to be confused with Dirty bomb.

A **salted bomb** is a nuclear weapon designed to function as a radiological weapon, producing enhanced quantities of radioactive fallout, rendering a large area uninhabitable.*[1] The term is derived both from the means of their manufacture, which involves the incorporation of additional elements to a standard atomic weapon, and from the expression "to salt the earth", meaning to render an area uninhabitable for generations. The idea originated with Hungarian-American physicist Leo Szilard, in February 1950. His intent was not to propose that such a weapon be built, but to show that nuclear weapon technology would soon reach the point where it could end human life on Earth.*[1]*[2] No intentionally salted bomb has ever been atmospherically tested,*[1] however the UK tested a 1 kiloton bomb incorporating a small amount of cobalt as an experimental radiochemical tracer at their Tadje testing site in Maralinga range, Australia, on September 14, 1957.*[3] Furthermore the triple "taiga" nuclear salvo test, as part of the preliminary March 1971 Pechora–Kama Canal project, produced substantial amounts of Co-60, with this fusion generated neutron activation product being responsible for about half of the gamma dose now (2011) at the test site.*[4]*[5]

A salted bomb should not be confused with a *dirty bomb*, which is an ordinary explosive bomb containing radioactive material which is spread over the area when the bomb explodes. A salted bomb is able to contaminate a much larger area than a dirty bomb.

11.1   Design

Salted versions of both fission and fusion weapons can be made by surrounding the core of the explosive device with a material containing an element that can be converted to a highly radioactive isotope by neutron bombardment.*[1] When the bomb explodes, the element absorbs neutrons released by the nuclear reaction, converting it to its radioactive form. The explosion scatters the resulting radioactive material over a wide area, leaving it uninhabitable far longer than an area affected by typical nuclear weapons. In a salted hydrogen bomb, the radiation case around the fusion fuel, which normally is made of some fissionable element, is replaced with a metallic salting element. Salted fission bombs can be made by replacing the neutron reflector between the fissionable core and the explosive layer with a metallic element. The energy yield from a salted weapon is usually lower than from an ordinary weapon of similar size as a consequence of these changes.

The radioactive isotope used for the fallout material would be a high intensity gamma ray emitter, with a half-life long enough that it remains lethal for an extended period. It would also have to have a chemistry that causes it to return to earth as fallout, rather than stay in the atmosphere after being vaporized in the explosion. Another consideration is biological: radioactive isotopes of elements normally taken up by animals as nutrition would pose a special threat to organisms that absorbed them, as their radiation would be delivered from within the body of the organism.

One example of a possible salted bomb would be a cobalt bomb, which would produce the radioactive isotope cobalt-60 (^{60}Co). Other radioactive isotopes that have been suggested for salted bombs include gold-198 (^{198}Au), tantalum-182 (^{182}Ta) and zinc-65 (^{65}Zn).*[2] Sodium-24 has also been proposed as a salting agent.*[6]

11.2   In popular culture

This concept is best known from the Soviet "Doomsday Machine" in the 1964 satirical Cold War film *Dr. Strangelove*. In the 1957 novel *On the Beach* by Nevil Shute, the death of all humanity is brought about by the detonation of cobalt bombs in the Northern Hemisphere. The 1970s movie *Beneath the Planet of the Apes* featured an atomic bomb that was hypothesized to use a cobalt casing. The use of a salted bomb is a component to the plot of Frank Miller's graphic novel series *The Dark Knight Returns* and 2008 TV programme *Ultimate Force Slow Bomb* episode.

Also, in the ABC show *The Whispers* season 1 episode 5, a "salted bomb" was referred to as a nuclear bomb laced with Arsenic, also known as "A.S. 33".

11.3 See also

- Doomsday device

- Fail-deadly

11.4 References

[1] Bhushan, K.; G. Katyal (2002). *Nuclear, Biological, and Chemical Warfare*. India: APH Publishing. pp. 75–77. ISBN 81-7648-312-5.

[2] Sublette, Carey (July 2007). "Types of nuclear weapons". *FAQ*. nuclearweaponarchive.org. Retrieved 2010-02-13.

[3] "British Nuclear Testing".

[4] Radiological investigations at the "Taiga" nuclear explosion site: Site description and in situ measurements V Ramzaev, V Repin, A Medvedev, E Khramtsov ··· - Journal of environmental ···, 2011 - Elsevier

[5] Radiological investigations at the "Taiga" nuclear explosion site, part II: man-made γ-ray emitting radionuclides in the ground and the resultant kerma rate in air V Ramzaev, V Repin, A Medvedev, E Khramtsov ··· - Journal of environmental ···, 2012 - Elsevier

[6] "Science: ty for Doomsday". *Time*. November 24, 1961.

Chapter 12

Psychological warfare

"Psywar" redirects here. For Norwegian black metal band Mayhem's song, see Psywar (Mayhem song).

Psychological warfare (**PSYWAR**), or the basic aspects of modern **psychological operations** (**PSYOP**), have been known by many other names or terms, including MISO, Psy Ops, Political Warfare, "Hearts and Minds", and propaganda.[1] The term is used "to denote any action which is practiced mainly by psychological methods with the aim of evoking a planned psychological reaction in other people".[2] Various techniques are used, and are aimed at influencing a target audience's value system, belief system, emotions, motives, reasoning, or behavior. It is used to induce confessions or reinforce attitudes and behaviors favorable to the originator's objectives, and are sometimes combined with black operations or false flag tactics. It is also used to destroy the morale of enemies through tactics that aim to depress troops' psychological states.[3][4] Target audiences can be governments, organizations, groups, and individuals, and is not just limited to soldiers. Civilians of foreign territories can also be targeted by technology and media so as to cause an effect in the government of their country.[5]

In *Propaganda: The Formation of Men's Attitudes*, Jacques Ellul discusses psychological warfare as a common peace policy practice between nations as a form of indirect aggression. This type of propaganda drains the public opinion of an opposing regime by stripping away its power on public opinion. This form of aggression is hard to defend against because no international court of justice is capable of protecting against psychological aggression since it cannot be legally adjudicated. "Here the propagandists is [sic] dealing with a foreign adversary whose morale he seeks to destroy by psychological means so that the opponent begins to doubt the validity of his beliefs and actions."[6][7]

12.1   History

12.1.1   Early

Mosaic of Alexander the Great on his campaign against the Persian Empire.

Since prehistoric times, warlords and chiefs have recognised the importance of inducing psychological terror in opponents. Facing armies would shout, hurl insults at each other and beat weapons together or on shields prior to an engagement, all designed to intimidate the enemy. Massacres and other atrocities were certainly first employed at this time to subdue enemy or rebellious populations or induce an enemy to abandon their struggle.

Currying favour with supporters was the other side of psychological warfare, and an early practitioner of such this was Alexander the Great, who successfully conquered large parts of Europe and the Middle East and held on to his territorial gains by co-opting local elites into the Greek administration and culture. Alexander left some of his men behind in each conquered city to introduce Greek culture and oppress dissident views. His soldiers were paid dowries to marry locals[8] in an effort to encourage assimilation.

Genghis Khan, leader of the Mongolian Empire in the 13th century AD employed less subtle techniques. Defeating the will of the enemy before having to attack and reaching a consented settlement was preferable to actually fighting.

The Mongol generals demanded submission to the Khan, and threatened the initially captured villages with complete destruction if they refused to surrender. If they had to fight to take the settlement, the Mongol generals fulfilled their threats and massacred the survivors. Tales of the encroaching horde spread to the next villages and created an aura of insecurity that undermined the possibility of future resistance.[9]

The Khan also employed tactics that made his numbers seem greater than they actually were. During night operations he ordered each soldier to light three torches at dusk to give the illusion of an overwhelming army and deceive and intimidate enemy scouts. He also sometimes had objects tied to the tails of his horses, so that riding on open and dry fields raised a cloud of dust that gave the enemy the impression of great numbers. His soldiers used arrows specially notched to whistle as they flew through the air, creating a terrifying noise.[10]

Another tactic favoured by the Mongols was catapulting severed human heads over city walls to frighten the inhabitants and spread disease in the besieged city's closed confines. This was especially used by the later Turko-Mongol chieftain.

The Muslim caliph Omar, in his battles against the Byzantine Empire, sent small reinforcements in the form of a continuous stream, giving the impression that a large force would accumulate eventually if not swiftly dealt with.

In the 6th century BCE Greek Bias of Priene successfully resisted the Lydian king Alyattes by fattening up a pair of mules and driving them out of the besieged city.[11] When Alyattes' envoy was then sent to Priene, Bias had piles of sand covered with corn to give the impression of plentiful resources.

This ruse appears to have been well known in medieval Europe: defenders in castles or towns under siege would throw food from the walls to show besiegers that provisions were plentiful. A famous example occurs in the 8th century legend of Lady Carcas, who supposedly persuaded the Franks to abandon a five-year siege by this means and gave her name to Carcassonne as a result.

12.1.2 Modern

First World War

The start of modern psychological operations in war is generally dated to the First World War. By that point, Western societies were increasingly educated and urbanized, and mass media was available in the form of large circulation newspapers and posters. It was also possible to transmit propaganda to the enemy via the use of airborne leaflets or

Lord Bryce led the commission of 1915 to document German atrocities committed against Belgian civilians.

through explosive delivery systems like modified artillery or mortar rounds.[12]

At the start of the war, the belligerents, especially the British and Germans, began distributing propaganda, both domestically and on the Western front. The British had several advantages that allowed them to succeed in the battle for world opinion; they had one of the world's most reputable news systems, with much experience in international and cross-cultural communication, and they controlled much of the undersea cable system then in operation. These capabilities were easily transitioned to the task of warfare.

The British also had a diplomatic service that kept up good relations with many nations around the world, in contrast to the reputation of the German services.[13] While German attempts to foment revolution in parts of the British Empire, such as Ireland and India, were ineffective, extensive experience in the Middle East allowed the British to successfully induce the Arabs to revolt against the Ottoman Empire.

In August 1914, David Lloyd George appointed Charles Masterman MP, to head a Propaganda Agency at Wellington House. A distinguished body of literary talent was enlisted for the task, with its members including Arthur Conan Doyle, Ford Madox Ford, G. K. Chesterton, Thomas Hardy, Rudyard Kipling and H. G. Wells. Over 1,160 pamphlets were published during the war and distributed to neutral countries, and eventually, to Germany. One of the first significant publications, the *Report on Alleged German Outrages* of 1915, had a great effect on general opinion across the world. The pamphlet documented atrocities, both actual and alleged, committed by the German army against Belgian civilians. A Dutch illustrator, Louis Raemaekers, provided the highly emotional drawings which appeared in the pamphlet.[14]

In 1917, the bureau was subsumed into the new Department of Information and branched out into telegraph communications, radio, newspapers, magazines and the cinema. In 1918, Viscount Northcliffe was appointed Director of Propaganda in Enemy Countries. The department was split between propaganda against Germany organized by H.G Wells and against the Austro-Hungarian Empire supervised by Wickham Steed and Robert William Seton-Watson; the attempts of the latter focused on the lack of ethnic cohesion in the Empire and stoked the grievances of minorities such as the Croats and Slovenes. It had a significant effect on the final collapse of the Austro-Hungarian Army at the Battle of Vittorio Veneto.[12]

Aerial leaflets were dropped over German trenches containing postcards from prisoners of war detailing their humane conditions, surrender notices and general propaganda against the Kaiser and the German generals. By the end of the war, MI7b had distributed almost 26 million leaflets. The Germans began shooting the leaflet-dropping pilots, prompting the British to develop unmanned leaflet balloons that drifted across no-man's land. At least one in seven of these leaflets were not handed in by the soldiers to their superiors, despite severe penalties for that offence. Even General Hindenburg admitted that "Unsuspectingly, many thousands consumed the poison", and POWs admitted to being disillusioned by the propaganda leaflets that depicted the use of German troops as mere cannon fodder. In 1915, the British began airdropping a regular leaflet newspaper *Le Courrier de l'Air* for civilians in German-occupied France and Belgium.[15]

At the start of the war, the French government took control of the media to suppress negative coverage. Only in 1916, with the establishment of the Maison de la Presse, did they begin to use similar tactics for the purpose of psychological warfare. One of its sections was the "Service de la Propagande aérienne" (Aerial Propaganda Service), headed by Professor Tonnelat and Jean-Jacques Waltz, an Alsatian artist code-named "*Hansi*". The French tended to distribute leaflets of images only, although the full publication of US President Woodrow Wilson's Fourteen Points, which had been heavily edited in the German newspapers, was distributed via airborne leaflets by the French.[16]

The Central Powers were slow to use these techniques; however, at the start of the war the Germans succeeded in inducing the Sultan of the Ottoman Empire to declare 'holy war', or Jihad, against the Western infidels. They also attempted to foment rebellion against the British Empire in places as far afield as Ireland, Afghanistan, and India. The Germans' greatest success was in giving the Russian revolutionary, Lenin, free transit on a sealed train from Switzerland to Finland after the overthrow of the Tsar. This soon paid off when the Bolshevik Revolution took Russia out of the war.[17]

World War II

An example of a World War II era leaflet meant to be dropped from an American B-17 over a German city. See the file description page for a translation.

Adolf Hitler was greatly influenced by the psychological tactics of warfare the British had employed during WWI, and attributed the defeat of Germany to the effects this propaganda had on the soldiers. He became committed to the use of mass propaganda to influence the minds of the German population in the decades to come. Joseph Goebbels was appointed as Propaganda Minister when Hitler came to power in 1933, and he portrayed Hitler as a messianic figure for the redemption of Germany. Hitler also coupled this with the resonating projections of his orations for effect.

Germany's *Fall Grün* plan of invasion of Czechoslovakia had a large part dealing with psychological warfare aimed both at the Czechoslovak civilians and government as well as, crucially, at Czechoslovak allies.[18] It became successful to the point that Germany gained support of UK and France through appeasement to occupy Czechoslovakia without having to fight an all-out war, sustaining only minimum losses in covert war before the Munich Agreement.

At the start of the Second World War, the British set up the Political Warfare Executive to produce and distribute propaganda. Through the use of powerful transmitters, broadcasts could be made across Europe. Sefton Delmer managed a successful black propaganda campaign through several radio stations which were designed to be popular with German troops while at the same time introducing news material that would weaken their morale under a veneer of authenticity. British Prime Minister Winston Churchill made use of radio broadcasts for propaganda against the Germans.

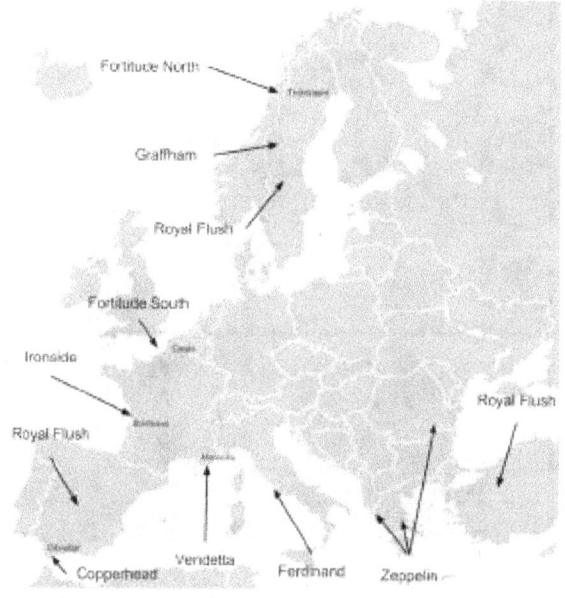

Map depicting the targets of all the subordinate plans of Operation Bodyguard.

During World War II, the British made extensive use of deception – developing many new techniques and theories. The main protagonists at this time were 'A' Force, set up in 1940 under Dudley Clarke, and the London Controlling Section, chartered in 1942 under the control of John Bevan.[19][20] Clarke pioneered many of the strategies of military deception. His ideas for combining fictional orders of battle, visual deception and double agents helped define Allied deception strategy during the war, for which he has been referred to as "the greatest British deceiver of WW2" .[21]

During the lead up to the Allied invasion of Normandy, many new tactics in psychological warfare were devised. The plan for Operation Bodyguard set out a general strategy to mislead German high command as to the exact date and location of the invasion. Planning began in 1943 under the auspices of the London Controlling Section (LCS). A draft strategy, referred to as Plan Jael, was presented to Allied high command at the Tehran Conference. Operation Fortitude was intended to convince the Germans of a greater

Allied military strength than existed, through fictional field armies, faked operations to prepare the ground for invasion and leaked information about the Allied order of battle and war plans.

Elaborate naval deceptions (Operations *Glimmer*, *Taxable* and *Big Drum*) were undertaken in the English Channel.[22] Small ships and aircraft simulated invasion fleets lying off Pas de Calais, Cap d'Antifer and the western flank of the real invasion force.[23] At the same time Operation *Titanic* involved the RAF dropping fake paratroopers to the east and west of the Normandy landings.

A dummy Sherman tank, used to deceive the Germans.

The deceptions were implemented with the use of double agents, radio traffic and visual deception. The British "Double Cross" anti-espionage operation had proven very successful from the outset of the war,[24] and the LCS was able to use double agents to send back misleading information about Allied invasion plans.[25] The use of visual deception, including mock tanks and other military hardware had been developed during the North Africa campaign. Mock hardware was created for *Bodyguard*; in particular, dummy landing craft were stockpiled to give the impression that the invasion would take place near Calais.

The Operation was a strategic success and the Normandy landings caught German defences unaware. Subsequent deception led Hitler into delaying reinforcement from the Calais region for nearly seven weeks.[26]

Vietnam War

The United States ran an extensive program of psychological warfare during the Vietnam War. The Phoenix Program had the dual aim of assassinating NLF personnel and terrorizing any potential sympathizers or passive supporters. Chieu Hoi program of the South Vietnam government promoted NLF defections.

When members of the PRG were assassinated, CIA and

"Viet Cong, beware!" – South Vietnam leaflets urging the defection of Viet Cong.

An American PSYOP leaflet disseminated during the Iraq War. It shows a caricature of Al-Qaeda in Iraq leader Abu Musab al-Zarqawi caught in a rat trap. The caption reads "This is your future, Zarqawi".

Special Forces operatives placed playing cards in the mouth of the deceased as a calling card. During the Phoenix Program, over 19,000 NLF supporters were killed.[27] The United States also used tapes of distorted human sounds and played them during the night making the Vietnamese soldiers think that the dead were back for revenge.

Recent operations

The CIA made extensive use of Contra soldiers to destabilize the Sandinista government in Nicaragua.[28] The CIA used psychological warfare techniques against the Panamanians by delivering unlicensed TV broadcasts. The CIA has extensively used propaganda broadcasts against the Cuban government through TV Marti, based in Miami, Florida. However, the Cuban government has been successful at jamming the signal of TV Marti.

In the Iraq War, the United States used the shock and awe campaign to psychologically maim, and break the will of the Iraqi Army to fight.

Social media has enabled the use of disinformation on a wide scale. Analysts have found evidence of doctored or misleading photographs spread by social media in the Syrian Civil War and 2014 Russian military intervention in Ukraine, possibly with state involvement.[29]

12.2   Methods

Most modern uses of the term psychological warfare, refers to the following military methods:

- Demoralization:
  - Distributing pamphlets that encourage desertion or supply instructions on how to surrender

- Shock and awe military strategy
  - Projecting repetitive and annoying sounds and music for long periods at high volume towards groups under siege like during Operation Nifty Package

- Propaganda radio stations, such as Lord Haw-Haw in World War II on the "Germany calling" station

- Renaming cities and other places when captured, such as the renaming of Saigon to Ho Chi Minh City after Vietnamese victory in the Vietnam War

- False flag events

- Use of loudspeaker systems to communicate with enemy soldiers

- Terrorism [30]

- The threat of chemical weapons [31]

Most of these techniques were developed during World War II or earlier, and have been used to some degree in every conflict since. Daniel Lerner was in the OSS (the predecessor to the American CIA) and in his book, attempts to analyze how effective the various strategies were. He concludes that there is little evidence that any of them were dramatically successful, except perhaps surrender instructions over loudspeakers when victory was imminent. It should be noted, though, that measuring the success or failure of psychological warfare is very hard, as the conditions are very far from being a controlled experiment.

Lerner also divides psychological warfare operations into three categories: [32]

- White propaganda (Omissions and Emphasis): Truthful and not strongly biased, where the source of information is acknowledged.

- Grey propaganda (Omissions, Emphasis and Racial/Ethnic/Religious Bias): Largely truthful, containing no information that can be proven wrong; the source is not identified.

- Black propaganda (Commissions of falsification): Inherently deceitful, information given in the product is attributed to a source that was not responsible for its creation.

Lerner points out that grey and black operations ultimately have a heavy cost, in that the target population sooner or later recognizes them as propaganda and discredits the source. He writes, "This is one of the few dogmas advanced by Sykewarriors that is likely to endure as an axiom

of propaganda: Credibility is a condition of persuasion. Before you can make a man do as you say, you must make him believe what you say." [32]:28 Consistent with this idea, the Allied strategy in World War II was predominantly one of truth (with certain exceptions).

12.3 By country

12.3.1 Soviet Union

Main articles: Zersetzung, Russian military deception, and Active measures

12.3.2 China

According to U.S. military analysts, attacking the enemy's mind is an important element of the People's Republic of China's military strategy. [33] This type of warfare is rooted in the Chinese Stratagems outlined by Sun Tzu in *The Art of War* and *Thirty-Six Stratagems*. In its dealings with its rivals, China is expected to utilize Marxism to mobilize communist loyalists, as well as flex its economic and military muscle to persuade other nations to act in China's interests. The Chinese government also tries to control the media to keep a tight hold on propaganda efforts for its people. [33]

12.3.3 Germany

In the German Bundeswehr, the **Zentrum Operative Information** and its subordinate **Batallion für Operative Information 950** are responsible for the PSYOP efforts (called **Operative Information** in German). Both the center and the battalion are subordinate to the new *Streitkräftebasis* (Joint Services Support Command, SKB) and together consist of about 1,200 soldiers specialising in modern communication and media technologies. One project of the German PSYOP forces is the radio station *Stimme der Freiheit* (Sada-e Azadi, Voice of Freedom), [34] heard by thousands of Afghans. Another is the publication of various newspapers and magazines in Kosovo and Afghanistan, where German soldiers serve with NATO.

12.3.4 United Kingdom

The British were one of the first major military powers to use psychological warfare in the First and Second World Wars. In current the British Armed Forces, PSYOPS are handled by the tri-service 15 Psychological Operations

Group. (See also MI5 and Secret Intelligence Service). The Psychological Operations Group comprises over 150 personnel, approximately 75 from the regular Armed Services and 75 from the Reserves. The Group supports deployed commanders in the provision of psychological operations in operational and tactical environments.[35][36]

The Group was established immediately after the 1991 Gulf War,[37] has since grown significantly in size to meet operational requirements,[38] and from 2015 it will be one of the sub-units of the 77th Brigade, formerly called the Security Assistance Group.[39] Stephen Jolly, the MOD's Director of Defence Communications and former Chair of the UK's National Security Communications Committee (2013–15), is thought to be the most senior serving psyops officer within British Defence.

In June 2015, NSA files published by Glenn Greenwald revealed details of the JTRIG group at British intelligence agency GCHQ covertly manipulating online communities.[40] This is in line with JTRIG's goal: to "destroy, deny, degrade [and] disrupt" enemies by "discrediting" them, planting misinformation and shutting down their communications.[41]

12.3.5   United States

See also: Psychological Operations (United States)
The term psychological warfare is believed to have mi-

U.S. Army soldier hands out a newspaper to a local in Mosul, Iraq.

grated from Germany to the United States in 1941.[42] During World War II, the United States Joint Chiefs of Staff defined psychological warfare broadly, stating "Psychological warfare employs *any* weapon to influence the mind of the enemy. The weapons are psychological only in the effect they produce and not because of the weapons themselves."[43] The U.S. Department of Defense currently defines psychological warfare as:

"The planned use of propaganda and other

U.S. Army loudspeaker team in action in Korea

psychological actions having the primary purpose of influencing the opinions, emotions, attitudes, and behavior of hostile foreign groups in such a way as to support the achievement of national objectives."[44]

This definition indicates that a critical element of the U.S. psychological operations capabilities includes propaganda and by extension counterpropaganda. Joint Publication 3-53 establishes specific policy to use public affairs mediums to counterpropaganda from foreign origins.[45]

The purpose of United States psychological operations is to induce or reinforce attitudes and behaviors favorable to US objectives. The Special Activities Division (SAD) is a division of the Central Intelligence Agency's National Clandestine Service, responsible for Covert Action and "Special Activities". These special activities include covert political influence (which includes psychological operations) and paramilitary operations.[46] SAD's political influence group is the only US unit allowed to conduct these operations covertly and is considered the primary unit in this area.[46]

Dedicated psychological operations units exist in the United States Army. The United States Navy also plans and executes limited PSYOP missions. United States PSYOP units and soldiers of all branches of the military are prohibited by law from targeting U.S. citizens with PSYOP within the borders of the United States (Executive Order S-1233, DOD Directive S-3321.1, and National Security Decision Directive 130). While United States Army PSYOP units may offer non-PSYOP support to domestic military missions, they can only target foreign audiences.

A U.S. Army field manual released in January 2013 states that "Inform and Influence Activities" are critical for describing, directing, and leading military operations. Several Army Division leadership staff are assigned to "planning, integration and synchronization of designated information-related capabilities." "[47]

12.4 See also

- Charles Douglas Jackson
- Demonizing the enemy
- Demoralization (warfare)
- Electromagnetic Weapon
- Information warfare
- Lawfare
- Media manipulation
- Military psychology
- Mind games
- Minor sabotage
- Political Warfare
- Propaganda: The Formation of Men's Attitudes
- Psychological manipulation
- Special Operations
- Taliban propaganda
- Unconventional Warfare
- Fear § Manipulation

NATO

- Able Archer 83

US specific:

- Information Operations Roadmap
- Military journalism
- NLF and PAVN battle tactics
- Psychological operations (United States)
- Special Activities Division
- Zarqawi PSYOP program

World War II:

- Political Warfare Executive
- Psychological Warfare Division

USSR

- Active measures

Related:

- Asymmetric warfare
- Fourth generation warfare

12.5 References

[1] "Forces.gc.ca". Journal.forces.gc.ca. Retrieved 2011-05-18.

[2] Szunyogh, Béla (1955). *Psychological warfare: an introduction to ideological propaganda and the techniques of psychological warfare*. United States: William-Frederick Press. p. 13. Retrieved 2015-02-11.

[3] Chekinov, S. C.; Bogdanov, S. A. *The Nature and Content of a New-Generation War* (PDF). United States: Military Thought. p. 16. ISSN 0869-5636. Retrieved 2015-02-11.

[4] Doob, Leonard W. "The Strategies Of Psychological Warfare." Public Opinion Quarterly 13.4 (1949): 635-644. SocINDEX with Full Text. Web. 20 Feb. 2015.

[5] Wall, Tyler (September 2010). *U.S Psychological Warfare and Civilian Targeting*. United States: Vanderbilt University. p. 289. Retrieved 2015-02-11.

[6] Ellul, Jacques (1973). *Propaganda: The Formation of Men's Attitudes*, p. xiii.Trans. Konrad Kellen & Jean Lerner. Vintage Books, New York. ISBN 978-0-394-71874-3.

[7] *The Psychology of Terrorism: Clinical aspects and responses - Google Books*. Books.google.com. Retrieved 2014-08-10.

[8] Lance B. Curke Ph.D., *The Wisdom of Alexander the Great: Enduring Leadership Lessons From the Man Who Created an Empire* (2004) p. 66

[9] David Nicolle, *The Mongol Warlords: Genghis Khan, Kublai Khan, Hulegu, Tamerlane* (2004) p. 21

[10] George H. Quester (2003). *Offense and Defense in the International System*. Transaction Publishers. p. 43. Retrieved 2016-03-19.

[11] Diogenes Laertius. *Lives and Opinions of the Eminent Philosophers*.

[12] "ALLIED PSYOP OF WWI". Retrieved 2012-12-17.

[13] Linebarger. Paul Myron Anthony (2006). *Psychological Warfare*. University of Chicago Press. Retrieved 2013-02-07.

[14] "The Battle for the Mind: German and British Propaganda in the First World War".

[15] Taylor. Philip M. (1999). *British Propaganda in the Twentieth Century: Selling Democracy*. Edinburgh University Press. Retrieved 2013-02-07.

[16] "ALLIED PSYOP OF WWI". Retrieved 2012-12-17.

[17] "GERMAN WWI PSYOP". Retrieved 2012-12-17.

[18] Hruška. Emil (2013), *Boj o pohraničí: Sudetoněmecký Freikorps v roce 1938* (1st ed.). Prague: Nakladatelství epocha, Pražská vydavatelská společnost, p. 9

[19] Latimer (2004), pg. 148–149

[20] Cruickshank (2004)

[21] Rankin. Nicholas (1 October 2008). *Churchill's Wizards: The British Genius for Deception. 1914–1945*. Faber and Faber. p. 178. ISBN 0-571-22195-5.

[22] Barbier, Mary (30 Oct 2007). *D-Day Deception: Operation Fortitude and the Normandy Invasion*. Greenwood Publishing Group. p. 70. ISBN 0275994791.

[23] Barbier, Mary (30 Oct 2007). *D-Day Deception: Operation Fortitude and the Normandy Invasion*. Greenwood Publishing Group. p. 108. ISBN 0275994791.

[24] Masterman. John C (1972) [1945]. *The Double-Cross System in the War of 1939 to 1945*. Australian National University Press. ISBN 978-0-7081-0459-0.

[25] Ambrose. Stephen E. (1981). "Eisenhower, the Intelligence Community, and the D-Day Invasion". *The Wisconsin Magazine of History*. Vol. 64 no. 4. Wisconsin Historical Society. p. 269. ISSN 0043-6534.

[26] Latimer. John (2001). *Deception in War*. New York: Overlook Press. p. 238. ISBN 978-1-58567-381-0.

[27] Janq Designs. "Special operation - Phoenix". Specialoperations.com. Archived from the original on May 12, 2011. Retrieved 2011-05-18.

[28] "Is the U.S. Organizing Salvador-Style Death Squads in Iraq?". *Democracy Now!*. 2005-01-10. Retrieved 2008-12-16.

[29] Rawlsey. Adam (1 November 2014). "Be Very Skeptical —A Lot of Your Open-Source Intel Is Fake". Medium. Retrieved 3 November 2014.

[30] Boaz, Gaynor (April 2004). "Terrorism as a strategy of psychological warfare". *Journal of Aggression, Maltreatment & Trauma*. Taylor and Francis. **9** (1–2): 33–4. doi:10.1300/J146v09n01_03.(subscription required)

[31] Romano Jr., James A.; King. James M. (2002). "Chemical warfare and chemical terrorism: psychological and performance outcomes". *Military Psychology*. American Psychological Association via PsycNET. **14** (2): 85–92. doi:10.1207/S15327876MP1402_2.(subscription required)

[32] Lerner. Daniel (1971) [1949]. *Psychological warfare against Nazi Germany: the Sykewar Campaign, D-Day to VE-Day*. Boston. Mass: MIT Press. ISBN 0-262-12045-3. Originally printed by George W. Stewart of New York. Alternative ISBN 0-262-62019-7

[33] "Chinese Military - Psychological Warfare". *ufl.edu*. Archived from the original on 15 April 2011.

[34] "Sada-e-azadi.net". Sada-e-azadi.net. Archived from the original on May 12, 2011. Retrieved 2011-05-18.

[35] "15 (UK) Psychological Operations Group". Ministry of Defence. Archived from the original on 2006-06-20. Retrieved 23 August 2008.

[36] "Psychological Ops Group". Royal Navy. Archived from the original on 2010-07-02. Retrieved 28 May 2013.

[37] Jolly. Stephen (October 2000). Minshall. David, ed. "Wearing the Stag's Head Badge: British Combat Propaganda since 1945". *Falling Leaf*. The Psywar Society (170): 86–89. ISSN 0956-2400.

[38] "15 (United Kingdom) Psychological Operations Group: Annual Report" (PDF). 15 (UK) PSYOPS Group. Retrieved 29 May 2011.

[39] Ewan MacAskill (31 January 2015). "British army creates team of Facebook warriors". The Guardian. Retrieved 31 January 2015.

[40] Greenwald. Glenn and Andrew Fishman. Controversial GCHQ Unit Engaged in Domestic Law Enforcement, Online Propaganda. Psychology Research. *The Intercept*. 2015-06-22.

[41] "Snowden Docs: British Spies Used Sex and 'Dirty Tricks'". NBC News. 7 February 2014. Retrieved 7 February 2014.

[42] WALL. TYLER. "U.S. Psychological Warfare And Civilian Targeting." Peace Review 22.3 (2010): 288-294. SocINDEX with Full Text. Web. 20 Feb. 2015.

[43] From "Overall Strategic Plan for the United States' Psychological Warfare." 1 March 1943. JCS Records, Strategic Issues, Reel 11. Quoted in Robert H. Keyserlingk (July 1990). *Austria in World War II*. McGill-Queen's University Press. p. 131. ISBN 0-7735-0800-7.

[44] Phil Taylor (1987). "Glossary of Relevant Terms & Acronyms Propaganda and Psychological Warfare Studies University of Leeds UK". University of Leeds UK. Archived from the original on 2013-06-22. Retrieved 2008-04-19.

[45] Garrison, WC (1999). "Information Operations and Counter-Propaganda: Making a Weapon of Public Affairs" (PDF). *Strategy Research Project, U.S. Army War College.* p. 12. Retrieved April 4, 2012.

[46] Executive Secrets: Covert Action and the Presidency. William J. Daugherty. University of Kentucky Press, 2004.

[47] "Pentagon gearing up to fight the PR war" *Washington Post.* February 6, 2013

12.5.1 Bibliography

- Fred Cohen. *Frauds, Spies, and Lies - and How to Defeat Them.* ISBN 1-878109-36-7 (2006). ASP Press.

- Fred Cohen. *World War 3 ... Information Warfare Basics.* ISBN 1-878109-40-5 (2006). ASP Press.

- Gagliano Giuseppe. *Guerra psicologia.Disinformazione e movimenti sociali.* Introduzione del Gen. Carlo Jean e di Alessandro Politi Editrice Aracne, Roma, 2012.

- Gagliano Giuseppe. *Guerra psicologia.Saggio sulle moderne tecniche militari,di guerra cognitiva e disinformazione.* Introduzione del Gen. Carlo Jean, Editrice Fuoco, Roma 2012.

- Paul M. A. Linebarger. *Psychological Warfare: International Propaganda and Communications.* ISBN 0-405-04755-X (1948). Revised second edition. Duell, Sloan and Pearce (1954).

12.6 External links

- Movie: *Psywar: The Real Battlefield is the Mind* by Metanoia films

-

- Paul Myron Anthony Linebarger. *Psychological Warfare* at Project Gutenberg

- The history of psychological warfare

- IWS Psychological Operations (PsyOps) / Influence Operations

- "Pentagon psychological warfare operation" , *USA Today.* December 15, 2005

- "U.S. Adapts Cold-War Idea to Fight Terrorists" , *New York Times,* March 18, 2008

- US Army PSYOPS Info - Detailed information about the US Army Psychological Operation Soldiers

- IWS —The Information Warfare Site

- U.S. —PSYOP producing mid-eastern kids comic book

- The Institute of Heraldry —Psychological Operations

- Psychological warfare

Chapter 13

Acute radiation syndrome

Radiation poisoning and radiation sickness redirect here. This page is about the short-term systemic health effects of a large dose. For other uses of these terms, see Radiation poisoning (disambiguation).

Acute radiation syndrome (**ARS**), also known as **radiation poisoning**, **radiation sickness**, or **radiation toxicity**, is a collection of health effects that are present within 24 hours of exposure to high amounts of ionizing radiation. The radiation causes cellular degradation due to damage to DNA and other key molecular structures within the cells in various tissues. This destruction, particularly because it affects the ability of cells to divide normally, in turn causes the symptoms. The symptoms can begin within one or two hours and may last for several months.[1][2] The terms refer to acute medical problems rather than ones that develop after a prolonged period.[3][4][5]

The onset and type of symptoms depends on the radiation exposure. Relatively smaller doses result in gastrointestinal effects, such as nausea and vomiting, and symptoms related to falling blood counts, and predisposition to infection and bleeding. Relatively larger doses can result in neurological effects and rapid death. Treatment of acute radiation syndrome is generally supportive with blood transfusions and antibiotics, with some more aggressive treatments, such as bone marrow transfusions, being required in extreme cases.[1]

Similar symptoms may appear months to years after exposure as chronic radiation syndrome when the dose rate is too low to cause the acute form.[6] Radiation exposure can also increase the probability of developing some other diseases, mainly different types of cancers. These diseases are sometimes referred to as radiation sickness, but they are never included in the term *acute radiation syndrome*.

13.1 Signs and symptoms

See also: Biological timeline of radiation poisoning
Classically acute radiation syndrome is divided into three

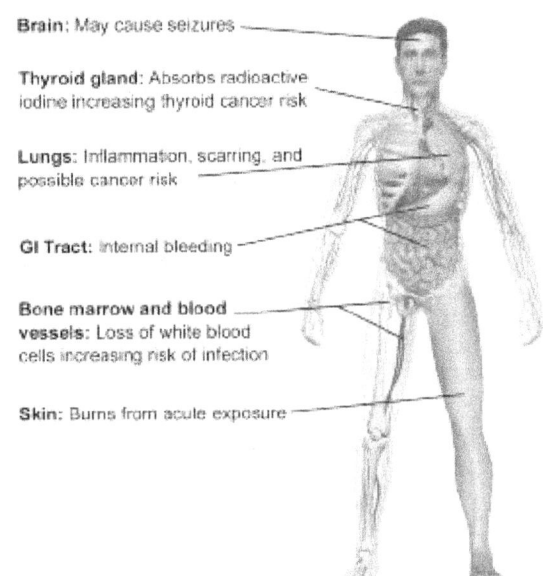

Brain: May cause seizures

Thyroid gland: Absorbs radioactive iodine increasing thyroid cancer risk

Lungs: Inflammation, scarring, and possible cancer risk

GI Tract: Internal bleeding

Bone marrow and blood vessels: Loss of white blood cells increasing risk of infection

Skin: Burns from acute exposure

Selected Risks from Radiation Sickness

Radiation Sickness

main presentations: hematopoietic, gastrointestinal, and neurological/vascular. These syndromes may or may not be preceded by a prodrome.[1] The speed of onset of symptoms is related to radiation exposure, with greater doses resulting in a shorter delay in symptom onset.[1] These presentations presume whole-body exposure and many of them are markers which are not valid if the entire body has not been exposed. Each syndrome requires that the tissue showing the syndrome itself be exposed. The hematopoietic syndrome requires exposure of the areas of bone marrow actively forming blood elements (i.e., the pelvis and sternum in adults). The neurovascular symptoms require exposure of the brain. The gastrointestinal syndrome is not seen if the stomach and intestines are not exposed to radiation.

1. Hematopoietic. This syndrome is marked by a drop in the number of blood cells, called aplastic anemia. This may result in infections due to a low amount of white blood cells, bleeding due to a lack of platelets, and anemia due to few red blood cells in the circulation.[1] These changes can be detected by blood tests after receiving a whole-body acute dose as low as 0.25 Gy, though they might never be felt by the patient if the dose is below 1 Gy. Conventional trauma and burns resulting from a bomb blast are complicated by the poor wound healing caused by hematopoietic syndrome, increasing mortality.

2. Gastrointestinal. This syndrome often follows absorbed doses of 6–30 Gy (600–3000 rad).[1] The signs and symptoms of this form of radiation injury include nausea, vomiting, loss of appetite, and abdominal pain.[7] Vomiting in this time-frame is a marker for whole body exposures that are in the fatal range above 4 Gy. Without exotic treatment such as bone marrow transplant, death with this dose is common.[1] The death is generally more due to infection than gastrointestinal dysfunction.

3. Neurovascular. This syndrome typically occurs at absorbed doses greater than 30 Gy (3000 rad), though it may occur at 10 Gy (1000 rad).[1] It presents with neurological symptoms such as dizziness, headache, or decreased level of consciousness, occurring within minutes to a few hours, and with an absence of vomiting. It is invariably fatal.[1]

The prodrome (early symptoms) of ARS typically includes nausea and vomiting, headaches, fatigue, fever, and a short period of skin reddening.[1] These symptoms may occur at radiation doses as low as 0.35 Gy (35 rad). These symptoms are common to many illnesses, and may not, by themselves, indicate acute radiation sickness.[1]

13.1.1 Whole-body absorbed dose effects chart

13.1.2 Skin changes

Main article: radiation burn

Cutaneous radiation syndrome (CRS) refers to the skin symptoms of radiation exposure.[5] Within a few hours after irradiation, a transient and inconsistent redness (associated with itching) can occur. Then, a latent phase may occur and last from a few days up to several weeks, when intense reddening, blistering, and ulceration of the irradiated site are visible. In most cases, healing occurs by regenerative means; however, very large skin doses can cause permanent hair loss, damaged sebaceous and sweat glands, atrophy, fibrosis (mostly Keloids), decreased or increased skin pigmentation, and ulceration or necrosis of the exposed tissue.[5] Notably, as seen at Chernobyl, when skin is irradiated with high energy beta particles, moist desquamation (peeling of skin) and similar early effects can heal, only to be followed by the collapse of the dermal vascular system after two months, resulting in the loss of the full thickness of the exposed skin.[9] This effect had been demonstrated previously with pig skin using high energy beta sources at the Churchill Hospital Research Institute, in Oxford.[10]

13.1.3 Cancer

Main article: radiation-induced cancer

According to the linear no-threshold model, any exposure to ionizing radiation, even at doses too low to produce any symptoms of radiation sickness, can induce cancer due to cellular and genetic damage. Under the assumption, survivors of acute radiation syndrome face an increased risk of developing cancer later in life. The probability of developing cancer is a linear function with respect to the effective radiation dose. In radiation-induced cancer, the speed at which the condition advances, the prognosis, the degree of pain, and every other feature of the disease are not believed to be functions of the radiation dosage.

However, some studies contradict the linear no-threshold model. These studies indicate that some low levels of radiation do not increase cancer risk at all, and that there may exist a threshold dosage of ionizing radiation below which exposure should be considered safe. Nonetheless the 'no safe amount' assumption is the basis of US and most national regulatory policies regarding "man-made" sources of radiation.

13.2 Cause

Radiation sickness is caused by exposure to a large dose of ionizing radiation (> ~0.1 Gy) over a short period of time. (> ~0.1 Gy/h) This might be the result of a nuclear explosion, a criticality accident, a radiotherapy accident as in Therac-25, a solar flare during interplanetary travel, misplacement of radioactive waste as in the 1987 Goiânia accident, human error in a nuclear reactor, or other possibilities. Acute radiation sickness due to ingestion of radioactive material is possible, but rare; examples include

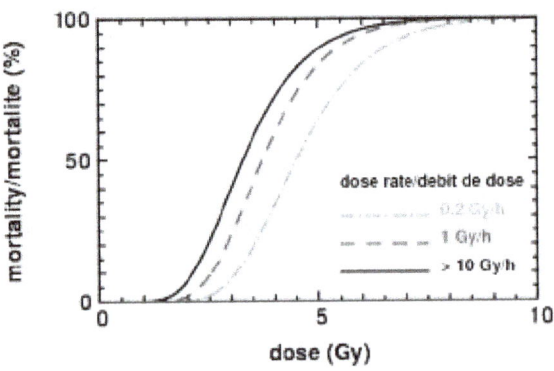

Both dose and dose rate contribute to the severity of acute radiation syndrome

the 1987 contamination of Leide das Neves Ferreira and the 2006 poisoning of Alexander Litvinenko.

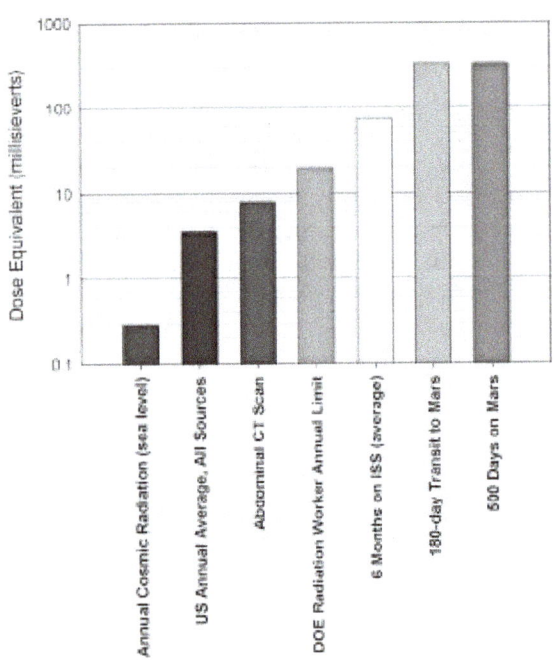

Comparison of Radiation Doses – includes the amount detected on the trip from Earth to Mars by the RAD on the MSL (2011 - 2013).[11]*[12]*[13]*[14]*

Alpha and beta radiation have low penetrating power and are unlikely to affect vital internal organs from outside the body. Any type of ionizing radiation can cause burns, but alpha and beta radiation can only do so if radioactive contamination or nuclear fallout is deposited on the individual's skin or clothing. Gamma and neutron radiation can travel much further distances and penetrate the body easily, so whole-body irradiation generally causes ARS before skin effects are evident. Local gamma irradiation can cause skin effects without any sickness. In the early twentieth century, radiographers would commonly calibrate their machines by irradiating their own hand and measuring the time to onset of erythema.*[15]

13.2.1 Spaceflight

During spaceflight, particularly flights beyond low Earth orbit, astronauts are exposed to both galactic cosmic radiation (GCR) and solar particle event (SPE) radiation. Evidence indicates past SPE radiation levels which would have been lethal for unprotected astronauts.*[16] GCR levels which might lead to acute radiation poisoning are less well understood.*[17]

13.3 Pathophysiology

The most commonly used predictor of acute radiation symptoms is the whole-body absorbed dose. Several related quantities, such as the equivalent dose, effective dose, and committed dose, are used to gauge long-term stochastic biological effects such as cancer incidence, but they are not designed to evaluate acute radiation syndrome.*[18] To help avoid confusion between these quantities, absorbed dose is measured in units of grays (in SI, unit symbol **Gy**) or rads (in CGS), while the others are measured in sieverts (in SI, unit symbol **Sv**) or rems (in CGS). 1 rad = 0.01 Gy and 1 rem = 0.01 Sv.*[19]

In most of the acute exposure scenarios that lead to radiation sickness, the bulk of the radiation is external whole-body gamma, in which case the absorbed, equivalent and effective doses are all equal. There are exceptions, such as the Therac-25 accidents and the 1958 Cecil Kelley criticality accident, where the absorbed doses in Gy or rad are the only useful quantities.

Radiotherapy treatments are typically prescribed in terms of the local absorbed dose, which might be 60 Gy or higher. The dose is fractionated (about 2 Gy per day for curative treatment), which allows for the normal tissues to undergo repair, allowing it to tolerate a higher dose than would otherwise be expected. The dose to the targeted tissue mass must be averaged over the entire body mass, most of which receives negligible radiation, to arrive at a whole-body absorbed dose that can be compared to the table above.

13.4 Diagnosis

Diagnosis is typically made based on a history of significant radiation exposure and suitable clinical findings.*[1]

An absolute lymphocyte count can give a rough estimate of radiation exposure.[1] Time from exposure to vomiting can also give estimates of exposure levels if they are less than 1000 rad.[1]

13.5 Prevention

See also: Radiation protection

The best prevention for radiation sickness is to minimize the exposure dose or to reduce the dose rate.

13.5.1 Distance

Increasing distance from the radiation source reduces the dose according to the inverse-square law for a point source. Distance can sometimes be effectively increased by means as simple as handling a source with forceps rather than fingers. This could reduce erythema to the fingers, but the extra few centimeters distance from the body will give little protection from acute radiation syndrome.

13.5.2 Time

The longer that humans are subjected to radiation the larger the dose will be. The advice in the nuclear war manual entitled "Nuclear War Survival Skills" published by Cresson Kearny in the U.S. was that if one needed to leave the shelter then this should be done as rapidly as possible to minimize exposure.

In chapter 12 he states that *"Quickly putting or dumping wastes outside is not hazardous once fallout is no longer being deposited. For example, assume the shelter is in an area of heavy fallout and the dose rate outside is 400 roentgen (R) per hour] enough to give a potentially fatal dose in about an hour to a person exposed in the open. If a person needs to be exposed for only 10 seconds to dump a bucket, in this 1/360 of an hour he will receive a dose of only about 1 R. Under war conditions, an additional 1-R dose is of little concern."*

In peacetime, radiation workers are taught to work as quickly as possible when performing a task which exposes them to radiation. For instance, the recovery of a lost radiography source should be done as quickly as possible.

13.5.3 Shielding

Matter attenuates radiation in most cases, so placing any mass (e.g., lead, dirt, sandbags, vehicles) between humans and the source will reduce the radiation dose. This is not always the case, however; care should be taken when constructing shielding for a specific purpose. For example, although high atomic number materials are very effective in shielding photons, using them to shield beta particles may cause higher radiation exposure due to the production of bremsstrahlung x-rays, and hence low atomic number materials are recommended. Also, using material with a high neutron activation cross section to shield neutrons will result in the shielding material itself becoming radioactive and hence more dangerous than if it were not present.

13.5.4 Reduction of incorporation into the human body

Where radioactive contamination is present, a gas mask, dust mask, or good hygiene practices may offer protection, depending on the nature of the contaminant. Potassium iodide (KI) tablets can reduce the risk of cancer in some situations due to slower uptake of ambient radioiodine. Although this doesn't protect any organ other than the thyroid gland, their effectiveness is still highly dependent on the time of ingestion which would protect the gland for the duration of a twenty-four-hour period. They do not prevent acute radiation syndrome as they provide no shielding from other environmental radionuclides.[20]

13.5.5 Fractionation of dose

If an intentional dose is broken up into a number of smaller doses, with time allowed for recovery between irradiations, the same total dose causes less cell death. Even without interruptions, a reduction in dose rate below 0.1 Gy/h also tends to reduce cell death.[18] This technique is routinely used in radiotherapy.

The human body contains many types of cells and a human can be killed by the loss of a single type of cells in a vital organ. For many short term radiation deaths (3 days to 30 days), the loss of two important types of cells that are constantly being regenerated causes death. The loss of cells forming blood cells (bone marrow) and the cells in the digestive system (microvilli which form part of the wall of the intestines) is fatal.

13.6 Management

Treatment is supportive with the use of antibiotics, blood products, colony stimulating factors, and stem cell transplant as clinically indicated.[1] Symptomatic measures may also be employed.[1]

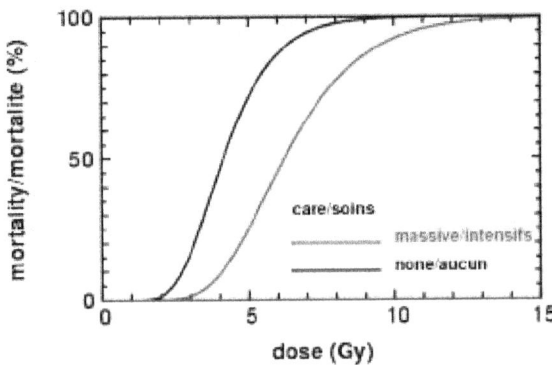

Effect of medical care on acute radiation syndrome

of Gram-negative aerobic bacilli (i.e., Enterobacteriace, Pseudomonas) that account for more than three quarters of the isolates causing sepsis. Because aerobic and facultative Gram-positive bacteria (mostly alpha-hemolytic streptococci) cause sepsis in about a quarter of the victims, coverage for these organisms may also be needed.[24]

A standardized management plane of febrile, neutropenic patients must be devised in each institution or agency. Empirical regimens must contain antibiotics broadly active against Gram-negative aerobic bacteria (quinolones: i.e., ciprofloxacin, levofloxacin, a third- or fourth-generation cephalosporin with pseudomonal coverage: e.g., cefepime, ceftazidime, or an aminoglycoside: i.e. gentamicin, amikacin).[25]

13.6.1 Antimicrobials

Main article: Treatment of infections after exposure to ionizing radiation

There is a direct relationship between the degree of the neutropenia that emerges after exposure to radiation and the increased risk of developing infection. Since there are no controlled studies of therapeutic intervention in humans, most of the current recommendations are based on animal research.

The treatment of established or suspected infection following exposure to radiation (characterized by neutropenia and fever) is similar to the one used for other febrile neutropenic patients. However, important differences between the two conditions exist. Individuals that develop neutropenia after exposure to radiation are also susceptible to irradiation damage in other tissues, such as the gastrointestinal tract, lungs and central nervous system. These patients may require therapeutic interventions not needed in other types of neutropenic patients. The response of irradiated animals to antimicrobial therapy can be unpredictable, as was evident in experimental studies where metronidazole[21] and pefloxacin[22] therapies were detrimental.

Antimicrobials that reduce the number of the strict anaerobic component of the gut flora (i.e., metronidazole) generally should not be given because they may enhance systemic infection by aerobic or facultative bacteria, thus facilitating mortality after irradiation.[23]

An empirical regimen of antimicrobials should be chosen based on the pattern of bacterial susceptibility and nosocomial infections in the affected area and medical center and the degree of neutropenia. Broad-spectrum empirical therapy (see below for choices) with high doses of one or more antibiotics should be initiated at the onset of fever. These antimicrobials should be directed at the eradication

13.7 History

Acute effects of ionizing radiation were first observed when Wilhelm Röntgen intentionally subjected his fingers to X-rays in 1895. He published his observations concerning the burns that developed, though he misattributed them to ozone, a free radical produced in air by X-rays. Other free radicals produced within the body are now understood to be more important. His injuries healed later.

The Radium Girls were female factory workers who contracted radiation poisoning from painting watch dials with self-luminous paint at the United States Radium factory in Orange, New Jersey, around 1917.

Ingestion of radioactive materials caused many radiation-induced cancers in the 1930s, but no one was exposed to high enough doses at high enough rates to bring on acute radiation syndrome. Marie Curie died of aplastic anemia caused by radiation, a possible early incident of acute radiation syndrome.

The atomic bombings of Hiroshima and Nagasaki resulted in high acute doses of radiation to a large number of Japanese, allowing for greater insight into its symptoms and dangers. Red Cross Hospital Surgeon, Terufumi Sasaki led intensive research into the syndrome in the weeks and months following the Hiroshima bombings. Dr Sasaki and his team were able to monitor the effects of radiation in patients of varying proximities to the blast itself, leading to the establishment of three recorded stages of the syndrome. Within 25–30 days of the explosion, the Red Cross surgeon noticed a sharp drop in white blood cell count and established this drop, along with symptoms of fever, as prognostic standards for Acute Radiation Syndrome.[26] Actress Midori Naka, who was present during the atomic bombing of Hiroshima, was the first incident of radiation poisoning to be extensively studied. Her death on August 24, 1945 was the first death ever to be officially certified as a result of acute radiation syndrome (or "Atomic bomb disease").

13.7.1 Notable incidents

Main articles: Nuclear and radiation accidents and Lists of nuclear disasters and radioactive incidents

There are two major databases that track radiation accidents: The American ORISE REAC/TS and the European IRSN ACCIRAD. REAC/TS shows 417 accidents occurring between 1944 and 2000, causing about 3000 cases of acute radiation syndrome, of which 127 were fatal.[27] ACCIRAD lists 580 accidents with 180 ARS fatalities for an almost identical period.[28] The two deliberate bombings are not included in either database, nor are any possible radiation-induced cancers from low doses. The detailed accounting is difficult because of confounding factors. ARS may be accompanied by conventional injuries such as steam burns, or may occur in someone with a pre-existing condition undergoing radiotherapy. There may be multiple causes for death, and the contribution from radiation may be unclear. Some documents may incorrectly refer to radiation-induced cancers as radiation poisoning, or may count all overexposed individuals as survivors without mentioning if they had any symptoms of ARS. The table below attempts to catalog some cases of ARS. Many of these incidents involved additional fatalities from other causes, such as cancer, which are excluded from this table.

13.8 Other animals

Thousands of scientific experiments have been performed to study acute radiation syndrome in animals.

There is a simple guide for predicting survival/death in mammals, including humans, following the acute effects of inhaling radioactive particles.[44]

13.9 See also

- Hibakusha – Japanese atomic bomb survivors
- List of civilian nuclear accidents
- List of military nuclear accidents
- Biological effects of ionizing radiation
- Orders of magnitude (radiation)
- 5-Androstenediol
- CBLB502

- Ex-Rad

13.10 References

[1] Donnelly EH, Nemhauser JB, Smith JM, et al. (June 2010). "Acute radiation syndrome: assessment and management". *South. Med. J.* **103** (6): 541–6. doi:10.1097/SMJ.0b013e3181ddd571. PMID 20710137.

[2] Xiao M, Whitnall MH (January 2009). "Pharmacological countermeasures for the acute radiation syndrome". *Curr Mol Pharmacol.* **2** (1): 122–33. doi:10.2174/1874467210902010122. PMID 20021452.

[3] "Acute Radiation Syndrome". Centers for Disease Control and Prevention. 2005-05-20. Archived from the original on 2015-12-04.

[4] "Acute Radiation Syndrome" (PDF). National Center for Environmental Health/Radiation Studies Branch. 2002-04-09. Retrieved 2009-06-22.

[5] "Acute Radiation Syndrome: A Fact Sheet for Physicians". Centers for Disease Control and Prevention. 2005-03-18.

[6] Reeves GI, Ainsworth EJ (May 1995). "Description of the chronic radiation syndrome in humans irradiated in the former Soviet Union". *Radiat. Res.* **142** (2): 242–3. doi:10.2307/3579035. PMID 7724741.

[7] Christensen DM, Iddins CJ, Sugarman SL (February 2014). "Ionizing radiation injuries and illnesses". *Emerg Med Clin North Am.* **32** (1): 245–65. doi:10.1016/j.emc.2013.10.002. PMID 24275177.

[8] "Radiation Exposure and Contamination". *Merck Manuals.* Retrieved 2 June 2013.

[9] The medical handling of skin lesions following high level accidental irradiation, IAEA Advisory Group Meeting, September 1987 Paris.

[10] Wells J; et al. (1982). "Non-Uniform Irradiation of Skin: Criteria for Limiting Non-Stochastic Effects". *Proceedings of the Third International Symposium of the Society for Radiological Protection _ Advances in Theory and Practice.* **2**: 537–542. ISBN 0-9508123-0-7.

[11] Kerr, Richard (31 May 2013). "Radiation Will Make Astronauts' Trip to Mars Even Riskier". *Science.* **340** (6136): 1031. doi:10.1126/science.340.6136.1031. PMID 23723213. Retrieved 31 May 2013.

[12] Zeitlin, C.; et al. (31 May 2013). "Measurements of Energetic Particle Radiation in Transit to Mars on the Mars Science Laboratory". *Science.* **340** (6136): 1080–1084. doi:10.1126/science.1235989. Retrieved 31 May 2013.

[13] Chang, Kenneth (30 May 2013). "Data Point to Radiation Risk for Travelers to Mars". New York Times. Retrieved 31 May 2013.

[14] Gelling, Cristy (June 29, 2013). "Mars trip would deliver big radiation dose; Curiosity instrument confirms expectation of major exposures". *Science News*. **183** (13): 8. Retrieved July 8, 2013.

[15] William, C. Inkret; Charles B. Meinhold; John C. Taschner (1995). "A Brief History of Radiation Protection Standards" (PDF). *Los Alamos Science* (23): 116–123. Retrieved 12 November 2012.

[16] "Superflares could kill unprotected astronauts". New Scientist. 21 March 2005.

[17] National Research Council (U.S.). Ad Hoc Committee on the Solar System Radiation Environment and NASA's Vision for Space Exploration (2006). *Space Radiation Hazards and the Vision for Space Exploration*. National Academies Press. ISBN 978-0-309-10264-3.

[18] "The 2007 Recommendations of the International Commission on Radiological Protection". *Annals of the ICRP*. ICRP publication 103. **37** (2–4). 2007. ISBN 978-0-7020-3048-2. Retrieved 17 May 2012.

[19] *The Effects of Nuclear Weapons*. Revised ed., US DOD 1962, p. 579

[20] "Radiation and its Health Effects". Nuclear Regulatory Commission. Retrieved 2013-11-19.

[21] Brook I, Ledney GD (1994). "Effect of antimicrobial therapy on the gastrointestinal bacterial flora, infection and mortality in mice exposed to different doses of irradiation". *Journal of Antimicrobial Chemotherapy*. **33**: 63–74. doi:10.1093/jac/33.1.63. ISSN 1460-2091.

[22] Patchen ML, Brook I, Elliott TB, Jackson WE (1993). "Adverse effects of pefloxacin in irradiated C3H/HeN mice: correction with glucan therapy". *Antimicrobial Agents and Chemotherapy*. **37** (9): 1882–9. doi:10.1128/AAC.37.9.1882. ISSN 0066-4804. PMC 188087. PMID 8239601.

[23] Brook I, Walker RI, MacVittie TJ (1988). "Effect of antimicrobial therapy on the bowel flora and bacterial infection in irradiated mice". *International Journal of Radiation Biology*. **53** (5): 709–18. doi:10.1080/09553008814551081. ISSN 1362-3095.

[24] Brook I, Ledney D (1992). "Quinolone therapy in the management of infection after irradiation". *Crit Rev Microbiol*: 18235–46.

[25] Brook I, Elliot TB, Ledney GD, Shomaker MO, Knudson GB (2004). "Management of postirradiation infection: lessons learned from animal models". *Military Medicine*. **169**: 194–7. ISSN 0026-4075.

[26] Carmichael, Ann G. (1991). *Medicine: A Treasury of Art and Literature*. New York: Harkavy Publishing Service. p. 376. ISBN 0-88363-991-2.

[27] Turai, István; Veress, Katalin (2001). "Radiation Accidents: Occurrence, Types, Consequences, Medical Management, and the Lessons to be Learned". *Central European Journal of Occupational and Environmental Medicine*. **7** (1): 3–14. Retrieved 1 June 2012.

[28] Chambrette, V.; Hardy, S.; Nenot, J. C. (2001). "Les accidents d'irradiation: Mise en place d'une base de données "ACCIRAD" à l'IPSN" (PDF). *Radioprotection*. **36** (4): 477–510. doi:10.1051/radiopro:2001105. Retrieved 13 June 2012.

[29] Goldfarb, Alex; Litvinenko, Marina (2007). *Death of a Dissident: The Poisoning of Alexander Litvinenko and the Return of the KGB*. Simon & Schuster UK. ISBN 978-1-4711-0301-8.

[30] Johnston, Wm. Robert. "K-19 submarine reactor accident, 1961". *Database of radiological incidents and related events*. Johnston's Archive. Retrieved 24 May 2012.

[31] Johnston, Wm. Robert. "K-27 submarine reactor accident, 1968". *Database of radiological incidents and related events*. Johnston's Archive. Retrieved 24 May 2012.

[32] "Lost Iridium-192 Source".

[33] Johnston, Wm. Robert. "K-431 submarine reactor accident, 1985". *Database of radiological incidents and related events*. Johnston's Archive. Retrieved 24 May 2012.

[34] The Radiological Accident in Goiania p. 2.

[35] Strengthening the Safety of Radiation Sources Archived 2009-06-08 at WebCite p. 15.

[36] Gusev, Igor; Guskova, Angelina; Mettler, Fred A. (12 December 2010). *Medical Management of Radiation Accidents* (Second ed.). CRC Press. pp. 299–303. ISBN 978-1-4200-3719-7.

[37] Bagla, Pallava (7 May 2010). "Radiation Accident a 'Wake-Up Call' For India's Scientific Community". *Science*. **328** (5979): 679. doi:10.1126/science.328.5979.679-a. PMID 20448162.

[38] International Atomic Energy Agency. "Investigation of an accidental Exposure of radiotherapy patients in Panama" (PDF).

[39] Johnston, Robert (September 23, 2007). "Deadliest radiation accidents and other events causing radiation casualties". Database of Radiological Incidents and Related Events.

[40] Patterson AJ (2007). "Ushering in the era of nuclear terrorism". *Critical Care Medicine*. **35** (3): 953–4. doi:10.1097/01.CCM.0000257229.97208.76. PMID 17421087.

[41] Acton JM, Rogers MB, Zimmerman PD (September 2007). "Beyond the Dirty Bomb: Re-thinking Radiological Terror". *Survival*. **49** (3): 151–168. doi:10.1080/00396330701564760.

[42] Sixsmith, Martin (2007). *The Litvinenko File: The Life and Death of a Russian Spy*. True Crime. p. 14. ISBN 0-312-37668-5.

[43] Bremer Mærli, Morten. "Radiological Terrorism: "Soft Killers"". *Bellona Foundation*.

[44] Wells J (1976). "A guide to the prognosis for survival in mammals following the acute effects of inhaled radioactive particles". *Journal of the Institution of Nuclear Engineers*. **17** (5): 126–131. ISSN 0368-2595.

13.11 Further reading

- Michihiko Hachiya, *Hiroshima Diary* (Chapel Hill: University of North Carolina, 1955), ISBN 0-8078-4547-7.

- John Hersey, *Hiroshima* (New York: Vintage, 1946, 1985 new chapter), ISBN 0-679-72103-7.

- Ibuse Masuji, *Black Rain* (1969) ISBN 0-87011-364-X

- Ernest J. Sternglass, *Secret Fallout: low-level radiation from Hiroshima to Three-Mile Island* (1981) ISBN 0-07-061242-0 (online)

- Norman Solomon, Harvey Wasserman *Killing Our Own: The Disaster of America's Experience with Atomic Radiation, 1945–1982*. New York: Dell, 1982. ISBN 0-385-28537-X, ISBN 0-385-28536-1, ISBN 0-440-04567-3 (online)

13.12 External links

- The Center for Disease Control's fact sheet on Acute Radiation Syndrome

- List of radiation accidents and other events causing radiation casualties

- The criticality accident in Sarov, International Atomic Energy Agency, 2001 —well documented account of the biological effects of a criticality accident

- Armed Forces Radiobiology Research Institute

- *This article incorporates public domain material from websites or documents of the Armed Forces Radiobiology Research Institute and the Center for Disease Control and Prevention*

Chapter 14

Radiation hormesis

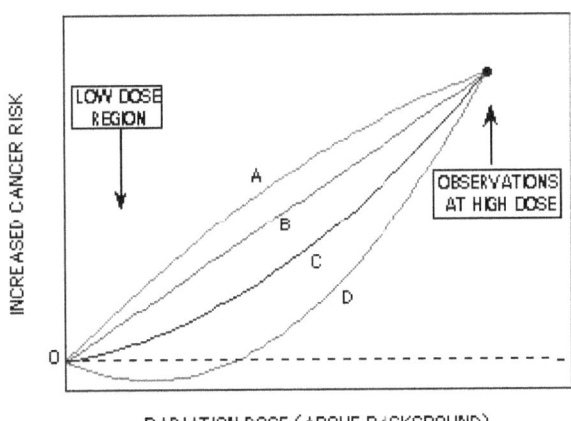

Alternative assumptions for the extrapolation of the cancer risk vs. radiation dose to low-dose levels, given a known risk at a high dose: supra-linearity (A), linear (B), linear-quadratic (C) and hormesis (D).

Radiation hormesis (also called **radiation homeostasis**) is the hypothesis that low doses of ionizing radiation (within the region of and just above natural background levels) are beneficial, stimulating the activation of repair mechanisms that protect against disease, that are not activated in absence of ionizing radiation. The reserve repair mechanisms are hypothesized to be sufficiently effective when stimulated as to not only cancel the detrimental effects of ionizing radiation but also inhibit disease not related to radiation exposure (see hormesis).[1][2][3][4] This counter-intuitive hypothesis has captured the attention of scientists and public alike in recent years.[5]

While the effects of high and acute doses of ionising radiation are easily observed and understood in humans (*e.g.* Japanese Atomic Bomb survivors), the effects of low-level radiation are very difficult to observe and highly controversial. This is because the baseline cancer rate is already very high and the risk of developing cancer fluctuates 40% because of individual life style and environmental effects.[6][7] obscuring the subtle effects of low-level radiation. An acute effective dose of 100 millisieverts may increase cancer risk by ~0.8%. However, children are partic-

ularly sensitive to radioactivity, with childhood leukemias and other cancers increasing even within natural and man-made background radiation levels (under 4 mSv cumulative with 1 mSv being an average annual dose from terrestrial and cosmic radiation excluding radon which primarily doses the lung).[8][9] There is also indication that exposures around this dose level will cause negative subclinical health impacts to neural development. Students born in regions of Sweden with higher Chernobyl fallout performed worse in secondary school, particularly in math. "Damage is accentuated within families (i.e., siblings comparison) and among children born to parents with low education..." who often don't have the resources to overcome this additional health challenge.[10]

Hormesis remains largely unknown to the public. Any policy change ought to consider hormesis first as a public health issue (versus an industrial regulatory issue). This would include the assessment of the public concern regarding exposure to small toxic doses. In addition, impact of hormesis policy change upon the management of industrial risks should be studied.[11] Government and regulatory bodies disagree on the existence of radiation hormesis and research points to the "severe problems and limitations" with the use of hormesis in general as the "principal dose-response default assumption in a risk assessment process charged with ensuring public health protection."[12]

Quoting results from a literature database research, the Académie des Sciences — Académie nationale de Médecine (French Academy of Sciences — National Academy of Medicine) stated in their 2005 report concerning the effects of low-level radiation that many laboratory studies have observed radiation hormesis.[13][14] However, they cautioned that it is not yet known if radiation hormesis occurs outside the laboratory, or in humans.[15]

Reports by the United States National Research Council and the National Council on Radiation Protection and Measurements and the United Nations Scientific Committee on the Effects of Atomic Radiation (UNSCEAR) argue[16] that there is no evidence for hormesis in humans and in the case of the National Research Council, that hormesis is outright rejected as a possibility despite population and scientific evidence.[17] Therefore, estimating Linear no-threshold model (LNT) continues to be the model generally used by regulatory agencies for human radiation exposure.

14.1 Proposed mechanism and ongoing debate

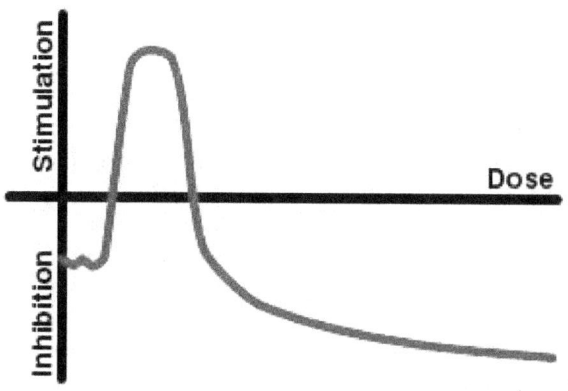

A very low dose of a chemical agent may trigger from an organism the opposite response to a very high dose.

Radiation hormesis proposes that radiation exposure comparable to and just above the natural background level of radiation is not harmful but beneficial, while accepting that much higher levels of radiation are hazardous. Proponents of radiation hormesis typically claim that radio-protective responses in cells and the immune system not only counter the harmful effects of radiation but additionally act to inhibit spontaneous cancer not related to radiation exposure. Radiation hormesis stands in stark contrast to the more generally accepted linear no-threshold model (LNT), which states that the radiation dose-risk relationship is linear across all doses, so that small doses are still damaging, albeit less so than higher ones. Opinion pieces on chemical and radiobiological hormesis appeared in the journals Nature[1] and Science[3] in 2003.

Assessing the risk of radiation at low doses (<100 mSv) and low dose rates (<0.1 mSv.min−1) is highly problematic and controversial.[18][19] While epidemiological studies on populations of people exposed to an acute dose of high level radiation such as Japanese Atomic Bomb Survivors (hibakusha (被爆者)) have robustly upheld the LNT (mean dose ~210 mSv),[20] studies involving low doses and low dose rates have failed to detect any increased cancer rate.[19] This is because the baseline cancer rate is already very high (~42 of 100 people will be diagnosed in their lifetime) and it fluctuates ~40% because of lifestyle and environmental effects,[7][21] obscuring the subtle effects of low level radiation. Epidemiological studies maybe capable of detecting elevated cancer rates as low as 1.2 to 1.3 *i.e.* 20% to 30% increase. But for low doses (1–100 mSv) the predicted elevated risks are only 1.001 to 1.04 and excess cancer cases, if present, cannot be detected due to confounding factors, errors and biases.[21][22]

In particular, variations in smoking prevalence or even accuracy in reporting smoking cause wide variation in excess cancer and measurement error bias. Thus, even a large study of many thousands of subjects with imperfect smoking prevalence information will fail to detect the effects of low level radiation than a smaller study that properly compensates for smoking prevalence.[23] Given the absence of direct epidemiological evidence, there is considerable debate as to whether the dose-response relationship <100 mSv is supralinear, linear (LNT), has a threshold or *sublinear i.e.* a hormetic response.

While most major consensus reports and government bodies currently adhere to LNT,[24] the 2005 French Academy of Sciences-National Academy of Medicine's report concerning the effects of low-level radiation rejected LNT as a scientific model of carcinogenic risk at low doses.[15]

"Using LNT to estimate the carcinogenic effect at doses of less than 20 mSv is not justified in the light of current radiobiologic knowledge."

They consider there to be several dose-effect relationships rather than only one, and that these relationships have many variables such as target tissue, radiation dose, dose rate and individual sensitivity factors. They request that further study is required on low doses (less than 100 mSv) and very low doses (less than 10 mSv) as well as the impact of tissue type and age. The Academy considers the LNT model is only useful for regulatory purposes as it simplifies the administrative task. Quoting results from literature research,[13][14] they furthermore claim that approximately 40% of laboratory studies on cell cultures and animals indicate some degree of chemical or radiobiological hormesis, and state:

"...its existence in the laboratory is beyond question and its mechanism of action appears well understood."

They go on to outline a growing body of research that illustrates that the human body is not a passive accumulator of radiation damage but it actively repairs the damage caused via a number of different processes, including:[15][19]

- Mechanisms that mitigate reactive oxygen species generated by ionizing radiation and oxidative stress.

- Apoptosis of radiation damaged cells that may undergo tumorigenesis is initiated at only few mSv.

- Cell death during meiosis of radiation damaged cells that were unsuccessfully repaired.

- The existence of a cellular signaling system that alerts neighboring cells of cellular damage.

- The activation of enzymatic DNA repair mechanisms around 10 mSv.

- Modern DNA microarray studies which show that numerous genes are activated at radiation doses well below the level that mutagenesis is detected.

- Radiation-induced tumorigenesis may have a threshold related to damage density, as revealed by experiments that employ blocking grids to thinly distribute radiation.

- A large increase in tumours in immunosuppressed individuals illustrates that the immune system efficiently destroys aberrant cells and nascent tumors.

Furthermore, increased sensitivity to radiation induced cancer in the inherited condition Ataxia-telangiectasia like disorder, illustrates the damaging effects of loss of the repair gene Mre11h resulting in the inability to fix DNA double-strand breaks.[25]

The BEIR-VII report argued that, "the presence of a true dose threshold demands totally error-free DNA damage response and repair." The specific damage they worry about is double strand breaks (DSBs) and they continue, "error-prone nonhomologous end joining (NHEJ) repair in postirradiation cellular response, argues strongly against a DNA repair-mediated low-dose threshold for cancer initiation".[26] Recent research observed that DSBs caused by CAT scans are repaired within 24-hours and DSBs maybe more efficiently repaired at low doses, suggesting the risk ionizing radiation at low doses may not by directly proportional to the dose.[27][28] However, it is not known if low dose ionizing radiation stimulates the repair of DSBs not caused by ionizing radiation i.e. a hormetic response.

Radon gas in homes is the largest source of radiation dose for most individuals and it is generally advised that the concentration be kept below 150 Bq/m^3 (4 pCi/L).[29] A recent retrospective case-control study of lung cancer risk showed substantial cancer rate reduction between 50 and 123 Bq per cubic meter relative to a group at zero to 25 Bq per cubic meter.[30] This study is cited as evidence for hormesis, but a single study all by itself cannot be regarded as definitive. Other studies into the effects of domestic radon exposure have not reported a hormetic effect; including for example the respected "Iowa Radon Lung Cancer

Study" of Field et al. (2000), which also used sophisticated radon exposure dosimetry.[31] In addition, Darby et al. (2005) argue that radon exposure is negatively correlated with the tendency to smoke and environmental studies need to accurately control for this: people living in urban areas where smoking rates are higher usually have lower levels of radon exposure due the increased prevalence of multi-story dwellings.[32] When doing so, they found a significant increase in lung cancer amongst smokers exposed to radon at doses as low as 100 to 199 Bq m^{-3} and warned that smoking greatly increases the risk posed by radon exposure i.e. reducing the prevalence of smoking would decrease deaths caused by radon.[32][33]

Furthermore, particle microbeam studies show that passage of even a single alpha particle (e.g. from radon and its progeny) through cell nuclei is highly mutagenic.[34] and that alpha radiation may have a higher mutagenic effect at low doses (even if a small fraction of cells are hit by alpha particles) than predicted by linear no-threshold model, a phenomenon attributed to bystander effect.[35] However, there is currently insufficient evidence at hand to suggest that the bystander effect promotes carcinogenesis in humans at low doses.[36]

In his October 4, 2016 Skeptoid podcast episode titled "Radiation Hormesis: Is It Good for You?," scientific skeptic author Brian Dunning evaluated the scientific literature concerning this issue. Regarding the scientific effort to replace the LTN model with a model that better reflects the facts concerning radiation exposure, Dunning concluded:

Nowhere will you find recommendations that radiation hormesis be made part of the model — at least, nowhere outside of the fringe sources. A dedicated literature search will indeed find claims that hormesis is real, but these effects are always found within the noise; and when you find references to these studies in mainstream sources, they always point out the flaws. To sum up radiation hormesis in one sentence, we can say that it is one claim of a pattern that some say can be found in the data, but that most dismiss because the data is simply far too noisy at that low level to support the drawing of any conclusions at all.

Most radiation researchers will probably agree that radiation hormesis is potentially plausible, but unproven and probably unprovable. Even if it exists, the data that can be interpreted to support it would show that the hypothetical protective effect is also too small to be detectable.[37]

14.2 Statements by leading nuclear bodies

Radiation hormesis has not been accepted by either the United States National Research Council.[17] or the National Council on Radiation Protection and Measurements.[38] In addition, the United Nations Scientific Committee on the Effects of Atomic Radiation (UNSCEAR) wrote in its most recent report:[39]

> Until the [...] uncertainties on low-dose response are resolved, the Committee believes that an increase in the risk of tumour induction proportionate to the radiation dose is consistent with developing knowledge and that it remains, accordingly, the most scientifically defensible approximation of low-dose response. However, a strictly linear dose response should not be expected in all circumstances.

This is a reference to the fact that very low doses of radiation have only marginal impacts on individual health outcomes. It is therefore difficult to detect the 'signal' of decreased or increased morbidity and mortality due to low-level radiation exposure in the 'noise' of other effects. The notion of radiation hormesis has been rejected by the National Research Council's (part of the National Academy of Sciences) 16-year-long study on the Biological Effects of Ionizing Radiation. "The scientific research base shows that there is no threshold of exposure below which low levels of ionizing radiation can be demonstrated to be harmless or beneficial. The health risks – particularly the development of solid cancers in organs – rise proportionally with exposure" says Richard R. Monson, associate dean for professional education and professor of epidemiology, Harvard School of Public Health, Boston.[40][41]

> The possibility that low doses of radiation may have beneficial effects (a phenomenon often referred to as "hormesis") has been the subject of considerable debate. Evidence for hormetic effects was reviewed, with emphasis on material published since the 1990 BEIR V study on the health effects of exposure to low levels of ionizing radiation. Although examples of apparent stimulatory or protective effects can be found in cellular and animal biology, the preponderance of available experimental information does not support the contention that low levels of ionizing radiation have a beneficial effect. The mechanism of any such possible effect remains obscure. At this time, the assumption that any stimulatory hormetic effects

from low doses of ionizing radiation will have a significant health benefit to humans that exceeds potential detrimental effects from radiation exposure at the same dose is unwarranted. —[41]

14.3 Studies of low level radiation

14.3.1 Very high natural background gamma radiation cancer rates at Kerala, India

Kerala's monazite sand (containing a third of the world's economically recoverable reserves of radioactive thorium) emits about 8 micro Sieverts per hour of gamma radiation, 80 times the dose rate equivalent in London, but a decade long study of 69,985 residents published in Health Physics in 2009: "showed no excess cancer risk from exposure to terrestrial gamma radiation. The excess relative risk of cancer excluding leukemia was estimated to be -0.13 Gy_1 (95% CI: -0.58, 0.46)", indicating no statistically significant positive or negative relationship between background radiation levels and cancer risk in this sample.[42]

14.3.2 Cultures

Studies in cell cultures can be useful for finding mechanisms for biological processes, but they also can be criticized for not effectively capturing the whole of the living organism.

A study by E.I. Azzam suggested that pre-exposure to radiation causes cells to turn on protection mechanisms.[43] A different study by de Toledo and collaborators, has shown that irradiation with gamma rays increases the concentration of glutathione, an antioxidant found in cells.[44]

In 2011, an *in vitro* study led by S.V. Costes showed in time-lapse images a strongly non-linear response of certain cellular repair mechanisms called radiation-induced foci (RIF). The study found that low doses of radiation prompted higher rates of RIF formation than high doses, and that after low-dose exposure RIF continued to form after the radiation had ended. Measured rates of RIF formation were 15 RIF/Gy at 2 Gy, and 64 RIF/Gy at .1 Gy.[28] These results suggest that low dose levels of ionizing radiation may not increase cancer risk directly proportional to dose and thus contradict the linear-no-threshold standard model.[45] Mina Bissell, a world-renowned breast cancer researcher and collaborator in this study stated "Our data show that at lower doses of ionizing radiation, DNA repair mechanisms work much better than at higher doses.

This non-linear DNA damage response casts doubt on the general assumption that any amount of ionizing radiation is harmful and additive." [45]

14.3.3   Animals

A study by Otsuka and collaborators found hormesis in animals. [46] Miyachi conducted a study on mice and found that a 200 mGy X-ray dose protects mice against both further X-ray exposure and ozone gas. [47] In another rodent study, Sakai and collaborators found that (1 mGy/hr) gamma irradiation prevents the development of cancer (induced by chemical means, injection of methylcholanthrene). [48]

In a 2006 paper, [49] a dose of 1 Gy was delivered to the cells (at constant rate from a radioactive source) over a series of lengths of time. These were between 8.77 and 87.7 hours, the abstract states for a dose delivered over 35 hours or more (low dose rate) no transformation of the cells occurred. Also for the 1 Gy dose delivered over 8.77 to 18.3 hours that the biological effect (neoplastic transformation) was about "1.5 times less than that measured at high dose rate in previous studies with a similar quality of [X-ray] radiation." Likewise it has been reported that fractionation of gamma irradiation reduces the likelihood of a neoplastic transformation. [50] Pre-exposure to fast neutrons and gamma rays from Cs-137 is reported to increase the ability of a second dose to induce a neoplastic transformation. [51]

Caution must be used in interpreting these results, as it noted in the BEIR VII report, these pre-doses can also increase cancer risk:

> In chronic low-dose experiments with dogs (75 mGy/d for the duration of life), vital hematopoietic progenitors showed increased radioresistance along with renewed proliferative capacity (Seed and Kaspar 1992). Under the same conditions, a subset of animals showed an increased repair capacity as judged by the unscheduled DNA synthesis assay (Seed and Meyers 1993). Although one might interpret these observations as an adaptive effect at the cellular level, the exposed animal population experienced a high incidence of myeloid leukemia and related myeloproliferative disorders. The authors concluded that "the acquisition of radioresistance and associated repair functions under the strong selective and mutagenic pressure of chronic radiation is tied temporally and causally to leukemogenic transformation by the radiation exposure" (Seed and Kaspar 1992).
> —BEIR VII report. [41]

However, 75 mGy/d cannot be accurately described as a low dose rate – it is equivalent to over 27 sieverts per year. The same study on dogs showed no increase in cancer nor reduction in life expectancy for dogs irradiated at 3 mGy/d. [52]

14.3.4   Humans

Effects of sunlight exposure

In an Australian study which analyzed the association between solar UV exposure and DNA damage, the results indicated that although the frequency of cells with chromosome breakage increased with increasing sun exposure, the misrepair of DNA strand breaks decreased as sun exposure was heightened. [53]

Effects of cobalt-60 exposure

The health of the inhabitants of radioactive apartment buildings in Taiwan has received prominent attention in popular treatments of radiation hormesis. In 1982, more than 20,000 tons of steel was accidentally contaminated with cobalt-60, and much of this radioactive steel was used to build apartments and exposed thousands of Taiwanese to gamma radiation levels of up to >1000 times background (average 47.7 mSv, maximum 2360 mSv excess cumulative dose) – it was not until 1992 that the radioactive contamination was discovered. A medical study published in 2004 claimed the cancer mortality rates in the exposed population were much lower than expected. [54] However, this initial study failed to control for age, comparing a much younger exposed population (mean age 17.2 years at initial exposure) with the much older general population of Taiwan (mean age approx. 34 years in 2004), a serious flaw. [55] [56] Older people have much higher cancer rates even in the absence of excess radiation exposure. However, Chen et al. did find a lower cancer incidence with time, still the opposite of what would be expected, even with a younger population.

A subsequent study by Hwang et al. (2006) found the incidence of "all cancers" in the irradiated population was 40% lower than expected (95 vs. 160.3 cases expected), except for leukaemia in men (6 vs. 1.8 cases expected) and thyroid cancer in women (6 vs. 2.8 cases expected), an increase only detected amongst those exposed before the age of 30. Hwang et al. proposed that the lower rate of "all cancers" might due to the exposed populations higher socioeconomic status and thus overall healthier lifestyle, but this was difficult to prove. Additionally, they cautioned that leukaemia was the first cancer type found to be elevated amongst the survivors of the Hiroshima and Nagasaki bombings, so it may be decades before any increase in more common cancer types is seen.[55]

Besides the excess risks of leukaemia and thyroid cancer, a later publication notes various DNA anomalies and other health effects among the exposed population:[57]

There have been several reports concerning the radiation effects on the exposed population, including cytogenetic analysis that showed increased micronucleus frequencies in peripheral lymphocytes in the exposed population, increases in acentromeric and single or multiple centromeric cytogenetic damages, and higher frequencies of chromosomal translocations, rings and dicentrics. Other analyses have shown persistent depression of peripheral leucocytes and neutrophils, increased eosinophils, altered distributions of lymphocyte subpopulations, increased frequencies of lens opacities, delays in physical development among exposed children, increased risk of thyroid abnormalities, and late consequences in hematopoietic adaptation in children.

14.4 Effects of no radiation

Given the uncertain effects of low-level and very-low-level radiation, there is a pressing need for quality research in this area. An expert panel convened at the 2006 Ultra-Low-Level Radiation Effects Summit at Carlsbad, New Mexico, proposed the construction of an Ultra-Low-Level Radiation laboratory.[58] The laboratory, if built, will investigate the effects of almost *no radiation* on laboratory animals and cell cultures, and it will compare these groups to control groups exposed to natural radiation levels. Precautions would be made, for example, to remove potassium-40 from the food of laboratory animals. The expert panel believes that the Ultra-Low-Level Radiation laboratory is the only experiment that can explore with authority and confidence the effects of low-level radiation; that it can confirm or discard the various radiobiological effects proposed at low radiation levels e.g. LNT, threshold and radiation hormesis.[59]

The first preliminary results of the effects of almost no-radiation on cell cultures was reported by two research groups in 2011 and 2012; researchers in the US studied cell cultures protected from radiation in a steel chamber 650 meters underground at the Waste Isolation Pilot Plant in Carlsbad, New Mexico[60] and researchers in Europe reported the effects of almost no-radiation on mouse cells (pKZ1 transgenic chromosomal inversion assay).[61]

14.5 See also

- Background radiation
- Dose fractionation
- Hormesis
- Radithor
- Linear no-threshold model
- Petkau effect
- Radioresistance
- Ramsar, Mazandaran

14.6 References

[1] Calabrese, Edward J; Baldwin, Linda A (2003). "Toxicology rethinks its central belief". *Nature*. **421** (6924): 691–2. Bibcode:2003Natur.421..691C. doi:10.1038/421691a. PMID 12610596.

[2] Feinendegen, L E (2005). "Evidence for beneficial low level radiation effects and radiation hormesis". *British Journal of Radiology*. **78** (925): 3–7. doi:10.1259/bjr/63353075. PMID 15673519.

[3] Kaiser, J. (2003). "HORMESIS: Sipping from a Poisoned Chalice". *Science*. **302** (5644): 376–9. doi:10.1126/science.302.5644.376. PMID 14563981.

[4] Wolff, Sheldon (1998). "The Adaptive Response in Radiobiology: Evolving Insights and Implications". *Environmental Health Perspectives*. **106**: 277–83. doi:10.2307/3433927. JSTOR 3433927. PMC 1533272. PMID 9539019.

[5] Allison, Wade (2009). *Radiation and Reason: The Impact of Science on a Culture of Fear*. York, England: York Publishing Services. p. 2. ISBN 0-9562756-1-3.

[6] "WHO Cancer Fact sheet N°297". Retrieved 2011-04-29.

[7] Parkin, D M; Boyd, L; Walker, L C (2011). "16. The fraction of cancer attributable to lifestyle and environmental factors in the UK in 2010". *British Journal of Cancer*. **105** (Suppl 2): S77–81. doi:10.1038/bjc.2011.489. PMC 3252065. PMID 22158327.

[8] Kendall; et al. (January 2013). "A record-based case-control study of natural background radiation and the incidence of childhood leukaemia and other cancers in Great Britain during 1980-2006". *Leukemia*. 27(1): 3–9. doi:10.1038/leu.2012.151.

[9] Spycher; et al. (23 Feb 2015). "Background Ionizing Radiation and the Risk of Childhood Cancer: A Census-Based Nationwide Cohort Study". *Environ Health Perspect*.

[10] Almond; et al. (2007). "Chernobyl's subclinical legacy: Prenatal exposure to radioactive fallout and school outcomes in Sweden." *Columbia University*.

[11] Poumadere, M. (2003). Hormesis: public health policy, organizational safety and risk communication. Human & experimental toxicology. 22(1). 39-41

[12] Kitchin; et al. (2005). ". A critique of the use of hormesis in risk assessment". *Human & Experimental Toxicology*. 24: 249–253. doi:10.1191/0960327105ht520oa.

[13] Calabrese, Edward J (2004). "Hormesis: From marginalization to mainstream". *Toxicology and Applied Pharmacology*. 197 (2): 125–36. doi:10.1016/j.taap.2004.02.007. PMID 15163548.

[14] Duport, P. (2003). "A database of cancer induction by low-dose radiation in mammals: Overview and initial observations". *International Journal of Low Radiation*. 1: 120–31. doi:10.1504/IJLR.2003.003488.

[15] Aurengo (2005-03-30). "Dose-effect relationships and estimation of the carcinogenic effects of low doses of ionizing radiation". Académie des Sciences & Académie nationale de Médecine. CiteSeerX 10.1.1.126.1681⊕.

[16] UNSCEAR 2000 REPORT Vol. II: Sources and Effects of Ionizing Radiation: Annex G: Biological effects at low radiation doses.

[17] http://books.nap.edu/catalog/11340.html Health Risks from Exposure to Low Levels of Ionizing Radiation: BEIR VII Phase 2

[18] Mullenders, Leon; Atkinson, Mike; Paretzke, Herwig; Sabatier, Laure; Bouffler, Simon (2009). "Assessing cancer risks of low-dose radiation". *Nature Reviews Cancer*. 9 (8): 596–604. doi:10.1038/nrc2677. PMID 19629073.

[19] Tubiana, M.; Feinendegen, L. E.; Yang, C.; Kaminski, J. M. (2009). "The Linear No-Threshold Relationship is Inconsistent with Radiation Biologic and Experimental Data". *Radiology*. 251 (1): 13–22. doi:10.1148/radiol.2511080671. PMC 2663584⊕. PMID 19332842.

[20] Samartzis, Dino; Nishi, N; Hayashi, M; Cologne, J; Cullings, HM; Kodama, K; Miles, EF; Funamoto, S; et al. (2011). "Exposure to Ionizing Radiation and Development of Bone Sarcoma: New Insights Based on Atomic-Bomb Survivors of Hiroshima and Nagasaki". *The Journal of Bone & Joint Surgery (American)*. 93 (11): 1008–15. doi:10.2106/JBJS.J.00256. PMID 21984980.

[21] Boice Jr, John D (2012). "Radiation epidemiology: A perspective on Fukushima". *Journal of Radiological Protection*. 32 (1): N33–40. doi:10.1088/0952-4746/32/1/N33. PMID 22395193.

[22] Boice, John D. (2010). "INVITED EDITORIAL. Uncertainties in studies of low statistical power Uncertainties in studies of low statistical power". *Journal of Radiological Protection*. 30 (2): 115–20. Bibcode:2010JRP....30..115B. doi:10.1088/0952-4746/30/2/E02. PMID 20548136.

[23] Lubin, Jay H.; Samet, Jonathan M.; Weinberg, Clarice (1990). "Design Issues in Epidemiologic Studies of Indoor Exposure to Rn and Risk of Lung Cancer". *Health Physics*. 59 (6): 807–17. doi:10.1097/00004032-199012000-00004. PMID 2228608.

[24] Hall, Eric J. (1998). "From Chimney Sweeps to Astronauts". *Health Physics*. 75 (4): 357–66. doi:10.1097/00004032-199810000-00001. PMID 9753358.

[25] Stewart, G; Maser, RS; Stankovic, T; Bressan, DA; Kaplan, MI; Jaspers, NG; Raams, A; Byrd, PJ; et al. (1999). "The DNA Double-Strand Break Repair Gene hMRE11 is Mutated in Individuals with an Ataxia-Telangiectasia-like Disorder". *Cell*. 99 (6): 577–87. doi:10.1016/S0092-8674(00)81547-0. PMID 10612394.

[26] BEIR VII, page 245

[27] Löbrich, Markus; Rief, Nicole; Kühne, Martin; Heckmann, Martina; Fleckenstein, Jochen; Rübe, Christian; Uder, Michael (2005). "In vivo formation and repair of DNA double-strand breaks after computed tomography examinations". *Proceedings of the National Academy of Sciences*. 102 (25): 8984–9. Bibcode:2005PNAS..102.8984L. doi:10.1073/pnas.0501895102. PMC 1150277⊕. PMID 15956203.

[28] Neumaier, T.; Swenson, J.; Pham, C.; Polyzos, A.; Lo, A. T.; Yang, P.; Dyball, J.; Asaithamby, A.; et al. (2012). "Evidence for formation of DNA repair centers and dose-response nonlinearity in human cells". *Proceedings of the National Academy of Sciences*. 109 (2): 443–8. Bibcode:2012PNAS..109..443N. doi:10.1073/pnas.1117849108. PMC 3258602⊕. PMID 22184222.

[29] "Surgeon General Releases National Health Advisory On Radon". US HHS Office of the Surgeon General. January 12, 2005. Retrieved 28 November 2008.

[30] Thompson, Richard E.; Nelson, Donald F.; Popkin, Joel H.; Popkin, Zenaida (2008). "Case-Control Study of Lung Cancer Risk from Residential Radon Exposure in Worcester County, Massachusetts". *Health Physics*. 94 (3): 228–41. doi:10.1097/01.HP.0000288561.53790.5f. PMID 18301096.

[31] Field, R. W.; Steck, D. J.; Smith, B. J.; Brus, C. P.; Fisher, E. L.; Neuberger, J. S.; Platz, C. E.; Robinson, R. A.; et al. (2000). "Residential Radon Gas Exposure and Lung Cancer: The Iowa Radon Lung Cancer Study". *American Journal of Epidemiology*. 151 (11): 1091–102. doi:10.1093/oxfordjournals.aje.a010153. PMID 10873134.

[32] Darby, S; Hill, D; Auvinen, A; Barros-Dios, JM; Baysson, H; Bochicchio, F; Deo, H; Falk, R; et al. (2005). "Radon in homes and risk of lung cancer: Collaborative analysis of individual data from 13 European case-control studies". *BMJ*. 330 (7485): 223. doi:10.1136/bmj.38308.477650.63. PMC 546066⊕. PMID 15613366.

[33] Méndez, David; Alshanqeety, Omar; Warner, Kenneth E.; Lantz, Paula M.; Courant, Paul N. (2011). "The Impact of Declining Smoking on Radon-Related Lung Cancer in the United States". *American Journal of Public Health.* **101** (2): 310–4. doi:10.2105/AJPH.2009.189225. PMC 3020207⊙. PMID 21228294.

[34] Hei, Tom K.; Wu, Li-Jun; Liu, Su-Xian; Vannais, Diane; Waldren, Charles A.; Randers-Pehrson, Gerhard (1997). "Mutagenic Effects of a Single and an Exact Number of α Particles in Mammalian Cells". *Proceedings of the National Academy of Sciences of the United States of America.* **94** (8): 3765–70. Bibcode:1997PNAS...94.3765H. doi:10.1073/pnas.94.8.3765. PMC 20515⊙. PMID 9108052.

[35] Zhou, Hongning; Randers-Pehrson, Gerhard; Waldren, Charles A.; Vannais, Diane; Hall, Eric J.; Hei, Tom K. (2000). "Induction of a bystander mutagenic effect of alpha particles in mammalian cells". *Proceedings of the National Academy of Sciences.* **97** (5): 2099–104. Bibcode:2000PNAS...97.2099Z. doi:10.1073/pnas.030420797. PMC 15760⊙. PMID 10681418.

[36] Blyth, Benjamin J.; Sykes, Pamela J. (2011). "Radiation-Induced Bystander Effects: What Are They, and How Relevant Are They to Human Radiation Exposures?". *Radiation Research.* **176** (2): 139–57. doi:10.1667/RR2548.1. PMID 21631286.

[37] Dunning, Brian. "Radiation Hormesis: Is It Good for You?". *Skeptoid.com.* Retrieved 5 October 2016.

[38] NCRP Report No. 136 —Evaluation of the Linear-Nonthreshold Dose-Response Model for Ionizing Radiation

[39] UNSCEAR 2000 REPORT Vol. II: Sources and Effects of Ionizing Radiation: Annex G: Biological effects at low radiation doses. page 160, paragraph 541.

[40] Vines, Vanee; Petty, Megan (2005-06-29). "Low Levels of Ionizing Radiation May Cause Harm". National Academy of Sciences. Retrieved 2010-01-27.

[41] *Health Risks from Exposure to Low Levels of Ionizing Radiation: BEIR VII Phase 2.* National Academies Press. 2006. ISBN 978-0-309-09156-5. Retrieved 2010-01-27.

[42] Nair, Raghu Ram K.; Rajan, Balakrishnan; Akiba, Suminori; Jayalekshmi, P; Nair, M Krishnan; Gangadharan, P; Koga, Taeko; Morishima, Hiroshige; et al. (2009). "Background Radiation and Cancer Incidence in Kerala, India— Karanagappally Cohort Study". *Health Physics.* **96** (1): 55–66. doi:10.1097/01.HP.0000327646.54923.11. PMID 19066487.

[43] Azzam, E.I.; Raaphorst, G. P.; Mitchel, R. E. J. (1994). "Radiation-Induced Adaptive Response for Protection against Micronucleus Formation and Neoplastic Transformation in C3H 10T1/2 Mouse Embryo Cells". *Radiation*

Research. Radiation Research. Vol. 138, No. 1. **138** (1): S28–S31. doi:10.2307/3578755. JSTOR 3578755. PMID 8146320.

[44] De Toledo, Sonia M.; Asaad, Nesrin; Venkatachalam, Perumal; Li, Ling; Howell, Roger W.; Spitz, Douglas R.; Azzam, Edouard I. (2006). "Adaptive Responses to Low-Dose/Low-Dose-Rate γ Rays in Normal Human Fibroblasts: The Role of Growth Architecture and Oxidative Metabolism". *Radiation Research.* **166** (6): 849–57. doi:10.1667/RR0640.1. PMID 17149977.

[45] http://www.healthcanal.com/public-health-safety/24865-New-Take-Impacts-Low-Dose-Radiation.html[]

[46] Otsuka, Kensuke; Koana, Takao; Tauchi, Hiroshi; Sakai, Kazuo (2006). "Activation of Antioxidative Enzymes Induced by Low-Dose-Rate Whole-Body γ Irradiation: Adaptive Response in Terms of Initial DNA Damage". *Radiation Research.* **166** (3): 474–8. doi:10.1667/RR0561.1. PMID 16953665.

[47] Miyachi, Y (2000). "Acute mild hypothermia caused by a low dose of X-irradiation induces a protective effect against mid-lethal doses of X-rays, and a low level concentration of ozone may act as a radiomimetic". *The British Journal of Radiology.* **73** (867): 298–304. doi:10.1259/bjr.73.867.10817047. PMID 10817047.

[48] Sakai, Kazuo; Iwasaki, Toshiyasu; Hoshi, Yuko; Nomura, Takaharu; Oda, Takeshi; Fujita, Kazuo; Yamada, Takeshi; Tanooka, Hiroshi (2002). "Suppressive effect of long-term low-dose rate gamma-irradiation on chemical carcinogenesis in mice". *International Congress Series.* **1236**: 487–490. doi:10.1016/S0531-5131(01)00861-5.

[49] Elmore, E.; Lao, X-Y.; Kapadia, R.; Redpath, J. L. (2006). "The Effect of Dose Rate on Radiation-Induced Neoplastic TransformationIn Vitroby Low Doses of Low-LET Radiation". *Radiation Research.* **166** (6): 832–8. doi:10.1667/RR0682.1. PMID 17149982.

[50] Hill, C.K.; Han, A.; Buonaguro, F.; Elkind, M.M. (1984). "Multifractionation of 60Co gamma-rays reduces neoplastic transformation in vitro". *Carcinogenesis.* **5** (2): 193–7. doi:10.1093/carcin/5.2.193. PMID 6697436.

[51] Cao, J.; Wells, R.L.; Elkind, M.M. (1992). "Enhanced Sensitivity to Neoplastic Transformation by137Cs γ-rays of Cells in the G2/M-phase Age Interval". *International Journal of Radiation Biology.* **62** (2): 191–9. doi:10.1080/09553009214552011. PMID 1355513.

[52] http://www.nuclearsafety.gc.ca/eng/pdfs/Presentations/Guest-Speakers/2013/20130625-Cuttler-CNSC-Fukushima-and-beneficial.pdf

[53] Nair-Shalliker, V.; Fenech, M.; Forder, P. M.; Clements, M. S.; Armstrong, B. K. (2012). "Sunlight and vitamin D affect DNA damage, cell division and cell death in human lymphocytes: A cross-sectional study in South Australia".

Mutagenesis. **27** (5): 609–14. doi:10.1093/mutage/ges026. PMID 22547344.

[54] Chen, W.L.; Luan, Y.C.; Shieh, M.C.; Chen, S.T.; Kung, H.T.; Soong, K.L.; Yeh, Y.C.; Chou, T.S.; Mong, S.H. (2004). "Is Chronic Radiation an Effective Prophylaxis Against Cancer?" (PDF). *Journal of the American Physicians and Surgeons.* **9** (1): 6–10.

[55] Hwang, S. -L.; Guo, H. -R.; Hsieh, W. -A.; Hwang, J. -S.; Lee, S. -D.; Tang, J. -L.; Chen, C. -C.; Chang, T. -C.; et al. (2006). "Cancer risks in a population with prolonged low dose-rate γ-radiation exposure in radiocontaminated buildings, 1983 – 2002". *International Journal of Radiation Biology.* **82** (12): 849–58. doi:10.1080/09553000601085980. PMID 17178625.

[56] Chen, C. Y.; Y. J. Chen (2011). *The Social Migration Effect Toward Population Aging-The Application of Perston's Rate of Change of a Population's Mean Age Improvement Model in Taiwan* (PDF). The 23rd Conference of the European Network for Housing Research. Retrieved 2012-05-09.

[57] Hwang, Su-Lun; Hwang, Jing-Shiang; Yang, Yi-Ta; Hsieh, Wanhua A.; Chang, Tien-Chun; Guo, How-Ran; Tsai, Mong-Hsun; Tang, Jih-Luh; et al. (2008). "Estimates of Relative Risks for Cancers in a Population after Prolonged Low-Dose-Rate Radiation Exposure: A Follow-up Assessment from 1983 to 2005". *Radiation Research.* **170** (2): 143–8. doi:10.1667/RR0732.1. PMID 18666807.

[58] "Ultra-Low-Level Radiation Effects Summit." January 2006. ORION International Technologies, Inc. (ORION) and sponsored by the U.S. Department of Energy's Waste Isolation Pilot Plant (WIPP) 03 Apr. 2008.

[59] http://www.orionint.com/ullre/report-2006.pdf[]

[60] Smith, Geoffrey Battle; Grof, Yair; Navarrette, Adrianne; Guilmette, Raymond A. (2011). "Exploring Biological Effects of Low Level Radiation from the Other Side of Background". *Health Physics.* **100** (3): 263–5. doi:10.1097/HP.0b013e318208cd44. PMID 21595063.

[61] Capece, D.; Fratini, E. (2012). "The use of pKZ1 mouse chromosomal inversion assay to study biological effects of environmental background radiation". *The European Physical Journal Plus.* **127** (4): 37. Bibcode:2012EPJP..127...37C. doi:10.1140/epjp/i2012-12037-7.

14.7 External links

- International Dose-Response Society. University of Massachusetts center for research on hormesis. Many papers on radiation hormesis.

- Health Risks from Exposure to Low Levels of Ionizing Radiation: BEIR VII Phase 2

- Radiation Hormesis Overview by T. D. Luckey, who wrote a book on the subject (Luckey, T. D. (1991). *Radiation Hormesis.* Boca Raton, FL: CRC Press. ISBN 0-8493-6159-1)

- Brenner, David J.; Doll, Richard; Goodhead, Dudley T.; Hall, Eric J.; Land, Charles E.; Little, John B.; Lubin, Jay H.; Preston, Dale L.; et al. (2003). "Cancer risks attributable to low doses of ionizing radiation: Assessing what we really know". *Proceedings of the National Academy of Sciences.* **100** (24): 13761–6. Bibcode:2003PNAS..10013761B. doi:10.1073/pnas.2235592100. JSTOR 3148861. PMC 283495. PMID 14610281.

- "Skeptoid #539: Radiation Hormesis: Is It Good for You?". *Skeptoid.*

Chapter 15

Radionuclide

Not to be confused with radionucleotide.

A **radionuclide** (**radioactive nuclide**, **radioisotope** or **radioactive isotope**) is an atom that has excess nuclear energy, making it unstable. This excess energy can be either emitted from the nucleus as gamma radiation, or create and emit from the nucleus a new particle (alpha particle or beta particle), or transfer this excess energy to one of its electrons, causing that electron to be ejected as a conversion electron. During those processes, the radionuclide is said to undergo radioactive decay.*[1] These emissions constitute ionizing radiation. The unstable nucleus is more stable following the emission, but will sometimes undergo further decay. Radioactive decay is a random process at the level of single atoms: it is impossible to predict when one particular atom will decay.*[2]*[3]*[4]*[5] However, for a collection of atoms of a single element the decay rate, and thus the half-life ($t_{1/2}$) for that collection can be calculated from their measured decay constants. The range of the half-lives of radioactive atoms have no known limits and span a time range of over 55 orders of magnitude.

Radionuclides occur naturally and are artificially produced in nuclear reactors, cyclotrons, particle accelerators or radionuclide generators. There are about 730 radionuclides with half-lives longer than 60 minutes (see list of nuclides). With the longest half lives are the 32 primordial radionuclides that have survived from the creation of the Solar System. Over 60 further radionuclides are detectable in nature, either as daughters of these, or through natural production on Earth by cosmic radiation. More than 2400 radionuclides have half-lives less than 60 minutes. Most of those are only produced artificially, and have very short half-lives. For comparison, there are about 254 stable nuclides.

All chemical elements have radionuclides. Even the lightest element, hydrogen, has a well-known radionuclide, tritium. Elements heavier than lead, and the elements technetium and promethium, exist only as radionuclides.

Unplanned exposure to radionuclides generally has a harmful effect on living organisms including humans, although low levels of exposure occur naturally without harm. The degree of harm will depend on the nature and extent of the radiation produced, the amount and nature of exposure (close contact, inhalation or ingestion), and the biochemical properties of the element; with increased risk of cancer the most usual consequence. However, radionuclides with suitable properties are used in nuclear medicine for both diagnosis and treatment. An imaging tracer made with radionuclides is called a radioactive tracer. A pharmaceutical drug made with radionuclides is called a radiopharmaceutical.

15.1 Origin

15.1.1 Natural

On Earth, naturally occurring radionuclides fall into three categories: primordial radionuclides, secondary radionuclides, and cosmogenic radionuclides.

- Radionuclides are produced in stellar nucleosynthesis and supernova explosions along with stable nuclides. Most decay quickly but can still be observed astronomically and can play a part in understanding astronomic processes. Primordial radionuclides, such as uranium and thorium, exist in the present time because their half-lives are so long (>100 million years) they have not yet completely decayed. Some radionuclides have half-lives so long (many times the age of the universe) that decay has only recently been detected, and for most practical purposes they can be considered stable, most notably bismuth-209: detection of this decay meant that bismuth was no longer considered stable. It is possible decay may be observed in other nuclides adding to this list of primordial radionuclides.

- Secondary radionuclides are radiogenic isotopes derived from the decay of primordial radionuclides. They have shorter half-lives than primordial radionuclides. They arise in the decay chain of the primordial isotopes thorium-232, uranium-238 and uranium-235.

Examples include the natural isotopes of polonium and radium.

- Cosmogenic isotopes, such as carbon-14, are present because they are continually being formed in the atmosphere due to cosmic rays.*[6]

Many of these radionuclides exist only in trace amounts in nature, including all cosmogenic nuclides. Secondary radionuclides will occur in proportion to their half-lives, so short-lived ones will be very rare. Thus polonium can be found in uranium ores at about 0.1 mg per metric ton (1 part in 10^{10}),.*[7]*[8] Further radionunclides may occur in nature in virtually undetectable amounts as a result of rare events such as spontaneous fission or uncommon cosmic ray interactions.

15.1.2   Nuclear fission

Radionuclides are produced as an unavoidable result of nuclear fission and thermonuclear explosions. The process of nuclear fission creates a wide range of fission products, most of which are radionuclides. Further radionuclides can be created from irradiation of the nuclear fuel (creating a range of actinides) and of the surrounding structures, yielding activation products. This complex mixture of radionuclides with different chemistries and radioactivity makes handling nuclear waste and dealing with nuclear fallout particularly problematic.

15.1.3   Synthetic

Synthetic radionuclides are deliberately synthesised using nuclear reactors, particle accelerators or radionuclide generators:

- As well as being extracted from nuclear waste, radioisotopes can be produced deliberately with nuclear reactors, exploiting the high flux of neutrons present. These neutrons activate elements placed within the reactor. A typical product from a nuclear reactor is iridium-192. The elements that have a large propensity to take up the neutrons in the reactor are said to have a high neutron cross-section.

- Particle accelerators such as cyclotrons accelerate particles to bombard a target to produce radionuclides. Cyclotrons accelerate protons at a target to produce positron-emitting radionuclides. e.g. fluorine-18.

- Radionuclide generators contain a parent radionuclide that decays to produce a radioactive daughter. The parent is usually produced in a nuclear reactor. A typical example is the technetium-99m generator used in

nuclear medicine. The parent produced in the reactor is molybdenum-99.

15.2   Uses

Radionuclides are used in two major ways: either for their radiation alone (irradiation, nuclear batteries) or for the combination of chemical properties and their radiation (tracers, biopharmaceuticals).

- In biology, radionuclides of carbon can serve as radioactive tracers because they are chemically very similar to the nonradioactive nuclides, so most chemical, biological, and ecological processes treat them in a nearly identical way. One can then examine the result with a radiation detector, such as a Geiger counter, to determine where the provided atoms were incorporated. For example, one might culture plants in an environment in which the carbon dioxide contained radioactive carbon; then the parts of the plant that incorporate atmospheric carbon would be radioactive. Radionuclides can be used to monitor processes such as DNA replication or amino acid transport.

- In nuclear medicine, radioisotopes are used for diagnosis, treatment, and research. Radioactive chemical tracers emitting gamma rays or positrons can provide diagnostic information about internal anatomy and the functioning of specific organs, including the human brain.*[9]*[10]*[11] This is used in some forms of tomography: single-photon emission computed tomography and positron emission tomography (PET) scanning and Cherenkov luminescence imaging. Radioisotopes are also a method of treatment in hemopoietic forms of tumors; the success for treatment of solid tumors has been limited. More powerful gamma sources sterilise syringes and other medical equipment.

- In food preservation, radiation is used to stop the sprouting of root crops after harvesting, to kill parasites and pests, and to control the ripening of stored fruit and vegetables.

- In industry, and in mining, radionuclides are used to examine welds, to detect leaks, to study the rate of wear, erosion and corrosion of metals, and for on-stream analysis of a wide range of minerals and fuels.

- In spacecraft and elsewhere, radionuclides are used to provide power and heat, notably through radioisotope thermoelectric generators (RTGs).

- In astronomy and cosmology radionuclides play a role in understanding stellar and planetary process.

- In particle physics, radionuclides help discover new physics (physics beyond the Standard Model) by measuring the energy and momentum of their beta decay products.[12]

- In ecology, radionuclides are used to trace and analyze pollutants, to study the movement of surface water, and to measure water runoffs from rain and snow, as well as the flow rates of streams and rivers.

- In geology, archaeology, and paleontology, natural radionuclides are used to measure ages of rocks, minerals, and fossil materials.

15.3 Examples

The following table lists properties of selected radionuclides illustrating the range of properties and uses.

Key: Z = no of protons; N = no of Neutrons; DM = Decay Mode; DE = Decay Energy; EC = Electron Capture

15.3.1 Household Smoke detectors

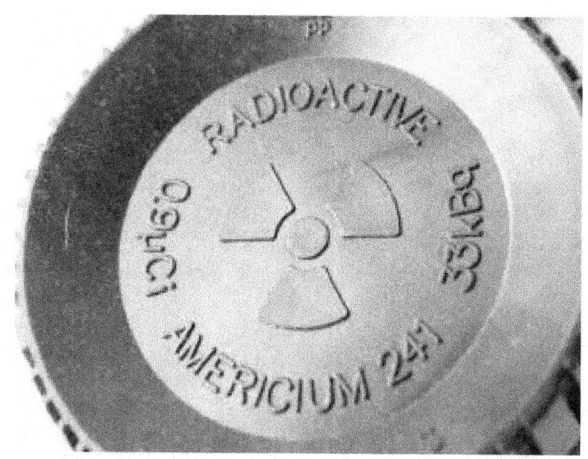

Americium-241 container in a smoke detector.

Radionuclides are present in many homes as they are used inside the most common household smoke detectors. The radionuclide used is americium-241, which is created by bombarding plutonium with neutrons in a nuclear reactor. It decays by emitting alpha particles and gamma radiation to become neptunium-237. Smoke detectors use a very small quantity of ^{241}Am (about 0.29 micrograms per smoke detector) in the form of americium dioxide. ^{241}Am is used as it emits alpha particles which ionise the air in the detector's ionization chamber. A small electric voltage is applied to the ionised air which gives rise to a small electric current. In the presence of smoke some of the ions are neutralized,

Americium-241 capsule as found in smoke detector. The circle of darker metal in the center is americium-241; the surrounding casing is aluminium.

thereby decreasing the current, which activates the detector's alarm.[13][14]

15.4 Impacts on organisms

Radionuclides that find their way into the environment may cause harmful effects as radioactive contamination. They can also cause damage if they are excessively used during treatment or in other ways exposed to living beings, by radiation poisoning. Potential health damage from exposure to radionuclides depends on a number of factors, and "can damage the functions of healthy tissue/organs. Radiation exposure can produce effects ranging from skin redness and hair loss, to radiation burns and acute radiation syndrome. Prolonged exposure can lead to cells being damaged and in turn lead to cancer. Signs of cancerous cells might not show up until years, or even decades, after exposure."[15]

15.5 Summary table for classes of nuclides, "stable"and radioactive

Following is a summary table for the total list of nuclides with half-lives greater than one hour. Ninety of these 905 nuclides are theoretically stable, except to proton-decay (which has never been observed). About 254 nuclides have never been observed to decay, and are classically considered stable.

The remaining 650 radionuclides have half-lives longer than 1 hour, and are well-characterized (see list of nuclides for a complete tabulation). They include 28 nuclides with measured half-lives longer than the estimated age of the universe (13.8 billion years[16]), and another 4 nuclides with half-lives long enough (> 100 million years) that they are radioactive primordial nuclides, and may be detected on Earth, having survived from their presence in interstellar dust since before the formation of the solar system, about 4.6 billion years ago. Another 60+ short-lived nuclides can be detected naturally as daughters of longer-lived nuclides or cosmic-ray products. The remaining known nuclides are known solely from artificial nuclear transmutation.

Numbers are not exact, and may change slightly in the future, as "stable nuclides" are observed to be radioactive with very long half-lives.

This is a summary table[17] for the 988 nuclides with half-lives longer than one hour (including those that are stable), given in list of nuclides.

15.6   List of commercially available radionuclides

This list covers common isotopes, most of which are available in very small quantities to the general public in most countries. Others that are not publicly accessible are traded commercially in industrial, medical, and scientific fields and are subject to government regulation. For a complete list of all known isotopes for every element (minus activity data), see List of nuclides and Isotope lists. For a table, see Table of nuclides.

15.6.1   Gamma emission only

15.6.2   Beta emission only

15.6.3   Alpha emission only

15.6.4   Multiple radiation emitters

15.7   See also

- List of nuclides shows all radionuclides with half-life > 1 hour

- Hyperaccumulators table – 3

- Radioactivity in biology

- Radiometric dating

- Radionuclide cisternogram

- Uses of radioactivity in oil and gas wells

15.8   Notes

[1] R.H. Petrucci, W.S. Harwood and F.G. Herring. *General Chemistry* (8th ed., Prentice-Hall 2002). p.1025–26

[2] "Decay and Half Life". Retrieved 2009-12-14.

[3] Stabin, Michael G. (2007). "3". *Radiation Protection and Dosimetry: An Introduction to Health Physics*. Springer. doi:10.1007/978-0-387-49983-3. ISBN 978-0387499826.

[4] Best, Lara; Rodrigues, George; Velker, Vikram (2013). "1.3". *Radiation Oncology Primer and Review*. Demos Medical Publishing. ISBN 978-1620700044.

[5] Loveland, W.; Morrissey, D.; Seaborg, G.T. (2006). *Modern Nuclear Chemistry*. Wiley-Interscience. p. 57. ISBN 0-471-11532-0.

[6] Eisenbud, Merril; Gesell, Thomas F (1997-02-25). *Environmental Radioactivity: From Natural, Industrial, and Military Sources*. p. 134. ISBN 9780122351549.

[7] Bagnall, K. W. (1962). "The Chemistry of Polonium". Advances in Inorganic Chemistry and Radiochemistry 4. New York: Academic Press. pp. 197–226. doi:10.1016/S0065-2792(08)60268-X. ISBN 0-12-023604-4. Retrieved June 14, 2012., p. 746

[8] Bagnall, K. W. (1962). "The Chemistry of Polonium". Advances in Inorganic Chemistry and Radiochemistry 4. New York: Academic Press., p. 198

[9] Ingvar, David H.; Lassen, Niels A. (1961). "Quantitative determination of regional cerebral blood-flow in man". *The Lancet*. **278** (7206): 806–807. doi:10.1016/s0140-6736(61)91092-3.

[10] Ingvar, David H.; Franzén, Göran (1974). "Distribution of cerebral activity in chronic schizophrenia". *The Lancet*. **304** (7895): 1484–1486. doi:10.1016/s0140-6736(74)90221-9.

[11] Lassen, Niels A.; Ingvar, David H.; Skinhøj, Erik (October 1978). "Brain Function and Blood Flow" (PDF). *Scientific American*. **239** (4): 62–71. doi:10.1038/scientificamerican1078-62.

[12] Severijns, Nathal; Beck, Marcus; Naviliat-Cuncic, Oscar (2006). "Tests of the standard electroweak model in nuclear beta decay". *Reviews of Modern Physics*. **78** (3): 991. arXiv:nucl-ex/0605029ə. Bibcode:2006RvMP...78..991S. doi:10.1103/RevModPhys.78.991.

[13] "Smoke Detectors and Americium". *world-nuclear.org*.

[14] Office of Radiation Protection – Am 241 Fact Sheet – Washington State Department of Health

[15] "Ionizing radiation, health effects and protective measures". World Health Organization. November 2012. Retrieved January 27, 2014.

[16] "Cosmic Detectives". The European Space Agency (ESA). 2013-04-02. Retrieved 2013-04-15.

[17] Table data is derived by counting members of the list: see WP:CALC. References for the list data itself are given below in the reference section in list of nuclides

15.9 References

- Carlsson, J.; Forssell Aronsson, E; Hietala, SO; Stigbrand, T; Tennvall, J; et al. (2003). "Tumour therapy with radionuclides: assessment of progress and problems". *Radiotherapy and Oncology*. **66** (2): 107–117. doi:10.1016/S0167-8140(02)00374-2. PMID 12648782.

- "Radioisotopes in Industry". *World Nuclear Association*.

- Martin, James (2006). *Physics for Radiation Protection: A Handbook*. p. 130. ISBN 3527406115.

15.10 Further reading

- Luig, H.; Kellerer, A. M.; Griebel, J. R. (2011). "Radionuclides, 1. Introduction". *Ullmann's Encyclopedia of Industrial Chemistry*. doi:10.1002/14356007.a22_499.pub2. ISBN 3527306730.

15.11 External links

- EPA – Radionuclides – EPA's Radiation Protection Program: Information.

- FDA – Radionuclides – FDA's Radiation Protection Program: Information.

- Interactive Chart of Nuclides – A chart of all nuclides

- National Isotope Development Center – U.S. Government source of radionuclides – production, research, development, distribution, and information

- The Live Chart of Nuclides – IAEA

Chapter 16

Radioactive waste

Radioactive waste is waste that contains radioactive material. Radioactive waste is usually a by-product of nuclear power generation and other applications of nuclear fission or nuclear technology, such as research and medicine. Radioactive waste is hazardous to most forms of life and the environment, and is regulated by government agencies in order to protect human health and the environment.

Radioactivity naturally decays over time, so radioactive waste has to be isolated and confined in appropriate disposal facilities for a sufficient period until it no longer poses a threat. The time radioactive waste must be stored for depends on the type of waste and radioactive isotopes. Current major approaches to managing radioactive waste have been segregation and storage for short-lived waste, near-surface disposal for low and some intermediate level waste, and deep burial or partitioning / transmutation for the high-level waste.

A summary of the amounts of radioactive waste and management approaches for most developed countries are presented and reviewed periodically as part of the International Atomic Energy Agency (IAEA) Joint Convention on the Safety of Spent Fuel Management and on the Safety of Radioactive Waste Management.[1]

16.1 Nature and significance of radioactive waste

Radioactive waste typically comprises a number of radionuclides: unstable configurations of elements that decay, emitting ionizing radiation which can be harmful to humans and the environment. Those isotopes emit different types and levels of radiation, which last for different periods of time.

16.1.1 Physics

Main article: Fission product yield
See also: Radioactive decay

The radioactivity of all radioactive waste diminishes with time. All radionuclides contained in the waste have a half-life—the time it takes for half of the atoms to decay into another nuclide—and eventually all radioactive waste decays into non-radioactive elements (i.e., stable nuclides). Certain radioactive elements (such as plutonium-239) will remain hazardous to humans and other creatures for hundreds or thousands of years. Other radionuclides remain radioactive for millions of years (though most of these products have so little activity as a result of their long half-lives that their radiation is lost in the background level). Thus, these wastes must be shielded for centuries and isolated from the living environment for millennia.[2] Since radioactive decay follows the half-life rule, the rate of decay is inversely proportional to the duration of decay. In other words, the radiation from a long-lived isotope like iodine-129 will be much less intense than that of a short-lived isotope like iodine-131.[3] The two tables show some of the major radioisotopes, their half-lives, and their radiation yield as a proportion of the yield of fission of uranium-235.

The energy and the type of the ionizing radiation emitted by a radioactive substance are also important factors in determining its threat to humans.[4] The chemical properties of the radioactive element will determine how mobile the substance is and how likely it is to spread into the environment and contaminate humans.[5] This is further complicated by the fact that many radioisotopes do not decay immediately to a stable state but rather to radioactive decay products within a decay chain before ultimately reaching a stable state.

16.1.2 Pharmacokinetics

Exposure to radioactive waste may cause serious harm or death. In humans, a dose of 1 sievert carries a 5.5% risk of developing cancer,[11] and this risk is assumed to be linearly proportional to dose even for low doses. Ionizing radiation causes deletions in chromosomes.[12] If a developing organism such as an unborn child is irradiated, it is possible a birth defect may be induced, but it is unlikely this defect will be in a gamete or a gamete-forming cell. The incidence of radiation-induced mutations in humans is small, as in most mammals, because of natural cellular-repair mechanisms, many just now coming to light. These mechanisms range from DNA, mRNA and protein repair, to internal lysosomic digestion of defective proteins, and even induced cell suicide—apoptosis[13]

Depending on the decay mode and the pharmacokinetics of an element (how the body processes it and how quickly), the threat due to exposure to a given activity of a radioisotope will differ. For instance iodine-131 is a short-lived beta and gamma emitter, but because it concentrates in the thyroid gland, it is more able to cause injury than caesium-137 which, being water soluble, is rapidly excreted in urine. In a similar way, the alpha emitting actinides and radium are considered very harmful as they tend to have long biological half-lives and their radiation has a high relative biological effectiveness, making it far more damaging to tissues per amount of energy deposited. Because of such differences, the rules determining biological injury differ widely according to the radioisotope, time of exposure and sometimes also the nature of the chemical compound which contains the radioisotope.

16.2 Sources of waste

Radioactive waste comes from a number of sources. In countries with nuclear power plants, nuclear armament, or nuclear fuel treatment plants, the majority of waste originates from the nuclear fuel cycle and nuclear weapons reprocessing, otherwise there is no waste of nuclear origin. Other sources include medical and industrial wastes, as well as naturally occurring radioactive materials (NORM) that can be concentrated as a result of the processing or consumption of coal, oil and gas, and some minerals, as discussed below.

16.2.1 Nuclear fuel cycle

Main articles: Nuclear fuel cycle and Spent nuclear fuel
See also: Nuclear power

Front end

Waste from the front end of the nuclear fuel cycle is usually alpha-emitting waste from the extraction of uranium. It often contains radium and its decay products.

Uranium dioxide (UO_2) concentrate from mining is not very radioactive – only a thousand or so times as radioactive as the granite used in buildings. It is refined from yellowcake (U_3O_8), then converted to uranium hexafluoride gas (UF_6). As a gas, it undergoes enrichment to increase the U-235 content from 0.7% to about 4.4% (LEU). It is then turned into a hard ceramic oxide (UO_2) for assembly as reactor fuel elements.[14]

The main by-product of enrichment is depleted uranium (DU), principally the U-238 isotope, with a U-235 content of ~0.3%. It is stored, either as UF_6 or as U_3O_8. Some is used in applications where its extremely high density makes it valuable such as anti-tank shells, even sailboat keels on at least one occasion.[15] It is also used with plutonium for making mixed oxide fuel (MOX) and to dilute, or downblend, highly enriched uranium from weapons stockpiles which is now being redirected to become reactor fuel.

Back end

See also: Nuclear reprocessing

The back end of the nuclear fuel cycle, mostly spent fuel rods, contains fission products that emit beta and gamma radiation, and actinides that emit alpha particles, such as uranium-234, neptunium-237, plutonium-238 and americium-241, and even sometimes some neutron emitters such as californium (Cf). These isotopes are formed in nuclear reactors.

It is important to distinguish the processing of uranium to make fuel from the reprocessing of used fuel. Used fuel contains the highly radioactive products of fission (see high level waste below). Many of these are neutron absorbers, called neutron poisons in this context. These eventually build up to a level where they absorb so many neutrons that the chain reaction stops, even with the control rods completely removed. At that point the fuel has to be replaced in the reactor with fresh fuel, even though there is still a substantial quantity of uranium-235 and plutonium present. In the United States, this used fuel is usually "stored", while in other countries such as Russia, the United Kingdom, France, Japan and India, the fuel is reprocessed to remove the fission products, and the fuel can then be reused, thus cutting costs, reducing health risks, saving time, and in general being far safer.[16] The fission products removed from the fuel are a concentrated form of high-level waste as are the chemicals used in the process. While these countries reprocess the fuel carrying out single plutonium cycles, India is the only country known to be planning multiple plutonium recycling schemes.[17]

Fuel composition and long term radioactivity

See also: Spent nuclear fuel and High level waste
Long-lived radioactive waste from the back end of the fuel

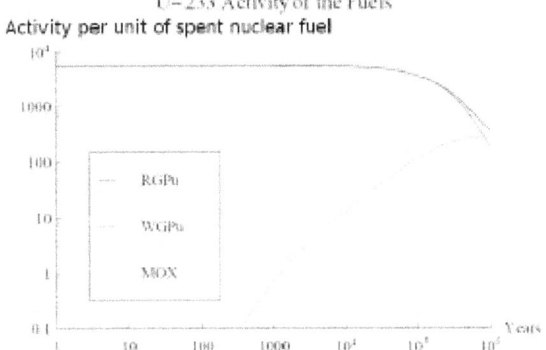

Activity of U-233 for three fuel types

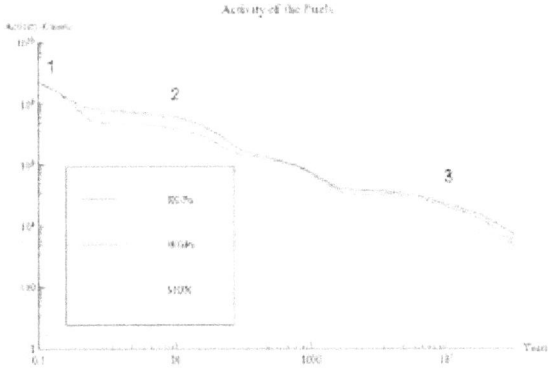

Total activity for three fuel types

cycle is especially relevant when designing a complete waste management plan for spent nuclear fuel (SNF). When looking at long term radioactive decay, the actinides in the SNF have a significant influence due to their characteristically long half-lives. Depending on what a nuclear reactor is fueled with, the actinide composition in the SNF will be different.

An example of this effect is the use of nuclear fuels with thorium. Th-232 is a fertile material that can undergo a neutron capture reaction and two beta minus decays, resulting in the production of fissile U-233. The SNF of a cycle with thorium will contain U-233. Its radioactive decay will strongly influence the long-term activity curve of the SNF around 1 million years. A comparison of the activity associated to U-233 for three different SNF types can be seen in the figure on the top right.

The burnt fuels are thorium with reactor-grade plutonium (RGPu), thorium with weapons-grade plutonium (WGPu)

and Mixed Oxide fuel (MOX). For RGPu and WGPu, the initial amount of U-233 and its decay around 1 million years can be seen. This has an effect in the total activity curve of the three fuel types. The absence of U-233 and its daughter products in the MOX fuel results in a lower activity in region 3 of the figure on the bottom right, whereas for RGPu and WGPu the curve is maintained higher due to the presence of U-233 that has not fully decayed.

The use of different fuels in nuclear reactors results in different SNF composition, with varying activity curves.

Proliferation concerns

See also: Nuclear proliferation and Reactor-grade plutonium

Since uranium and plutonium are nuclear weapons materials, there have been proliferation concerns. Ordinarily (in spent nuclear fuel), plutonium is reactor-grade plutonium. In addition to plutonium-239, which is highly suitable for building nuclear weapons, it contains large amounts of undesirable contaminants: plutonium-240, plutonium-241, and plutonium-238. These isotopes are extremely difficult to separate, and more cost-effective ways of obtaining fissile material exist (e.g. uranium enrichment or dedicated plutonium production reactors).[18]

High-level waste is full of highly radioactive fission products, most of which are relatively short-lived. This is a concern since if the waste is stored, perhaps in deep geological storage, over many years the fission products decay, decreasing the radioactivity of the waste and making the plutonium easier to access. The undesirable contaminant Pu-240 decays faster than the Pu-239, and thus the quality of the bomb material increases with time (although its quantity decreases during that time as well). Thus, some have argued, as time passes, these deep storage areas have the potential to become "plutonium mines", from which material for nuclear weapons can be acquired with relatively little difficulty. Critics of the latter idea have pointed out the difficulty of recovering useful material from sealed deep storage areas makes other methods preferable. Specifically, the high radioactivity and heat (80 C in surrounding rock) greatly increases the difficulty of mining a storage area, and the enrichment methods required have high capital costs.[19]

Pu-239 decays to U-235 which is suitable for weapons and which has a very long half-life (roughly 10^9 years). Thus plutonium may decay and leave uranium-235. However, modern reactors are only moderately enriched with U-235 relative to U-238, so the U-238 continues to serve as a denaturation agent for any U-235 produced by plutonium decay.

One solution to this problem is to recycle the plutonium and use it as a fuel e.g. in fast reactors. In pyrometallurgical fast reactors, the separated plutonium and uranium are contaminated by actinides and cannot be used for nuclear weapons.

16.2.2 Nuclear weapons decommissioning

Waste from nuclear weapons decommissioning is unlikely to contain much beta or gamma activity other than tritium and americium. It is more likely to contain alpha-emitting actinides such as Pu-239 which is a fissile material used in bombs, plus some material with much higher specific activities, such as Pu-238 or Po.

In the past the neutron trigger for an atomic bomb tended to be beryllium and a high activity alpha emitter such as polonium; an alternative to polonium is Pu-238. For reasons of national security, details of the design of modern bombs are normally not released to the open literature.

Some designs might contain a radioisotope thermoelectric generator using Pu-238 to provide a long lasting source of electrical power for the electronics in the device.

It is likely that the fissile material of an old bomb which is due for refitting will contain decay products of the plutonium isotopes used in it, these are likely to include U-236 from Pu-240 impurities, plus some U-235 from decay of the Pu-239; due to the relatively long half-life of these Pu isotopes, these wastes from radioactive decay of bomb core material would be very small, and in any case, far less dangerous (even in terms of simple radioactivity) than the Pu-239 itself.

The beta decay of Pu-241 forms Am-241; the in-growth of americium is likely to be a greater problem than the decay of Pu-239 and Pu-240 as the americium is a gamma emitter (increasing external-exposure to workers) and is an alpha emitter which can cause the generation of heat. The plutonium could be separated from the americium by several different processes; these would include pyrochemical processes and aqueous/organic solvent extraction. A truncated PUREX type extraction process would be one possible method of making the separation. Naturally occurring uranium is not fissile because it contains 99.3% of U-238 and only 0.7% of U-235.

16.2.3 Legacy waste

Due to historic activities typically related to radium industry, uranium mining, and military programs, there are numerous sites that contain or are contaminated with radioactivity. In the United States alone, the Department of Energy states there are "millions of gallons of radioactive waste" as well as "thousands of tons of spent nuclear fuel and ma-

terial" and also "huge quantities of contaminated soil and water."[20] Despite copious quantities of waste, the DOE has stated a goal of cleaning all presently contaminated sites successfully by 2025.[20] The Fernald, Ohio site for example had "31 million pounds of uranium product", "2.5 billion pounds of waste", "2.75 million cubic yards of contaminated soil and debris", and a "223 acre portion of the underlying Great Miami Aquifer had uranium levels above drinking standards."[20] The United States has at least 108 sites designated as areas that are contaminated and unusable, sometimes many thousands of acres.[20][21] DOE wishes to clean or mitigate many or all by 2025, using the recently developed method of geomelting, however the task can be difficult and it acknowledges that some may never be completely remediated. In just one of these 108 larger designations, Oak Ridge National Laboratory, there were for example at least "167 known contaminant release sites" in one of the three subdivisions of the 37,000-acre (150 km^2) site.[20] Some of the U.S. sites were smaller in nature, however, cleanup issues were simpler to address, and DOE has successfully completed cleanup, or at least closure, of several sites.[20]

16.2.4 Medical

Radioactive medical waste tends to contain beta particle and gamma ray emitters. It can be divided into two main classes. In diagnostic nuclear medicine a number of short-lived gamma emitters such as technetium-99m are used. Many of these can be disposed of by leaving it to decay for a short time before disposal as normal waste. Other isotopes used in medicine, with half-lives in parentheses, include:

- Y-90, used for treating lymphoma (2.7 days)

- I-131, used for thyroid function tests and for treating thyroid cancer (8.0 days)

- Sr-89, used for treating bone cancer, intravenous injection (52 days)

- Ir-192, used for brachytherapy (74 days)

- Co-60, used for brachytherapy and external radiotherapy (5.3 years)

- Cs-137, used for brachytherapy, external radiotherapy (30 years)

16.2.5 Industrial

Industrial source waste can contain alpha, beta, neutron or gamma emitters. Gamma emitters are used in radiography while neutron emitting sources are used in a range of applications, such as oil well logging.[22]

16.2.6 Naturally occurring radioactive material (NORM)

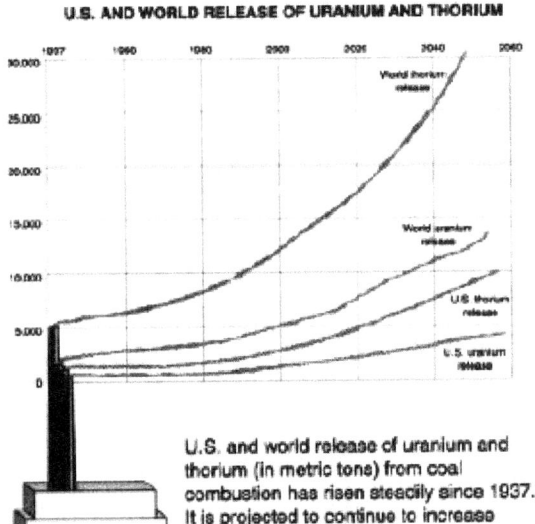

U.S. AND WORLD RELEASE OF URANIUM AND THORIUM

U.S. and world release of uranium and
thorium (in metric tons) from coal
combustion has risen steadily since 1937.
It is projected to continue to increase
through 2040 and beyond.

Annual release of uranium and thorium radioisotopes from coal combustion, predicted by ORNL to cumulatively amount to 2.9 million tons over the 1937–2040 period, from the combustion of an estimated 637 billion tons of coal worldwide.[23]

Substances containing natural radioactivity are known as NORM. After human processing that exposes or concentrates this natural radioactivity (such as mining bringing coal to the surface or burning it to produce concentrated ash), it becomes technologically enhanced naturally occurring radioactive material (TENORM).*[24] A lot of this waste is alpha particle-emitting matter from the decay chains of uranium and thorium. The main source of radiation in the human body is potassium–40 (^{40}K), typically 17 milligrams in the body at a time and 0.4 milligrams/day intake.*[25] Most rocks, due to their components, have a low level of radioactivity. Usually ranging from 1 millisievert (mSv) to 13 mSv annually depending on location, average radiation exposure from natural radioisotopes is 2.0 mSv per person a year worldwide.*[26] This makes up the majority of typical total dosage (with mean annual exposure from other sources amounting to 0.6 mSv from medical tests averaged over the whole populace, 0.4 mSv from cosmic rays, 0.005 mSv from the legacy of past atmospheric nuclear testing, 0.005 mSv occupational exposure, 0.002 mSv from the Chernobyl disaster, and 0.0002 mSv from the nuclear fuel cycle).*[26]

TENORM is not regulated as restrictively as nuclear reactor waste, though there are no significant differences in the radiological risks of these materials.*[27]

Coal

Coal contains a small amount of radioactive uranium, barium, thorium and potassium, but, in the case of pure coal, this is significantly less than the average concentration of those elements in the Earth's crust. The surrounding strata, if shale or mudstone, often contain slightly more than average and this may also be reflected in the ash content of 'dirty' coals.*[23]*[28] The more active ash minerals become concentrated in the fly ash precisely because they do not burn well.*[23] The radioactivity of fly ash is about the same as black shale and is less than phosphate rocks, but is more of a concern because a small amount of the fly ash ends up in the atmosphere where it can be inhaled.*[29] According to U.S. NCRP reports, population exposure from 1000-MWe power plants amounts to 490 person-rem/year for coal power plants, 100 times as great as nuclear power plants (4.8 person-rem/year). (The exposure from the complete nuclear fuel cycle from mining to waste disposal is 136 person-rem/year; the corresponding value for coal use from mining to waste disposal is "probably unknown".)*[23]

Oil and gas

Residues from the oil and gas industry often contain radium and its decay products. The sulfate scale from an oil well can be very radium rich, while the water, oil and gas from a well often contain radon. The radon decays to form solid radioisotopes which form coatings on the inside of pipework. In an oil processing plant the area of the plant where propane is processed is often one of the more contaminated areas of the plant as radon has a similar boiling point to propane.*[30]

16.3 Classification of radioactive waste

Classifications of radioactive waste varies by country. The IAEA, which publishes the Radioactive Waste Safety Standards (RADWASS), also plays a significant role.*[31]

16.3.1 Uranium tailings

Main article: Uranium tailings
Uranium tailings are waste by-product materials left over from the rough processing of uranium-bearing ore. They are not significantly radioactive. Mill tailings are sometimes referred to as **11(e)2 wastes**, from the section of the Atomic Energy Act of 1946 that defines them. Uranium mill tailings typically also contain chemically hazardous heavy metal such as lead and arsenic. Vast mounds

Removal of very low-level waste

Spent fuel flasks are transported by railway in the United Kingdom. Each flask is constructed of 14 in (360 mm) thick solid steel and weighs in excess of 50 tons

of uranium mill tailings are left at many old mining sites, especially in Colorado, New Mexico, and Utah.

See also: Uranium Mill Tailings Remedial Action

16.3.2 Low-level waste

Low level waste (LLW) is generated from hospitals and industry, as well as the nuclear fuel cycle. Low-level wastes include paper, rags, tools, clothing, filters, and other materials which contain small amounts of mostly short-lived radioactivity. Materials that originate from any region of an Active Area are commonly designated as LLW as a precautionary measure even if there is only a remote possibility of being contaminated with radioactive materials. Such LLW typically exhibits no higher radioactivity than one would expect from the same material disposed of in a non-active area, such as a normal office block.

Some high-activity LLW requires shielding during handling and transport but most LLW is suitable for shallow land burial. To reduce its volume, it is often compacted or incinerated before disposal. Low-level waste is divided into four classes: **class A, class B, class C**, and **Greater Than Class C (GTCC)**.

16.3.3 Intermediate-level waste

Intermediate-level waste (ILW) contains higher amounts of radioactivity and in general require shielding, but not cooling.[32] Intermediate-level wastes includes resins, chemical sludge and metal nuclear fuel cladding, as well as contaminated materials from reactor decommissioning. It may be solidified in concrete or bitumen for disposal. As a general rule, short-lived waste (mainly non-fuel materials

from reactors) is buried in shallow repositories, while long-lived waste (from fuel and fuel reprocessing) is deposited in geological repository. U.S. regulations do not define this category of waste; the term is used in Europe and elsewhere.

16.3.4 High-level waste

High-level waste (HLW) is produced by nuclear reactors. The exact definition of HLW differs internationally. After a nuclear fuel rod serves one fuel cycle and is removed from the core, it is considered HLW.[33] Fuel rods contain fission products and transuranic elements generated in the reactor core. Spent fuel is highly radioactive and often hot. HLW accounts for over 95 percent of the total radioactivity produced in the process of nuclear electricity generation. The amount of HLW worldwide is currently increasing by about 12,000 metric tons every year, which is the equivalent to about 100 double-decker buses or a two-story structure with a footprint the size of a basketball court.[34] A 1000-MW nuclear power plant produces about 27 tonnes of spent nuclear fuel (unreprocessed) every year.[35] In 2010, there was very roughly estimated to be stored some 250,000 tons of nuclear HLW,[36] that does not include amounts that have escaped into the environment from accidents or tests. Japan estimated to hold 17,000 tons of HLW in storage in 2015.[37] HLW have been shipped to other countries to be stored or reprocessed, and in some cases, shipped back as active fuel.

The ongoing controversy over high-level radioactive waste disposal is a major constraint on the nuclear power's global expansion.[38] Most scientists agree[39] that the main proposed long-term solution is deep geological burial, either in a mine or a deep borehole. However, almost six decades after commercial nuclear energy began, no government has succeeded in opening such a repository for civilian high-level nuclear waste,*[38] although Finland is in the advanced stage of the construction of such facility, the Onkalo spent nuclear fuel repository. Reprocessing or recycling spent nuclear fuel options already available or under active development still generate waste and so are not a total solution, but can reduce the sheer quantity of waste, and there are many such active programs worldwide. Deep geological burial remains the only responsible way to deal with high-level nuclear waste.*[40] The Morris Operation is currently the only de facto high-level radioactive waste storage site in the United States.

16.3.5   Transuranic waste

Transuranic waste (TRUW) as defined by U.S. regulations is, without regard to form or origin, waste that is contaminated with alpha-emitting transuranic radionuclides with half-lives greater than 20 years and concentrations greater than 100 nCi/g (3.7 MBq/kg), excluding high-level waste. Elements that have an atomic number greater than uranium are called transuranic ("beyond uranium"). Because of their long half-lives, TRUW is disposed more cautiously than either low- or intermediate-level waste. In the U.S., it arises mainly from weapons production, and consists of clothing, tools, rags, residues, debris and other items contaminated with small amounts of radioactive elements (mainly plutonium).

Under U.S. law, transuranic waste is further categorized into "contact-handled" (CH) and "remote-handled" (RH) on the basis of the radiation dose rate measured at the surface of the waste container. CH TRUW has a surface dose rate not greater than 200 mrem per hour (2 mSv/h), whereas RH TRUW has a surface dose rate of 200 mrem/h (2 mSv/h) or greater. CH TRUW does not have the very high radioactivity of high-level waste, nor its high heat generation, but RH TRUW can be highly radioactive, with surface dose rates up to 1,000,000 mrem/h (10,000 mSv/h). The U.S. currently disposes of TRUW generated from military facilities at the Waste Isolation Pilot Plant (WIPP) in a deep salt formation in New Mexico.*[41]

16.4   Prevention of waste

A theoretical way to reduce waste accumulation is to phase out current reactors in favour of Generation IV Reactors or Liquid Fluoride Thorium Reactors, which output less waste per power generated. Fast reactors can theoretically consume some existing waste. The UK's Nuclear Decommissioning Authority published a position paper in 2014 on the progress on approaches to the management of separated

plutonium, which summarises the conclusions of the work that NDA shared with UK government.*[42]

16.5   Management of waste

Modern medium to high level transport container for nuclear waste

See also: High-level radioactive waste management, List of nuclear waste treatment technologies, and Environmental effects of nuclear power

Of particular concern in nuclear waste management are two long-lived fission products, Tc-99 (half-life 220,000 years) and I-129 (half-life 15.7 million years), which dominate spent fuel radioactivity after a few thousand years. The most troublesome transuranic elements in spent fuel are Np-237 (half-life two million years) and Pu-239 (half-life 24,000 years).*[43] Nuclear waste requires sophisticated treatment and management to successfully isolate it from interacting with the biosphere. This usually necessitates treatment, followed by a long-term management strategy involving storage, disposal or transformation of the waste into a non-toxic form.*[44] Governments around the world are considering a range of waste management and disposal options, though there has been limited progress toward long-term waste management solutions.*[45]

In second half of 20th century, several methods of disposal of radioactive waste were investigated by nuclear nations,*[46] which are :

- "Long term above ground storage", not implemented.

- "Disposal in outer space" (for instance, inside the Sun), not implemented - as it would be currently too expensive.

- "Deep borehole disposal", not implemented.

- "Rock-melting" , not implemented.

- "Disposal at subduction zones" , not implemented.

- **"Ocean disposal"**, done by the USSR, the United Kingdom,[47] Switzerland, the United States, Belgium, France, The Netherlands, Japan, Sweden, Russia, Germany, Italy and South Korea. (1954–93) This is no longer permitted by international agreements.

- "Sub seabed disposal", not implemented, not permitted by international agreements.

- "Disposal in ice sheets" , rejected in Antarctic Treaty

- **"Direct injection"**, done by USSR and USA.

In the USA, waste management policy completely broke down with the ending of work on the incomplete Yucca Mountain Repository.[48] At present there are 70 nuclear power plant sites where spent fuel is stored. A Blue Ribbon Commission was appointed by President Obama to look into future options for this and future waste. A deep geological repository seems to be favored.[48]

16.5.1 Initial treatment of waste

Vitrification

Long-term storage of radioactive waste requires the stabilization of the waste into a form which will neither react nor degrade for extended periods. It is theorized that one way to do this might be through vitrification.[49] Currently at Sellafield the high-level waste (PUREX first cycle raffinate) is mixed with sugar and then calcined. Calcination involves passing the waste through a heated, rotating tube. The purposes of calcination are to evaporate the water from the waste, and de-nitrate the fission products to assist the stability of the glass produced.[50]

The 'calcine' generated is fed continuously into an induction heated furnace with fragmented glass.[51] The resulting glass is a new substance in which the waste products are bonded into the glass matrix when it solidifies. As a melt, this product is poured into stainless steel cylindrical containers ("cylinders") in a batch process. When cooled, the fluid solidifies ("vitrifies") into the glass. After being formed, the glass is highly resistant to water.[52]

After filling a cylinder, a seal is welded onto the cylinder head. The cylinder is then washed. After being inspected for external contamination, the steel cylinder is stored, usually in an underground repository. In this form, the waste products are expected to be immobilized for thousands of years.[53]

The glass inside a cylinder is usually a black glossy substance. All this work (in the United Kingdom) is done using

hot cell systems. Sugar is added to control the ruthenium chemistry and to stop the formation of the volatile RuO_4 containing radioactive ruthenium isotopes. In the West, the glass is normally a borosilicate glass (similar to Pyrex), while in the former Soviet bloc it is normal to use a phosphate glass.[54] The amount of fission products in the glass must be limited because some (palladium, the other Pt group metals, and tellurium) tend to form metallic phases which separate from the glass. Bulk vitrification uses electrodes to melt soil and wastes, which are then buried underground.[55] In Germany a vitrification plant is in use; this is treating the waste from a small demonstration reprocessing plant which has since been closed down.[50][56]

Ion exchange

It is common for medium active wastes in the nuclear industry to be treated with ion exchange or other means to concentrate the radioactivity into a small volume. The much less radioactive bulk (after treatment) is often then discharged. For instance, it is possible to use a ferric hydroxide floc to remove radioactive metals from aqueous mixtures.[57] After the radioisotopes are absorbed onto the ferric hydroxide, the resulting sludge can be placed in a metal drum before being mixed with cement to form a solid waste form.[58] In order to get better long-term performance (mechanical stability) from such forms, they may be made from a mixture of fly ash, or blast furnace slag, and Portland cement, instead of normal concrete (made with Portland cement, gravel and sand).

Synroc

The Australian Synroc (synthetic rock) is a more sophisticated way to immobilize such waste, and this process may eventually come into commercial use for civil wastes (it is currently being developed for US military wastes). Synroc was invented by Prof Ted Ringwood (a geochemist) at the Australian National University.[59] The Synroc contains pyrochlore and cryptomelane type minerals. The original form of Synroc (Synroc C) was designed for the liquid high level waste (PUREX raffinate) from a light water reactor. The main minerals in this Synroc are hollandite ($BaAl_2Ti_6O_{16}$), zirconolite ($CaZrTi_2O_7$) and perovskite ($CaTiO_3$). The zirconolite and perovskite are hosts for the actinides. The strontium and barium will be fixed in the perovskite. The caesium will be fixed in the hollandite.

16.5.2 Long term management of waste

See also: Economics of new nuclear power plants § Waste disposal

The time frame in question when dealing with radioactive waste ranges from 10,000 to 1,000,000 years.[60] according to studies based on the effect of estimated radiation doses.[61] Researchers suggest that forecasts of health detriment for such periods should be examined critically.[62] [63] Practical studies only consider up to 100 years as far as effective planning[64] and cost evaluations[65] are concerned. Long term behavior of radioactive wastes remains a subject for ongoing research projects in geoforecasting.[66]

Above-ground disposal

Dry cask storage typically involves taking waste from a spent fuel pool and sealing it (along with an inert gas) in a steel cylinder, which is placed in a concrete cylinder which acts as a radiation shield. It is a relatively inexpensive method which can be done at a central facility or adjacent to the source reactor. The waste can be easily retrieved for reprocessing.[67]

Geologic disposal

The process of selecting appropriate deep final repositories for high level waste and spent fuel is now under way in several countries with the first expected to be commissioned some time after 2010. The basic concept is to locate a large, stable geologic formation and use mining technology to excavate a tunnel, or large-bore tunnel boring machines (similar to those used to drill the Channel Tunnel from England to France) to drill a shaft 500 metres (1,600 ft) to 1,000 metres (3,300 ft) below the surface where rooms or vaults can be excavated for disposal of high-level radioactive waste. The goal is to permanently isolate nuclear waste from the human environment. Many people remain uncomfortable with the immediate stewardship cessation of this disposal system, suggesting perpetual management and monitoring would be more prudent.

Because some radioactive species have half-lives longer than one million years, even very low container leakage and radionuclide migration rates must be taken into account.[69] Moreover, it may require more than one half-life until some nuclear materials lose enough radioactivity to cease being lethal to living things. A 1983 review of the Swedish radioactive waste disposal program by the National Academy of Sciences found that country's estimate of several hundred thousand years—perhaps up to one million years—being necessary for waste isolation "fully justified."[70]

Ocean floor disposal of radioactive waste has been suggested by the finding that deep waters in the North Atlantic

On Feb. 14, 2014, at the Waste Isolation Pilot Plant, radioactive materials leaked from a damaged storage drum (see photo). Analysis of several accidents, by DOE, have shown lack of a "safety culture" at the facility.[68]

Ocean do not present an exchange with shallow waters for about 140 years based on oxygen content data recorded over a period of 25 years.[71] They include burial beneath a stable abyssal plain, burial in a subduction zone that would slowly carry the waste downward into the Earth's mantle,[72][73] and burial beneath a remote natural or human-made island. While these approaches all have merit and would facilitate an international solution to the problem of disposal of radioactive waste, they would require an amendment of the Law of the Sea.[74]

Article 1 (Definitions), 7., of the 1996 Protocol to the Convention on the Prevention of Marine Pollution by Dumping of Wastes and Other Matter, (the London Dumping Convention) states:

""Sea" means all marine waters other than the internal waters of States, as well as the seabed and the subsoil thereof; it does not include subseabed repositories accessed only from land."

The proposed land-based subductive waste disposal method disposes of nuclear waste in a subduction zone accessed from land and therefore is not prohibited by international agreement. This method has been described as the most viable means of disposing of radioactive waste.*[75] and as the state-of-the-art as of 2001 in nuclear waste disposal technology.*[76] Another approach termed Remix & Return*[77] would blend high-level waste with uranium mine and mill tailings down to the level of the original radioactivity of the uranium ore, then replace it in inactive uranium mines. This approach has the merits of providing jobs for miners who would double as disposal staff, and of facilitating a cradle-to-grave cycle for radioactive materials, but would be inappropriate for spent reactor fuel in the absence of reprocessing, due to the presence in it of highly toxic radioactive elements such as plutonium.

Deep borehole disposal is the concept of disposing of high-level radioactive waste from nuclear reactors in extremely deep boreholes. Deep borehole disposal seeks to place the waste as much as 5 kilometres (3.1 mi) beneath the surface of the Earth and relies primarily on the immense natural geological barrier to confine the waste safely and permanently so that it should never pose a threat to the environment. The Earth's crust contains 120 trillion tons of thorium and 40 trillion tons of uranium (primarily at relatively trace concentrations of parts per million each adding up over the crust's $3 * 10^{19}$ ton mass), among other natural radioisotopes.*[78]*[79]*[80] Since the fraction of nuclides decaying per unit of time is inversely proportional to an isotope's half-life, the relative radioactivity of the lesser amount of human-produced radioisotopes (thousands of tons instead of trillions of tons) would diminish once the isotopes with far shorter half-lives than the bulk of natural radioisotopes decayed.

In January 2013, Cumbria county council rejected UK central government proposals to start work on an underground storage dump for nuclear waste near to the Lake District National Park. "For any host community, there will be a substantial community benefits package and worth hundreds of millions of pounds" said Ed Davey, Energy Secretary, but nonetheless, the local elected body voted 7–3 against research continuing, after hearing evidence from independent geologists that "the fractured strata of the county was impossible to entrust with such dangerous material and a hazard lasting millennia."*[81]*[82]

Transmutation

Main article: Nuclear transmutation

There have been proposals for reactors that consume nuclear waste and transmute it to other, less-harmful nuclear waste. In particular, the Integral Fast Reactor was a proposed nuclear reactor with a nuclear fuel cycle that produced no transuranic waste and in fact, could consume transuranic waste. It proceeded as far as large-scale tests, but was then canceled by the US Government. Another approach, considered safer but requiring more development, is to dedicate subcritical reactors to the transmutation of the left-over transuranic elements.

An isotope that is found in nuclear waste and that represents a concern in terms of proliferation is Pu-239. The estimated world total of plutonium in the year 2000 was of 1,645 metric tons, of which 210 metric tons had been separated by reprocessing. The large stock of plutonium is a result of its production inside uranium-fueled reactors and of the reprocessing of weapons-grade plutonium during the weapons program. An option for getting rid of this plutonium is to use it as a fuel in a traditional Light Water Reactor (LWR). Several fuel types with differing plutonium destruction efficiencies are under study.

Transmutation was banned in the US in April 1977 by President Carter due to the danger of plutonium proliferation.*[83] but President Reagan rescinded the ban in 1981.*[84] Due to the economic losses and risks, construction of reprocessing plants during this time did not resume. Due to high energy demand, work on the method has continued in the EU. This has resulted in a practical nuclear research reactor called Myrrha in which transmutation is possible. Additionally, a new research program called ACTINET has been started in the EU to make transmutation possible on a large, industrial scale. According to President Bush's Global Nuclear Energy Partnership (GNEP) of 2007, the US is now actively promoting research on transmutation technologies needed to markedly reduce the problem of nuclear waste treatment.*[85]

There have also been theoretical studies involving the use of fusion reactors as so called "actinide burners" where a fusion reactor plasma such as in a tokamak, could be "doped" with a small amount of the "minor" transuranic atoms which would be transmuted (meaning fissioned in the actinide case) to lighter elements upon their successive bombardment by the very high energy neutrons produced by the fusion of deuterium and tritium in the reactor. A study at MIT found that only 2 or 3 fusion reactors with parameters similar to that of the International Thermonuclear Experimental Reactor (ITER) could transmute the entire annual minor actinide production from all of the light water reactors presently operating in the United States fleet while simultaneously generating approximately 1 gigawatt of power from each reactor.[86]

Re-use of waste

Main article: Nuclear reprocessing

Another option is to find applications for the isotopes in nuclear waste so as to re-use them.[87] Already, caesium-137, strontium-90 and a few other isotopes are extracted for certain industrial applications such as food irradiation and radioisotope thermoelectric generators. While re-use does not eliminate the need to manage radioisotopes, it reduces the quantity of waste produced.

The Nuclear Assisted Hydrocarbon Production Method,[88] Canadian patent application 2,659,302, is a method for the temporary or permanent storage of nuclear waste materials comprising the placing of waste materials into one or more repositories or boreholes constructed into an unconventional oil formation. The thermal flux of the waste materials fracture the formation and alters the chemical and/or physical properties of hydrocarbon material within the subterranean formation to allow removal of the altered material. A mixture of hydrocarbons, hydrogen, and/or other formation fluids is produced from the formation. The radioactivity of high-level radioactive waste affords proliferation resistance to plutonium placed in the periphery of the repository or the deepest portion of a borehole.

Breeder reactors can run on U-238 and transuranic elements, which comprise the majority of spent fuel radioactivity in the 1,000–100,000-year time span.

Space disposal

Space disposal is attractive because it removes nuclear waste from the planet. It has significant disadvantages, such as the potential for catastrophic failure of a launch vehicle, which could spread radioactive material into the atmosphere and around the world. A high number of launches would be required because no individual rocket would be able to carry very much of the material relative to the total amount that needs to be disposed of. This makes the proposal impractical economically and it increases the risk of at least one or more launch failures.[89] To further complicate matters, international agreements on the regulation of such a program would need to be established.[90] Costs and inadequate reliability of modern rocket launch systems for space disposal has been one of the motives for interest in non-rocket space launch systems such as mass drivers, space elevators, and other proposals.

16.5.3   National management plans

See also: High-level radioactive waste management

Most countries are considerably ahead of the United States in developing plans for high-level radioactive waste disposal. Sweden and Finland are furthest along in committing to a particular disposal technology, while many others reprocess spent fuel or contract with France or Great Britain to do it, taking back the resulting plutonium and high-level waste. "An increasing backlog of plutonium from reprocessing is developing in many countries... It is doubtful that reprocessing makes economic sense in the present environment of cheap uranium."[91]

In many European countries (e.g., Britain, Finland, the Netherlands, Sweden and Switzerland) the risk or dose limit for a member of the public exposed to radiation from a future high-level nuclear waste facility is considerably more stringent than that suggested by the International Commission on Radiation Protection or proposed in the United States. European limits are often more stringent than the standard suggested in 1990 by the International Commission on Radiation Protection by a factor of 20, and more stringent by a factor of ten than the standard proposed by the US Environmental Protection Agency (EPA) for Yucca Mountain nuclear waste repository for the first 10,000 years after closure.[92]

The U.S. EPA's proposed standard for greater than 10,000 years is 250 times more permissive than the European limit.[92] The U.S. EPA proposed a legal limit of a maximum of 3.5 millisieverts (350 millirem) each annually to local individuals after 10,000 years, which would be up to several percent of the exposure currently received by some populations in the highest natural background regions on Earth, though the U.S. DOE predicted that received dose would be much below that limit.[93] Over a timeframe of thousands of years, after the most active short half-life radioisotopes decayed, burying U.S. nuclear waste would increase the radioactivity in the top 2000 feet of rock and soil in the United States (10 million km^2) by $\approx$ 1 part in 10 million over the cumulative amount of natural radioisotopes in such a volume, but the vicinity of the site would have a far higher concentration of artificial radioisotopes underground than such an average.[94]

Mongolia

After serious opposition had arisen about plans and negotiations between Mongolia with Japan and the United States of America to build nuclear-waste facilities in Mongolia, Mongolia stopped all negotiations in September 2011. These negotiations had started after U.S. Deputy Secretary of Energy Daniel B. Poneman visited Mongolia in September, 2010. Talks took place in Washington DC between officials of Japan, the United States and Mongolia in February 2011. After this the United Arab Emirates (UAE), which wanted to buy nuclear fuel from Mongolia, joined in the negotiations. The talks were kept secret, and although The *Mainichi Daily News* reported on them in May, Mongolia officially denied the existence of these negotiations. However, alarmed by this news, Mongolian citizens protested against the plans, and demanded the government withdraw the plans and disclose information. The Mongolian President Tsakhiagiin Elbegdorj issued a presidential order on September 13 banning all negotiations with foreign governments or international organizations on nuclear-waste storage plans in Mongolia.*[95] The Mongolian government has accused the newspaper of distributing false claims around the world. After the presidential order, the Mongolian president fired the individual who was supposedly involved in these conversations.

16.5.4   Illegal dumping

Main article: Toxic waste dumping by the 'Ndrangheta

Authorities in Italy are investigating a 'Ndrangheta mafia clan accused of trafficking and illegally dumping nuclear waste. According to a whistleblower, a manager of the Italy's state energy research agency Enea paid the clan to get rid of 600 drums of toxic and radioactive waste from Italy, Switzerland, France, Germany, and the US, with Somalia as the destination, where the waste was buried after buying off local politicians. Former employees of Enea are suspected of paying the criminals to take waste off their hands in the 1980s and 1990s. Shipments to Somalia continued into the 1990s, while the 'Ndrangheta clan also blew up shiploads of waste, including radioactive hospital waste, sending them to the sea bed off the Calabrian coast.*[96] According to the environmental group Legambiente, former members of the 'Ndrangheta have said that they were paid to sink ships with radioactive material for the last 20 years.*[97]

16.6   Accidents involving radioactive waste

Main article: Nuclear and radiation accidents

A few incidents have occurred when radioactive material was disposed of improperly, shielding during transport was defective, or when it was simply abandoned or even stolen from a waste store.*[98] In the Soviet Union, waste stored in Lake Karachay was blown over the area during a dust storm after the lake had partly dried out.*[99] At Maxey Flat, a low-level radioactive waste facility located in Kentucky, containment trenches covered with dirt, instead of steel or cement, collapsed under heavy rainfall into the trenches and filled with water. The water that invaded the trenches became radioactive and had to be disposed of at the Maxey Flat facility itself. In other cases of radioactive waste accidents, lakes or ponds with radioactive waste accidentally overflowed into the rivers during exceptional storms. In Italy, several radioactive waste deposits let material flow into river water, thus contaminating water for domestic use.*[100] In France, in the summer of 2008 numerous incidents happened;*[101] in one, at the Areva plant in Tricastin, it was reported that during a draining operation, liquid containing untreated uranium overflowed out of a faulty tank and about 75 kg of the radioactive material seeped into the ground and, from there, into two rivers nearby;*[102] in another case, over 100 staff were contaminated with low doses of radiation.*[103]

Scavenging of abandoned radioactive material has been the cause of several other cases of radiation exposure, mostly in developing nations, which may have less regulation of dangerous substances (and sometimes less general education about radioactivity and its hazards) and a market for scavenged goods and scrap metal. The scavengers and those who buy the material are almost always unaware that the material is radioactive and it is selected for its aesthetics or scrap value.*[104] Irresponsibility on the part of the radioactive material's owners, usually a hospital, university or military, and the absence of regulation concerning radioactive waste, or a lack of enforcement of such regulations, have been significant factors in radiation exposures. For an example of an accident involving radioactive scrap originating from a hospital see the Goiânia accident.*[104]

Transportation accidents involving spent nuclear fuel from power plants are unlikely to have serious consequences due to the strength of the spent nuclear fuel shipping casks.*[105]

On 15 December 2011 top government spokesman Osamu Fujimura of the Japanese government admitted that nuclear substances were found in the waste of Japanese nuclear facilities. Although Japan did commit itself in 1977 to these inspections in the safeguard agreement with the IAEA, the reports were kept secret for the inspectors of the International Atomic Energy Agency. Japan did start discussions with the IAEA about the large quantities of enriched uranium and plutonium that were discovered in nuclear waste cleared away by Japanese nuclear operators. At the press conference Fujimura said: "Based on investigations so far, most nuclear substances have been properly managed as waste, and from that perspective, there is no problem in safety management." But according to him, the matter was at that moment still being investigated.*[106]

16.7 Associated hazard warning signs

- The trefoil symbol used to indicate ionising radiation.

- 2007 ISO radioactivity danger symbol intended for IAEA Category 1, 2 and 3 sources defined as dangerous sources capable of death or serious injury.*[107]

- The dangerous goods transport classification sign for radioactive materials

16.8 See also

- Depleted uranium
- Ducrete
- Environmental remediation
- Environmental racism
- Geomelting
- Hot cell
- Human Interference Task Force
- *Into Eternity (film)*

- Lists of nuclear disasters and radioactive incidents
- Mixed waste (radioactive/hazardous)
- Microbial corrosion
- Nuclear decommissioning
- Personal protective equipment
- Radiation protection
- Radioactive contamination
- Radioactive scrap metal
- Radioecology
- Taylor Wilson's nuclear waste-fired small reactor
- Toxic waste
- Waste management

16.9 References

[1] "The Joint Convention". IAEA.

[2] Nuclear Information and Resource Service, Radioactive Waste Project. Retrieved September 2007.

[3] "What about Iodine-129 – Half-Life is 15 Million Years". *Berkeley Radiological Air and Water Monitoring Forum*. University of California. 28 March 2011. Retrieved 1 December 2012.

[4] Attix, Frank (1986). *Introduction to Radiological Physics and Radiation Dosimetry*. New York: Wiley-VCH. pp. 2–15,468,474. ISBN 978-0-471-01146-0.

[5] Anderson, Mary; Woessner, William (1992). *Applied Groundwater Modeling*. San Diego, CA: Academic Press Inc. pp. 325–327. ISBN 0-12-059485-4.

[6] Plus radium (element 88). While actually a sub-actinide, it immediately precedes actinium (89) and follows a three-element gap of instability after polonium (84) where no isotopes have half-lives of at least four years (the longest-lived isotope in the gap is radon-222 with a half life of less than four *days*). Radium's longest lived isotope, at 1,600 years, thus merits the element's inclusion here.

[7] Specifically from thermal neutron fission of U-235, e.g. in a typical nuclear reactor.

[8] Milsted, J.; Friedman, A. M.; Stevens, C. M. (1965). "The alpha half-life of berkelium-247; a new long-lived isomer of berkelium-248". *Nuclear Physics*. **71** (2): 299. doi:10.1016/0029-5582(65)90719-4.
"The isotopic analyses disclosed a species of mass 248 in constant abundance in three samples analysed over a period

of about 10 months. This was ascribed to an isomer of Bk^{248} with a half-life greater than 9 y. No growth of Cf^{248} was detected, and a lower limit for the β^- half-life can be set at about 10^4 y. No alpha activity attributable to the new isomer has been detected; the alpha half-life is probably greater than 300 y."

[9] This is the heaviest isotope with a half-life of at least four years before the "Sea of Instability".

[10] Excluding those "classically stable" isotopes with half-lives significantly in excess of ^{232}Th: e.g., while ^{113m}Cd has a half-life of only fourteen years, that of ^{114}Cd is nearly eight quadrillion years.

[11] "The 2007 Recommendations of the International Commission on Radiological Protection". *Annals of the ICRP*. ICRP publication 103. **37** (2–4). 2007. ISBN 978-0-7020-3048-2.

[12] Gofman, John W. *Radiation and human health*. San Francisco: Sierra Club Books. 1981. 787.

[13] Sancar, A. et al *Molecular mechanisms of mammalian DNA repair and the DNA damage checkpoints*. Washington D.C.: NIH PubMed.gov. 2004.

[14] Cochran, Robert (1999). *The Nuclear Fuel Cycle: Analysis and Management*. La Grange Park, IL: American Nuclear Society. pp. 52–57. ISBN 0-89448-451-6.

[15] "Global Defence News and Defence Headlines – IHS Jane's 360".

[16] Hashem, Heba. "Recycling spent nuclear fuel: the ultimate solution for the US?". *nuclearenergyinsider.com*. Nuclear Energy Insider. Retrieved 2015-07-29.

[17] "Continuous Plutonium Recycling In India: Improvements in Reprocessing Technology".

[18] World Nuclear Association (March 2009). "Plutonium". Retrieved 2010-03-18.

[19] Lyman, Edwin S. (December 1994). "A Perspective on the Proliferation Risks of Plutonium Mines". Nuclear Control Institute. Retrieved 2015-11-25.

[20] U.S. Department of Energy Environmental Management – "Department of Energy Five Year Plan FY 2007-FY 2011 Volume II." Retrieved 8 April 2007.

[21] American Scientist Jan/Feb 2007

[22] "Nuclear Logging". Retrieved 2009-07-07.

[23] Gabbard, Alex (1993). "Coal Combustion". *ORNL Review*. **26** (3–4). Archived from the original on February 5, 2007.

[24] "TENORM Sources | Radiation Protection | US EPA". Epa.gov. 2006-06-28. Retrieved 2013-08-01.

[25] Idaho State University. Radioactivity in Nature

[26] United Nations Scientific Committee on the Effects of Atomic Radiation. Sources and Effects of Ionizing Radiation. UNSCEAR 2008

[27] "Regulation of TENORM". Tenorm.com. Retrieved 2013-08-01.

[28] Cosmic origins of Uranium. uic.com.au (November 2006)

[29] U.S. Geological Survey. Radioactive Elements in Coal and Fly Ash: Abundance, Forms, and Environmental Significance, *Fact Sheet* FS-163-1997, October 1997. Retrieved September 2007.

[30] Survey & Identification of NORM Contaminated Equipment. enprotec-inc.com.

[31] *Classification of Radioactive Waste*. IAEA, Vienna (1994)

[32] Janicki, Mark (26 November 2013). "Iron boxes for ILW transport and storage". Nuclear Engineering International. Retrieved 4 December 2013.

[33] Rogner, H. (2010). "NUCLEAR POWER AND SUSTAINABLE DEVELOPMENT". *Journal of International Affairs*. **64**: 149.

[34] "Myths and Realities of Radioactive Waste". February 2016.

[35] "Radioactive Waste Management". World Nuclear Association. July 2015. Retrieved 2015-08-25.

[36] Geere, Duncan. (2010-09-20) Where do you put 250,000 tonnes of nuclear waste? (Wired UK). Wired.co.uk. Retrieved on 2015-12-15.

[37] Humber, Yuriy (2015-07-10). "Japan's 17,000 Tons of Nuclear Waste in Search of a Home". *Bloomberg*.

[38] Findlay, Trevor (2010). "Nuclear Energy to 2030 and its Implications for Safety, Security and Nonproliferation: Overview" (PDF). *Nuclear energy futures project*.

[39] "Radioactive Waste Management | Nuclear Waste Disposal". World Nuclear Association. July 2015. Retrieved 2015-08-25.

[40] Biello, David (Jul 29, 2011). "Presidential Commission Seeks Volunteers to Store U.S. Nuclear Waste". *Scientific American*.

[41] Why Wipp?. wipp.energy.gov

[42] "Progress on approaches to management of separated plutonium". *Nuclear Decommissioning Authority*. 2014-01-20.

[43] Vandenbosch, p. 21.

[44] Ojovan, M. I. and Lee, W.E. (2014) *An Introduction to Nuclear Waste Immobilisation*. Elsevier, Amsterdam. ISBN 9780080993928

[45] See, for example, Paul Brown, 'Shoot it at the sun. Send it to Earth's core. What to do with nuclear waste?'. *The Guardian*, 14 April 2004.

[46] World Nuclear Association "Storage and Disposal Options" retrieved 2011-11-14

[47] "Ministers admit nuclear waste was dumped in sea". *The Independent*. London, 1997-07-01.

[48] *Blue Ribbon Commission on America's Nuclear Future: Executive Summary*, January 2012.

[49] Ojovan, M. I. and Lee, W.E. (2005) *An Introduction to Nuclear Waste Immobilisation*, Elsevier, Amsterdam, p. 315

[50] National Research Council (1996). *Nuclear Wastes: Technologies for Separation and Transmutation*. Washington DC: National Academy Press.

[51] "Laboratory-scale vitrification and leaching of Hanford high-level waste for the purpose of simulant and glass property models validation". Retrieved 2009-07-07.

[52] Ojovan, M.I.; et al. (2006). "Corrosion of nuclear waste glasses in non-saturated conditions: Time-Temperature behaviour" (PDF). Retrieved 2008-06-30.

[53] OECD Nuclear Energy Agency (1994). *The Economics of the Nuclear Fuel Cycle*. Paris: OECD Nuclear Energy Agency.

[54] Ojovan, Michael I.; Lee, William E. (2010). "Glassy Wasteforms for Nuclear Waste Immobilization". *Metallurgical and Materials Transactions A*. **42** (4): 837. Bibcode:2011MMTA...42..837O. doi:10.1007/s11661-010-0525-7.

[55] "Waste Form Release Calculations for the 2005 Integrated Disposal Facility Performance Assessment" (PDF). *PNNL-15198*. Pacific Northwest National Laboratory. July 2005. Retrieved 2006-11-08.

[56] Hensing, I. & Schultz, W. (1995). *Economic Comparison of Nuclear Fuel Cycle Options*. Cologne: Energiewirtschaftlichen Instituts.

[57] Brünglinghaus, Marion. "Waste processing". Euronuclear.org. Retrieved 2013-08-01.

[58] Wilmarth, W.R. et al. (2004) Removal of Silicon from High Level Waste Streams via Ferric Flocculation. srs.gov.

[59] World Nuclear Association, Synroc, *Nuclear Issues Briefing Paper* 21. Retrieved January 2009.

[60] National Research Council (1995). *Technical Bases for Yucca Mountain Standards*. Washington, D.C.: National Academy Press. cited in "The Status of Nuclear Waste Disposal". The American Physical Society. January 2006. Retrieved 2008-06-06.

[61] "Public Health and Environmental Radiation Protection Standards for Yucca Mountain, Nevada; Proposed Rule" (PDF). Environmental Protection Agency. 2005-08-22. Retrieved 2008-06-06.

[62] Peterson, Per; William Kastenberg; Michael Corradini. "Nuclear Waste and the Distant Future". *Issues in Science and Technology*. Washington, DC: National Academy of Sciences (Summer 2006).

[63] "Issues relating to safety standards on the geological disposal of radioactive waste" (PDF). International Atomic Energy Agency. 2001-06-22. Retrieved 2008-06-06.

[64] "IAEA Waste Management Database: Report 3 – L/ILW-LL" (PDF). International Atomic Energy Agency. 2000-03-28. Retrieved 2008-06-06.

[65] "Decommissioning costs of WWER-440 nuclear power plants" (PDF). International Atomic Energy Agency. November 2002. Retrieved 2008-06-06.

[66] International Atomic Energy Agency. *Spent Fuel and High Level Waste: Chemical Durability and Performance under Simulated Repository Conditions*, IAEA-TECDOC-1563, October 2007.

[67] "Fact Sheet on Dry Cask Storage of Spent Nuclear Fuel". NRC. May 7, 2009. Retrieved 2011-06-25.

[68] Cameron L. Tracy, Megan K. Dustin & Rodney C. Ewing, Policy: Reassess New Mexico's nuclear-waste repository, *Nature*, 13 January 2016.

[69] Vandenbosch, p. 10.

[70] Yates, Marshall (July 6, 1989). "DOE waste management criticized: On-site storage urged". *Public Utilities Fortnightly*. **124**: 33.

[71] Hoare, J.P. (1968) *Electrochemistry of Oxygen*, Interscience Publishers

[72] Hafemeister, David W. (2007). *Physics of societal issues: calculations on national security, environment, and energy*. Berlin: Springer. p. 187. ISBN 0387689095.

[73] Shipman, J.T.; Wison J.D.; Todd A. (2007). *An Introduction to Physical Science* (10 ed.). Cengage Learning. p. 279. ISBN 978-0-618-93596-3.

[74] "Dumping and Loss overview". *Oceans in the Nuclear Age*. Retrieved March 23, 2011.

[75] Utah Nuclear Waste Summary, by Tricia Jack, Jordan Robertson, Center for Public Policy & Administration, University of Utah

[76] Rao, K. R. (25 December 2001). "Radioactive waste: The problem and its management" (PDF). *Current Science*. **81** (12).

[77] Remix & Return: A Complete Low-Level Nuclear Waste Solution. scientiapress.com

[78] Sevior M. (2006). "Considerations for nuclear power in Australia". *International Journal of Environmental Studies*. **63** (6): 859–872. doi:10.1080/00207230601047255.

[79] Thorium Resources In Rare Earth Elements. uiuc.edu

[80] American Geophysical Union. Fall Meeting 2007. abstract #V33A-1161. Mass and Composition of the Continental Crust

[81] Wainwright, Martin (30 January 2013). "Cumbria rejects underground nuclear storage dump". *The Guardian*. London. Retrieved 1 February 2013.

[82] Macalister, Terry (31 January 2013). "Cumbria sticks it to the nuclear dump lobby – despite all the carrots on offer". *The Guardian*. London. Retrieved 1 February 2013.

[83] Review of the SONIC Proposal to Dump High-Level Nuclear Waste at Piketon. Southern Ohio Neighbors Group

[84] National Policy Analysis #396: The Separations Technology and Transmutation Systems (STATS) Report: Implications for Nuclear Power Growth and Energy Sufficiency – February 2002. Nationalcenter.org. Retrieved on 2015-12-15.

[85] Global Nuclear Energy Partnership Statement of Principles. gnep.energy.gov (2007-09-16)

[86] Freidberg, Jeffrey P. "Department of Nuclear Engineering: Reports to the President 2000–2001". Web.mit.edu. Retrieved 2013-08-01.

[87] Milton R. (January 17, 1978) Nuclear By-Products : A Resource for the Future. heritage.org

[88] "節素でブチ断食｜成功させる秘訣は代苔ドリンクにあった！". Nuclearhydrocarbons.com. Retrieved 2013-08-01.

[89] National Research Council (U.S.). Committee on Disposition of High-Level Radioactive Waste Through Geological Isolation (2001). *Disposition of high-level waste and spent nuclear fuel: the continuing societal and technical challenges*. National Academies Press. p. 122. ISBN 978-0-309-07317-2.

[90] "Managing nuclear waste: Options considered". *DOE Factsheets*. Department of Energy: Office of Civilian Radioactive Waste Management. Yucca Mountain Project. November 2003. Archived from the original on 2009-05-15.

[91] Vandenbosch, p. 247.

[92] Vandenbosch, p. 248

[93] U.S. Federal Register. 40 CFR Part 197. Environmental Protection Agency. Public Health and Environmental Radiation Protection Standards for Yucca Mountain, Nevada: Final Rule

[94] Cohen, Bernard L. (1998). "Perspectives on the High Level Waste Disposal Problem". *Interdisciplinary Science Reviews*. **23**: 193–203.

[95] The Mainichi Daily News (15 October 2011)Mongolia abandons nuclear waste storage plans, and informs Japan of decision

[96] From cocaine to plutonium: mafia clan accused of trafficking nuclear waste, The Guardian, October 9, 2007

[97] Mafia sank boat with radioactive waste: official. AFP. September 14, 2009

[98] Strengthening the safety of radiation sources & the security of radioactive materials: timely action, by Abel J. González, IAEA Bulletin, 41/3/1999

[99] GlobalSecurity.org. Chelyabinsk-65/Ozersk. Retrieved September 2007.

[100] Report RAI.it. L'Eredità (in Italian), 2 November 2008

[101] Reuters UK, New incident at French nuclear plant. Retrieved March 2009.

[102] "'It feels like a sci-fi film' – accidents tarnish nuclear dream". *The Guardian*. London. 25 July 2008.

[103] Reuters UK, Too many French nuclear workers contaminated. Retrieved March 2009.

[104] International Atomic Energy Agency. *The radiological accident in Goiânia*, 1988. Retrieved September 2007.

[105] "Nuclear Flask Train Crash Test – BBC News 1984". YouTube. 1984-07-17. Retrieved 2013-08-01.

[106] The Mainichi Daily News (December 15, 2011) Gov't admits nuclear substances found in waste, unreported to IAEA

[107] "New Symbol Launched to Warn Public About Radiation Dangers". International Atomic Energy Agency. 2007.

16.10   Cited sources

- Vandenbosch, Robert & Vandenbosch, Susanne E. (2007). *Nuclear waste stalemate*. Salt Lake City: University of Utah Press. ISBN 0874809037.

16.11   External links

- Alsos Digital Library – Radioactive Waste (annotated bibliography)

- Euridice European Interest Group in charge of Hades URL operation (link)

- Ondraf/Niras, the waste management authority in Belgium (link)

- Critical Hour: Three Mile Island, The Nuclear Legacy, And National Security (PDF)

- Environmental Protection Agency – Yucca Mountain (documents)

- Grist.org – How to tell future generations about nuclear waste (article)

- International Atomic Energy Agency – Internet Directory of Nuclear Resources (links)

- Nuclear Files.org – Yucca Mountain (documents)

- Nuclear Regulatory Commission – Radioactive Waste (documents)

- Nuclear Regulatory Commission – Spent Fuel Heat Generation Calculation (guide)

- Radwaste Solutions (magazine)

- UNEP Earthwatch – Radioactive Waste (documents and links)

- World Nuclear Association – Radioactive Waste (briefing papers)

- Worries can't be buried as nuclear waste piles up. Los Angeles Times. January 21, 2008

- RadWaste.org

- Radioactivity.eu.com

Chapter 17

Enriched uranium

Enriched uranium is a type of uranium in which the percent composition of uranium-235 has been increased through the process of isotope separation. Natural uranium is 99.284% ^{238}U isotope, with ^{235}U only constituting about 0.711% of its weight. ^{235}U is the only nuclide existing in nature (in any appreciable amount) that is fissile with thermal neutrons.[1]

Enriched uranium is a critical component for both civil nuclear power generation and military nuclear weapons. The International Atomic Energy Agency attempts to monitor and control enriched uranium supplies and processes in its efforts to ensure nuclear power generation safety and curb nuclear weapons proliferation.

During the Manhattan Project enriched uranium was given the codename **oralloy**, a shortened version of Oak Ridge alloy, after the location of the plants where the uranium was enriched. The term oralloy is still occasionally used to refer to enriched uranium. There are about 2,000 tonnes (t, Mg) of highly enriched uranium in the world,[2] produced mostly for nuclear weapons, naval propulsion, and smaller quantities for research reactors.

The ^{238}U remaining after enrichment is known as depleted uranium (DU), and is considerably less radioactive than even natural uranium, though still very dense and extremely hazardous in granulated form – such granules are a natural by-product of the shearing action that makes it useful for armor-penetrating weapons and radiation shielding. At present, 95 percent of the world's stocks of depleted uranium remain in secure storage.

17.1 Grades

Uranium as it is taken directly from the Earth is not suitable as fuel for most nuclear reactors and requires additional processes to make it usable. Uranium is mined either underground or in an open pit depending on the depth in which it is found. After the uranium ore is mined, it must go through a milling process to extract the uranium from the ore. This is accomplished by a combination of chemical processes with the end product being concentrated uranium oxide, which is known as "yellowcake", contains roughly 60% uranium whereas the ore typically contains less than 1% uranium and as little as 0.1% uranium (Henderson 2000). After the milling process is complete, the uranium must next undergo a process of conversion, "to either uranium dioxide, which can be used as the fuel for those types of reactors that do not require enriched uranium, or into uranium hexafluoride, which can be enriched to produce fuel for the majority of types of reactors". Naturally-occurring uranium is made of a mixture of U-235 and U-238. The U-235 is fissile meaning it is easily split with neutrons while the remainder is U-238, but in nature, more than 99% of the extracted ore is U-238. Most nuclear reactors require enriched uranium, which is uranium with higher concentrations of U-235 ranging between 3.5% and 4.5%. There are two commercial enrichment processes: gaseous diffusion and gas centrifugation. Both enrichment processes involve the use of uranium hexafluoride and produce enriched uranium oxide.

17.1.1 Reprocessed uranium (RepU)

Main article: Reprocessed uranium

Reprocessed uranium (RepU) is a product of nuclear fuel cycles involving nuclear reprocessing of spent fuel. RepU recovered from light water reactor (LWR) spent fuel typically contains slightly more U-235 than natural uranium, and therefore could be used to fuel reactors that customarily use natural uranium as fuel, such as CANDU reactors. It also contains the undesirable isotope uranium-236, which undergoes neutron capture, wasting neutrons (and requiring higher U-235 enrichment) and creating neptunium-237, which would be one of the more mobile and troublesome radionuclides in deep geological repository disposal of nuclear waste.

17.1.2   Low-enriched uranium (LEU)

Low-enriched uranium (LEU) has a lower than 20% concentration of ^{235}U; for instance, in commercial light water reactors (LWR), the most prevalent power reactors in the world, uranium is enriched to 3 to 5% ^{235}U. Fresh LEU used in research reactors is usually enriched 12% to 19.75% U-235, the latter concentration being used to replace HEU fuels when converting to LEU.[3]

17.1.3   Highly enriched uranium (HEU)

Highly enriched uranium (HEU) has a 20% or higher concentration of ^{235}U. The fissile uranium in nuclear weapon primaries usually contains 85% or more of ^{235}U known as weapons-grade, though theoretically for an implosion design, a minimum of 20% could be sufficient (called weapon(s)-usable) although it would require hundreds of kilograms of material and "would not be practical to design";[4][5] even lower enrichment is hypothetically possible, but as the enrichment percentage decreases the critical mass for unmoderated fast neutrons rapidly increases, with for example, an infinite mass of 5.4% ^{235}U being required.[4] For criticality experiments, enrichment of uranium to over 97% has been accomplished.[6]

The very first uranium bomb, Little Boy dropped by the United States on Hiroshima in 1945, used 64 kilograms of 80% enriched uranium. Wrapping the weapon's fissile core in a neutron reflector (which is standard on all nuclear explosives) can dramatically reduce the critical mass. Because the core was surrounded by a good neutron reflector, at explosion it comprised almost 2.5 critical masses. Neutron reflectors, compressing the fissile core via implosion, fusion boosting, and "tamping", which slows the expansion of the fissioning core with inertia, allow nuclear weapon designs that use less than what would be one bare-sphere critical mass at normal density. The presence of too much of the ^{238}U isotope inhibits the runaway nuclear chain reaction that is responsible for the weapon's power. The critical mass for 85% highly enriched uranium is about 50 kilograms (110 lb), which at normal density would be a sphere about 17 centimetres (6.7 in) in diameter.

Later US nuclear weapons usually use plutonium-239 in the primary stage, but the jacket or tamper secondary stage, which is compressed by the primary nuclear explosion often uses HEU with enrichment between 40% and 80%[7] along with the fusion fuel lithium deuteride. For the secondary of a large nuclear weapon, the higher critical mass of less-enriched uranium can be an advantage as it allows the core at explosion time to contain a larger amount of fuel. The ^{238}U is not fissile but still fissionable by fusion neutrons.

HEU is also used in fast neutron reactors, whose cores re-

quire about 20% or more of fissile material, as well as in naval reactors, where it often contains at least 50% ^{235}U, but typically does not exceed 90%. The Fermi-1 commercial fast reactor prototype used HEU with 26.5% ^{235}U. Significant quantities of HEU are used in the production of medical isotopes, for example molybdenum-99 for technetium-99m generators.[8]

17.2   Enrichment methods

Isotope separation is difficult because two isotopes of the same elements have very nearly identical chemical properties, and can only be separated gradually using small mass differences. (^{235}U is only 1.26% lighter than ^{238}U.) This problem is compounded by the fact that uranium is rarely separated in its atomic form, but instead as a compound (^{235}UF$_6$ is only 0.852% lighter than ^{238}UF$_6$.) A cascade of identical stages produces successively higher concentrations of ^{235}U. Each stage passes a slightly more concentrated product to the next stage and returns a slightly less concentrated residue to the previous stage.

There are currently two generic commercial methods employed internationally for enrichment: gaseous diffusion (referred to as *first* generation) and gas centrifuge (*second* generation), which consumes only 2% to 2.5%[9] as much energy as gaseous diffusion, with centrifuges being at least a "factor of 20" more efficient.[10] Some work is being done that would use nuclear resonance; however there is no reliable evidence that any nuclear resonance processes have been scaled up to production.

17.2.1   Diffusion techniques

Gaseous diffusion

Main article: Gaseous diffusion

Gaseous diffusion is a technology used to produce enriched uranium by forcing gaseous uranium hexafluoride (*hex*) through semi-permeable membranes. This produces a slight separation between the molecules containing ^{235}U and ^{238}U. Throughout the Cold War, gaseous diffusion played a major role as a uranium enrichment technique, and as of 2008 accounted for about 33% of enriched uranium production,[11] but in 2011 was deemed an obsolete technology that is steadily being replaced by the later generations of technology as the diffusion plants reach their ends-of-life.[12] In 2013, the Paducah facility in the US ceased operating, it was the last commercial ^{235}U gaseous diffusion plant in the world.[13]

Thermal diffusion

Thermal diffusion utilizes the transfer of heat across a thin liquid or gas to accomplish isotope separation. The process exploits the fact that the lighter ^{235}U gas molecules will diffuse toward a hot surface, and the heavier ^{238}U gas molecules will diffuse toward a cold surface. The S-50 plant at Oak Ridge, Tennessee was used during World War II to prepare feed material for the EMIS process. It was abandoned in favor of gaseous diffusion.

17.2.2 Centrifuge techniques

Gas centrifuge

Main article: Gas centrifuge
The gas centrifuge process uses a large number of rotating cylinders in series and parallel formations. Each cylinder's rotation creates a strong centripetal force so that the heavier gas molecules containing ^{238}U move tangentially toward the outside of the cylinder and the lighter gas molecules rich in ^{235}U collect closer to the center. It requires much less energy to achieve the same separation than the older gaseous diffusion process, which it has largely replaced and so is the current method of choice and is termed *second generation*. It has a separation factor per stage of 1.3 relative to gaseous diffusion of 1.005,[11] which translates to about one-fiftieth of the energy requirements. Gas centrifuge techniques produce about 54% of the world's enriched uranium.

Zippe centrifuge

The Zippe centrifuge is an improvement on the standard gas centrifuge, the primary difference being the use of heat. The bottom of the rotating cylinder is heated, producing convection currents that move the ^{235}U up the cylinder, where it can be collected by scoops. This improved centrifuge design is used commercially by Urenco to produce nuclear fuel and was used by Pakistan in their nuclear weapons program.

17.2.3 Laser techniques

Laser processes promise lower energy inputs, lower capital costs and lower tails assays, hence significant economic advantages. Several laser processes have been investigated or are under development. Separation of Isotopes by Laser Excitation (SILEX) is well advanced and licensed for commercial operation in 2012.

Atomic vapor laser isotope separation (AVLIS)

Atomic vapor laser isotope separation employs specially tuned lasers[14] to separate isotopes of uranium using selective ionization of hyperfine transitions. The technique uses lasers tuned to frequencies that ionize ^{235}U atoms and no others. The positively charged ^{235}U ions are then attracted to a negatively charged plate and collected.

Molecular laser isotope separation (MLIS)

Molecular laser isotope separation uses an infrared laser directed at UF_6, exciting molecules that contain a ^{235}U atom. A second laser frees a fluorine atom, leaving uranium pentafluoride, which then precipitates out of the gas.

Separation of Isotopes by Laser Excitation (SILEX)

Separation of isotopes by laser excitation is an Australian development that also uses UF_6. After a protracted development process involving U.S. enrichment company USEC acquiring and then relinquishing commercialization rights to the technology, GE Hitachi Nuclear Energy (GEH) signed a commercialization agreement with Silex Systems in 2006.[15] GEH has since built a demonstration test loop and announced plans to build an initial commercial facility.[16] Details of the process are classified and restricted by intergovernmental agreements between United States, Australia, and the commercial entities. SILEX has been projected to be an order of magnitude more efficient than existing production techniques but again, the exact figure is classified.[11] In August, 2011 Global Laser Enrichment, a subsidiary of GEH, applied to the U.S. Nuclear Regulatory Commission (NRC) for a permit to build a commercial plant.[17] In September 2012, the NRC issued a license for GEH to build and operate a commercial SILEX enrichment plant, although the company had not yet decided whether the project would be profitable enough to begin construction, and despite concerns that the technology could contribute to nuclear proliferation.[18]

17.2.4 Other techniques

Aerodynamic processes

Aerodynamic enrichment processes include the Becker jet nozzle techniques developed by E. W. Becker and associates using the LIGA process and the vortex tube separation process. These aerodynamic separation processes depend upon diffusion driven by pressure gradients, as does the gas centrifuge. They in general have the disadvantage of requiring complex systems of cascading of individual separating elements to minimize energy consumption. In effect, aerodynamic processes can be considered as non-rotating centrifuges. Enhancement of the centrifugal forces is achieved by dilution of UF_6 with hydrogen or helium as a carrier gas achieving a much higher flow velocity for the gas than could be obtained using pure uranium hexafluoride. The Uranium Enrichment Corporation of South Africa (UCOR) developed and deployed the continuous Helikon vortex separation cascade for high production rate low enrichment and the substantially different semi-batch Pelsakon low production rate high enrichment cascade both using a particular vortex tube separator design, and both embodied in industrial plant.[19] A demonstration plant was built in Brazil by NUCLEI, a consortium led by Industrias Nucleares do Brasil that used the separation nozzle process. However all methods have high energy consumption and substantial requirements for removal of waste heat; none are currently still in use.

Electromagnetic isotope separation

Main article: Calutron

In the electromagnetic isotope separation process (EMIS), metallic uranium is first vaporized, and then ionized to positively charged ions. The cations are then accelerated and subsequently deflected by magnetic fields onto their respective collection targets. A production-scale mass spectrometer named the Calutron was developed during World War II that provided some of the ^{235}U used for the Little Boy nuclear bomb, which was dropped over Hiroshima in 1945. Properly the term 'Calutron' applies to a multistage device arranged in a large oval around a powerful electromagnet. Electromagnetic isotope separation has been largely abandoned in favour of more effective methods.

Chemical methods

One chemical process has been demonstrated to pilot plant stage but not used. The French CHEMEX process exploited a very slight difference in the two isotopes' propensity to change valency in oxidation/reduction, utilising immiscible aqueous and organic phases. An ion-exchange process was developed by the Asahi Chemical Company in Japan that applies similar chemistry but effects separation on a proprietary resin ion-exchange column.

Plasma separation

Plasma separation process (PSP) describes a technique that makes use of superconducting magnets and plasma physics. In this process, the principle of ion cyclotron resonance is used to selectively energize the ^{235}U isotope in a plasma

containing a mix of ions. The French developed their own version of PSP, which they called RCI. Funding for RCI was drastically reduced in 1986, and the program was suspended around 1990, although RCI is still used for stable isotope separation.

17.3 Separative work unit

"Separative work" – the amount of separation done by an enrichment process – is a function of the concentrations of the feedstock, the enriched output, and the depleted tailings; and is expressed in units that are so calculated as to be proportional to the total input (energy / machine operation time) and to the mass processed. Separative work is *not* energy. The same amount of separative work will require different amounts of energy depending on the efficiency of the separation technology. Separative work is measured in *Separative work units* SWU, kg SW, or kg UTA (from the German *Urantrennarbeit* – literally *uranium separation work*)

- 1 SWU = 1 kg SW = 1 kg UTA

- 1 kSWU = 1 tSW = 1 t UTA

- 1 MSWU = 1 ktSW = 1 kt UTA

Further information: Separative work units

17.4 Cost issues

In addition to the separative work units provided by an enrichment facility, the other important parameter to be considered is the mass of natural uranium (NU) that is needed to yield a desired mass of enriched uranium. As with the number of SWUs, the amount of feed material required will also depend on the level of enrichment desired and upon the amount of ^{235}U that ends up in the depleted uranium. However, unlike the number of SWUs required during enrichment, which increases with decreasing levels of ^{235}U in the depleted stream, the amount of NU needed will decrease with decreasing levels of ^{235}U that end up in the DU.

For example, in the enrichment of LEU for use in a light water reactor it is typical for the enriched stream to contain 3.6% ^{235}U (as compared to 0.7% in NU) while the depleted stream contains 0.2% to 0.3% ^{235}U. In order to produce one kilogram of this LEU it would require approximately 8 kilograms of NU and 4.5 SWU if the DU stream was allowed to have 0.3% ^{235}U. On the other hand, if the depleted stream had only 0.2% ^{235}U, then it would require just 6.7 kilograms of NU, but nearly 5.7 SWU of enrichment. Because the amount of NU required and the number of SWUs required during enrichment change in opposite directions, if NU is cheap and enrichment services are more expensive, then the operators will typically choose to allow more ^{235}U to be left in the DU stream whereas if NU is more expensive and enrichment is less so, then they would choose the opposite.

- Uranium enrichment calculator designed by the WISE Uranium Project

17.5 Downblending

The opposite of enriching is downblending; surplus HEU can be downblended to LEU to make it suitable for use in commercial nuclear fuel.

The HEU feedstock can contain unwanted uranium isotopes: ^{234}U is a minor isotope contained in natural uranium; during the enrichment process, its concentration increases but remains well below 1%. High concentrations of ^{236}U are a byproduct from irradiation in a reactor and may be contained in the HEU, depending on its manufacturing history. HEU reprocessed from nuclear weapons material production reactors (with an ^{235}U assay of approx. 50%) may contain ^{236}U concentrations as high as 25%, resulting in concentrations of approximately 1.5% in the blended LEU product. ^{236}U is a neutron poison; therefore the actual ^{235}U concentration in the LEU product must be raised accordingly to compensate for the presence of ^{236}U.

The blendstock can be NU, or DU, however depending on feedstock quality, SEU at typically 1.5 wt% ^{235}U may used as a blendstock to dilute the unwanted byproducts that may be contained in the HEU feed. Concentrations of these isotopes in the LEU product in some cases could exceed ASTM specifications for nuclear fuel, if NU, or DU were used. So, the HEU downblending generally cannot contribute to the waste management problem posed by the existing large stockpiles of depleted uranium.

A major downblending undertaking called the Megatons to Megawatts Program converts ex-Soviet weapons-grade HEU to fuel for U.S. commercial power reactors. From 1995 through mid-2005, 250 tonnes of high-enriched uranium (enough for 10,000 warheads) was recycled into low-enriched-uranium. The goal is to recycle 500 tonnes by 2013. The decommissioning programme of Russian nuclear warheads accounted for about 13% of total world requirement for enriched uranium leading up to 2008.[11]

The United States Enrichment Corporation has been involved in the disposition of a portion of the 174.3 tonnes of highly enriched uranium (HEU) that the U.S. government declared as surplus military material in 1996. Through the U.S. HEU Downblending Program, this HEU material, taken primarily from dismantled U.S. nuclear warheads, was recycled into low-enriched uranium (LEU) fuel, used by nuclear power plants to generate electricity.[20]

- A uranium downblending calculator designed by the WISE Uranium Project

17.6 Global enrichment facilities

The following countries are known to operate enrichment facilities: Argentina, Brazil, China, France, Germany, India, Iran, Japan, the Netherlands, North Korea, Pakistan, Russia, the United Kingdom, and the United States.[21][22] Belgium, Iran, Italy, and Spain hold an investment interest in the French Eurodif enrichment plant, with Iran's holding entitling it to 10% of the enriched uranium output. Countries that had enrichment programs in the past include Libya and South Africa, although Libya's facility was never operational.[23] Australia has developed a laser enrichment process known as SILEX, which it intends to pursue through financial investment in a U.S. commercial venture by General Electric.[24] It has also been claimed that Israel has a uranium enrichment program housed at the Negev Nuclear Research Center site near Dimona.[25]

17.7 See also

- Areva
- List of laser articles
- MOX fuel
- Nuclear fuel bank
- Nuclear power
- Uranium market
- Uranium mining

17.8 References

[1] OECD Nuclear Energy Agency (2003). *Nuclear Energy Today*. OECD Publishing. p. 25. ISBN 9789264103283.

[2] Thomas B. Cochran (Natural Resources Defense Council) (12 June 1997). "Safeguarding Nuclear Weapon-Usable Materials in Russia" (PDF). Proceedings of international forum on illegal nuclear traffic.

[3] Alexander Glaser (6 November 2005). "About the Enrichment Limit for Research Reactor Conversion : Why 20%?" (PDF). Princeton University. Retrieved 18 April 2014.

[4] Forsberg, C. W.; Hopper, C. M.; Richter, J. L.; Vantine, H. C. (March 1998). "Definition of Weapons-Usable Uranium-233" (PDF). *ORNL/TM-13517*. Oak Ridge National Laboratories. Retrieved 30 October 2013.

[5] Sublette, Carey (4 October 1996). "Nuclear Weapons FAQ, Section 4.1.7.1: Nuclear Design Principles – Highly Enriched Uranium". *Nuclear Weapons FAQ*. Retrieved 2 October 2010.

[6] Mosteller, R.D. (1994). "Detailed Reanalysis of a Benchmark Critical Experiment: Water-Reflected Enriched-Uranium Sphere" (PDF). *Los Alamos technical paper* (LA–UR–93-4097): 2. Retrieved 19 December 2007. The enrichment of the pin and of one of the hemispheres was 97.67 w/o, while the enrichment of the other hemisphere was 97.68 w/o.

[7] "Nuclear Weapons FAQ". Nuclearweaponarchive.org. Retrieved 26 January 2013.

[8] Frank N. Von Hippel; Laura H. Kahn (December 2006). "Feasibility of Eliminating the Use of Highly Enriched Uranium in the Production of Medical Radioisotopes". *Science & Global Security*. **14** (2 & 3): 151–162. doi:10.1080/08929880600993071. Retrieved 26 March 2010.

[9] "Uranium Enrichment". *world-nuclear.org*.

[10] *Economic Perspective for Uranium Enrichment* (PDF). The throughput per centrifuge unit is very small compared to that of a diffusion unit so small, in fact, that it is not compensated by the higher enrichment per unit. To produce the same amount of reactor-grade fuel requires a considerably larger number (approximately 50,000 to 500,000) of centrifuge units than diffusion units. This disadvantage, however, is outweighed by **the considerably lower (by a factor of 20) energy consumption per SWU** for the gas centrifuge

[11] "Lodge Partners Mid-Cap Conference 11 April 2008" (PDF). Silex Ltd. 11 April 2008.

[12] Rod Adams (24 May 2011). "McConnell asks DOE to keep using 60 year old enrichment plant to save jobs". Atomic Insights. Retrieved 26 January 2013.

[13] "Paducah enrichment plant to be closed. *The 1950s facility is the last remaining gaseous diffusion uranium enrichment plant in the world.*".

[14] F. J. Duarte and L.W. Hillman (Eds.), *Dye Laser Principles* (Academic, New York, 1990) Chapter 9.

[15] Archived 23 July 2015 at the Wayback Machine.

[16] "GE Hitachi Nuclear Energy Selects Wilmington, N.C. as Site for Potential Commercial Uranium Enrichment Facility". Business Wire. 30 April 2008. Retrieved 30 September 2012.

[17] Broad, William J. (20 August 2011). "Laser Advances in Nuclear Fuel Stir Terror Fear". *The New York Times*. Retrieved 21 August 2011.

[18] "Uranium Plant Using Laser Technology Wins U.S. Approval". *New York Times*. September 2012.

[19] Smith, Michael; Jackson A G M (2000). "Dr". *South African Institution of Chemical Engineers – Conference 2000*: 280–289.

[20] Archived 6 April 2001 at the Wayback Machine.

[21] Arjun Makhijani; Lois Chalmers; Brice Smith (15 October 2004). *Uranium enrichment* (PDF). Institute for Energy and Environmental Research. Retrieved 21 November 2009.

[22] Australia's uranium - Greenhouse friendly fuel for an energy hungry world (PDF). *Standing Committee on Industry and Resources* (Report). The Parliament of the Commonwealth of Australia. November 2006. p. 730. Retrieved 3 April 2015.

[23] BBC (1 September 2006). "Q&A: Uranium enrichment". *BBC News*. Retrieved 3 January 2010.

[24] "Laser enrichment could cut cost of nuclear power". The Sydney Morning Herald. 26 May 2006.

[25] "Israel's Nuclear Weapons Program". Nuclear Weapon Archive. 10 December 1997. Retrieved 7 October 2007.

17.9 External links

- Annotated bibliography on enriched uranium from the Alsos Digital Library for Nuclear Issues

- Silex Systems Ltd

- Uranium Enrichment, World Nuclear Association

- Overview and history of U.S. HEU production

- News Resource on Uranium Enrichment

- Nuclear Chemistry-Uranium Enrichment

- A busy year for SWU (a 2008 review of the commercial enrichment marketplace), Nuclear Engineering International, 1 September 2008

- *Uranium Enrichment and Nuclear Weapon Proliferation*, by Allan S. Krass, Peter Boskma, Boelie Elzen and Wim A. Smit, 296 pp., published for SIPRI by Taylor and Francis Ltd, London, 1983

- Poliakoff, Martyn (2009). "How do you enrich Uranium?". *The Periodic Table of Videos*. University of Nottingham.

- Gilinsky, V.; Hoehn, W. (December 1969). "The Military Significance of Small Uranium Enrichment Facilities Fed with Low-Enrichment Uranium (Redacted)". *Defense Technical Information Center*. RAND Corporation.

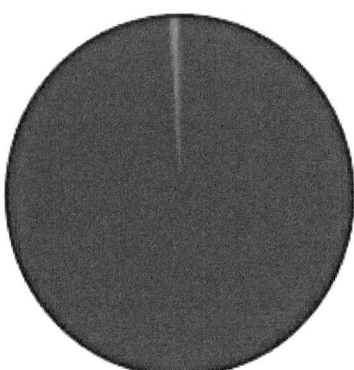

Natural uranium
> 99.2% U-238
0.72% U-235

A drum of yellowcake (a mixture of uranium precipitates)

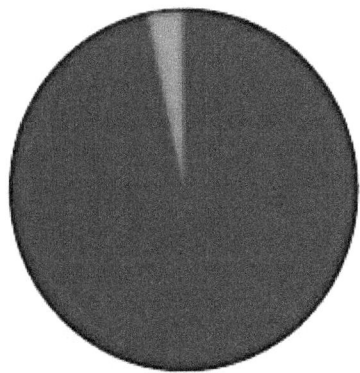

Low-enriched uranium
(reactor grade)
3-4% U-235

A billet of highly enriched uranium metal

Highly enriched uranium
(weapons grade)
90% U-235

A cascade of gas centrifuges at a U.S. enrichment plant

*Proportions of uranium-238 (blue) and uranium-235 (red) found
naturally versus enriched grades*

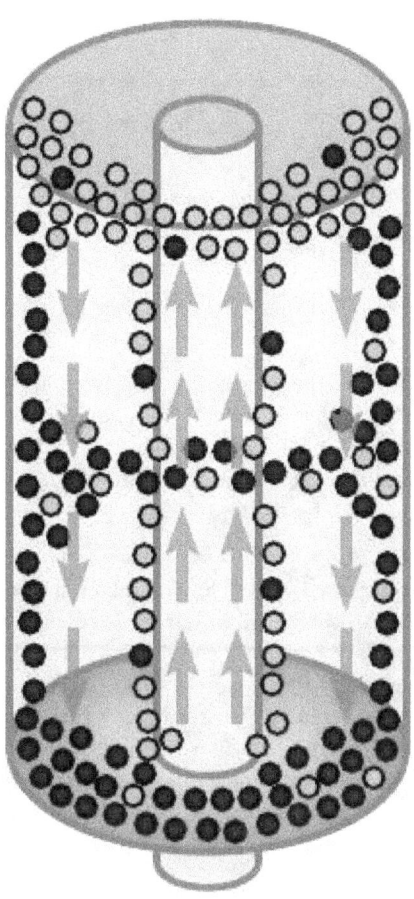

Diagram of the principles of a Zippe-type gas centrifuge with U-238 represented in dark blue and U-235 represented in light blue

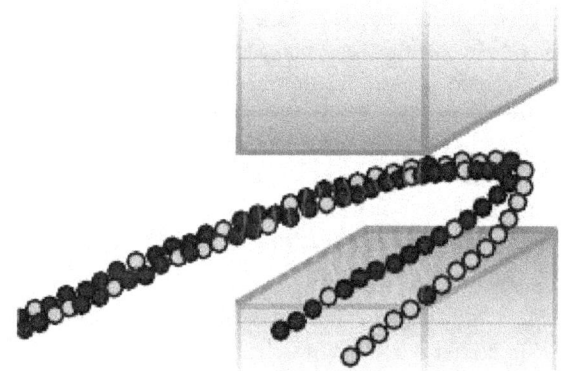

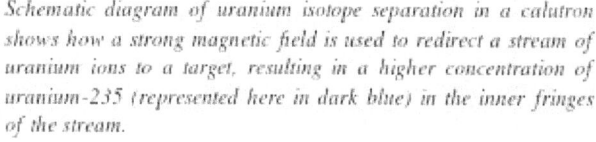

Schematic diagram of uranium isotope separation in a calutron shows how a strong magnetic field is used to redirect a stream of uranium ions to a target, resulting in a higher concentration of uranium-235 (represented here in dark blue) in the inner fringes of the stream.

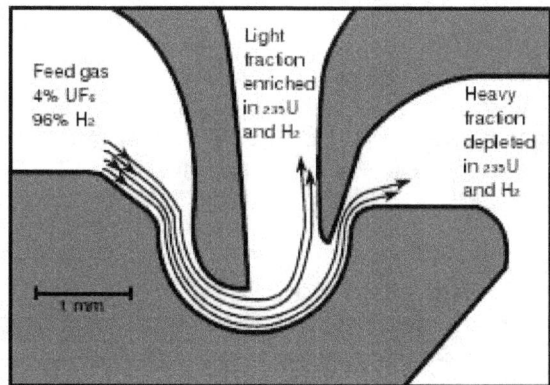

Schematic diagram of an aerodynamic nozzle. Many thousands of these small foils would be combined in an enrichment unit.

Chapter 18

Uranium-238

For the German submarine, see German submarine U-238.

Uranium-238 (^{238}U or **U-238**) is the most common isotope of uranium found in nature and is used as a fuel in nuclear power plants. It can capture a slow neutron and after two beta decays become fissile plutonium-239. ^{238}U is fissionable by fast neutrons, and cannot support a chain reaction because inelastic scattering reduces neutron energy below the range where fast fission of one or more next-generation nuclei is probable. Doppler broadening of U-238's neutron absorption resonances, increasing absorption as fuel temperature increases, is also an essential negative feedback mechanism for reactor control.

Around 99.284% of natural uranium is uranium-238, which has a half-life of 1.41×10^{17} seconds (4.468×10^{9} years, or 4.468 billion years).[1] Depleted uranium has an even higher concentration of the ^{238}U isotope, and even low-enriched uranium (LEU), while having a higher proportion of the uranium-235 isotope (in comparison to depleted uranium), is still mostly ^{238}U. Reprocessed uranium is also mainly ^{238}U, with about as much uranium-235 as natural uranium, a comparable proportion of uranium-236, and much smaller amounts of other isotopes of uranium such as uranium-234, uranium-233, and uranium-232.[2]

18.1 Nuclear energy applications

In a fission nuclear reactor, uranium-238 can be used to generate ^{239}Pu, which itself can be used in a nuclear weapon or as a nuclear-reactor fuel supply. In a typical nuclear reactor, up to one-third of the generated power does come from the fission of ^{239}Pu, which is not supplied as a fuel to the reactor, but rather, produced from ^{238}U.

18.1.1 Breeder reactors

^{238}U is not usable directly as nuclear fuel, though it can produce energy via "fast" fission. In this process, a neutron that has a kinetic energy in excess of 1 MeV can cause the nucleus of ^{238}U to split in two. Depending on design, this process can contribute some one to ten percent of all fission reactions in a reactor, but too few of the about 1.7 neutrons produced in each fission have enough speed to continue a chain reaction.

^{238}U can be used as a source material for creating plutonium-239, which can in turn be used as nuclear fuel. Breeder reactors carry out such a process of transmutation to convert the fertile isotope ^{238}U into fissile Pu-239. It has been estimated that there is anywhere from 10,000 to five billion years worth of ^{238}U for use in these power plants.[3] Breeder technology has been used in several experimental nuclear reactors.[4]

By December 2005, the only breeder reactor producing power was the 600-megawatt BN-600 reactor at the Beloyarsk Nuclear Power Station in Russia. Russia has planned to build another unit, BN-800, at the Beloyarsk nuclear power plant. Also, Japan's Monju breeder reactor is planned to be started, having been shut down since 1995, and both China and India have announced plans to build nuclear breeder reactors.

The breeder reactor as its name implies creates even larger quantities of Pu-239 than the fission nuclear reactor.

The Clean And Environmentally Safe Advanced Reactor (CAESAR), a nuclear reactor concept that would use steam as a moderator to control delayed neutrons, will potentially be able to burn ^{238}U as fuel once the reactor is started with LEU fuel. This design is still in the early stages of development.

18.1.2 Radiation shielding

^{238}U is also used as a radiation shield – its alpha radiation is easily stopped by the non-radioactive casing of the shielding and the uranium's high atomic weight and high number of electrons are highly effective in absorbing gamma rays and x-rays. It is not as effective as ordinary water for stopping fast neutrons. Both metallic depleted uranium and depleted uranium dioxide are used for radiation shielding. Uranium is about five times better as a gamma ray shield than lead, so a shield with the same effectiveness can be packed into a thinner layer.

DUCRETE, a concrete made with uranium dioxide aggregate instead of gravel, is being investigated as a material for dry cask storage systems to store radioactive waste.

18.1.3 Downblending

The opposite of enriching is downblending. Surplus highly enriched uranium can be downblended with depleted uranium or natural uranium to turn it into low enriched uranium suitable for use in commercial nuclear fuel.

^{238}U from depleted uranium and natural uranium is also used with recycled Pu-239 from nuclear weapons stockpiles for making mixed oxide fuel (MOX), which is now being redirected to become fuel for nuclear reactors. This dilution, also called downblending, means that any nation or group that acquired the finished fuel would have to repeat the very expensive and complex chemical separation of uranium and plutonium process before assembling a weapon.

18.2 Nuclear weapons

Most modern nuclear weapons utilize ^{238}U as a "tamper" material (see nuclear weapon design). A tamper which surrounds a fissile core works to reflect neutrons and to add inertia to the compression of the Pu-239 charge. As such, it increases the efficiency of the weapon and reduces the critical mass required. In the case of a thermonuclear weapon ^{238}U can be used to encase the fusion fuel, the high flux of very energetic neutrons from the resulting fusion reaction causes ^{238}U nuclei to split and adds more energy to the "yield" of the weapon. Such weapons are referred to as *fission-fusion-fission* weapons after the three consecutive stages of the explosion. An example of such a weapon is Castle Bravo.

The larger portion of the total explosive yield in this design comes from the final fission stage fueled by ^{238}U, producing enormous amounts of radioactive fission products. For example, an estimated 77% of the 10.4-megaton yield of the Ivy Mike thermonuclear test in 1952 came from fast fission of the depleted uranium tamper. Because depleted uranium has no critical mass, it can be added to thermonuclear bombs in almost unlimited quantity. The Soviet Union's test of the "Tsar Bomba" in 1961 produced "only" 50 megatons of explosive power, over 90% of which came from fusion, because the ^{238}U final stage had been replaced with lead. Had ^{238}U been used instead, the yield of the "Tsar

Bomba" could have been well-above 100 megatons, and it would have produced nuclear fallout equivalent to one third of the global total that had been produced up to that time.

18.3 Radioactivity and decay

Alpha-decay of Uranium accompanies with low energy gamma-radiation – 49.5 keV [5]

^{238}U radiates alpha-particles and decays (by way of thorium−234 and protactinium−234) into uranium-234. ^{234}U has a half-life of 245,500 years. The relation between ^{238}U and ^{234}U gives an indication of the age of sediments that are between 100,000 years and 1,200,000 years in age. [6]

The Voyager spacecrafts carry small amounts of initially pure ^{238}U on the covers of their golden records to facilitate dating in the same manner.

^{238}U occasionally decays by spontaneous fission or double beta decay with probabilities of 5×10^{-7} and 2×10^{-12} per alpha decay, respectively. [7]

18.4 Radium series (or uranium series)

The 4n+2 chain of ^{238}U is commonly called the "radium series" (sometimes "uranium series"). Beginning with naturally occurring uranium−238, this series includes the following elements: astatine, bismuth, lead, polonium, protactinium, radium, radon, thallium, and thorium. All are present, at least transiently, in any uranium-containing sample, whether metal, compound, or mineral.

The mean lifetime of ^{238}U is 1.41×10^{17} seconds divided by 0.693 (or multiplied by 1.443), i.e. ca. 2×10^{17} seconds, so 1 mole of ^{238}U emits 3×10^{6} alpha particles per second, producing the same number of thorium-234 (Th-234) atoms. In a closed system an equilibrium would be reached, with all amounts except for lead-206 and ^{238}U in fixed ratios, in slowly decreasing amounts. The amount of Pb-206 will increase accordingly while that of ^{238}U decreases; all steps in the decay chain have this same rate of 3×10^{6} decayed particles per second per mole ^{238}U.

Thorium-234 has a mean lifetime of 3×10^{6} seconds, so there is equilibrium if one mole of ^{238}U contains 9×10^{12} atoms of thorium-234, which is 1.5×10^{-11} mole (the ratio of the two half-lives). Similarly, in an equilibrium in a closed system the amount of each decay product, except the end product lead, is proportional to its half-life.

As already touched upon above, when starting with pure

^{238}U, within a human timescale the equilibrium applies for the first three steps in the decay chain only. Thus, for one mole of ^{238}U, 3×10^{6} times per second one alpha and two beta particles and gamma ray are produced, together 6.7 MeV, a rate of 3 μW. Extrapolated over 2×10^{17} seconds this is 600 gigajoules, the total energy released in the first three steps in the decay chain.

18.5   See also

- Depleted uranium

18.6   References

[1] Mcclain, D.E.; A.C. Miller; J.F. Kalinich (December 20, 2007). "Status of Health Concerns about Military Use of Depleted Uranium and Surrogate Metals in Armor-Penetrating Munitions" (pdf). NATO. Retrieved November 14, 2010.

[2] Nuclear France: Materials and sites. "Uranium from reprocessing".

[3] Facts from Cohen. Formal.stanford.edu (2007-01-26). Retrieved on 2010-10-24.

[4] Advanced Nuclear Power Reactors | Generation III+ Nuclear Reactors. World-nuclear.org. Retrieved on 2010-10-24.

[5] Komura, K.; Yamamoto, M.; Ueno, K. (1990). "Abundance of low-energy gamma rays in the decay of 238U, 234U, 230Th, 227Ac, 226Ra and 214Pb". Nuclear Instruments and Methods in Physics Research Section A. 295 (3): 461–465. Bibcode:1990NIMPA.295..461K. doi:10.1016/0168-9002(90)90727-N.

[6] Encyclopædia Britannica (14 November 2007). "uranium-234–uranium-238 dating".

[7] Table of Isotopes (1998). nucleardata.nuclear.lu.se

18.7   External links

- NLM Hazardous Substances Databank – Uranium, Radioactive
- Simulation of U238 using the Monte Carlo method

Chapter 19

Plutonium

This article is about the radioactive element. For other uses, see Plutonium (disambiguation).

Plutonium is a transuranic radioactive chemical element with symbol **Pu** and atomic number 94. It is an actinide metal of silvery-gray appearance that tarnishes when exposed to air, and forms a dull coating when oxidized. The element normally exhibits six allotropes and four oxidation states. It reacts with carbon, halogens, nitrogen, silicon and hydrogen. When exposed to moist air, it forms oxides and hydrides that can expand the sample up to 70% in volume, which in turn flake off as a powder that is pyrophoric. It is radioactive and can accumulate in bones, which makes the handling of plutonium dangerous.

Plutonium was first produced and isolated on December 14, 1940 by Dr. Glenn T. Seaborg, Joseph W. Kennedy, Edwin M. McMillan, and Arthur C. Wahl by deuteron bombardment of uranium-238 in the 60-inch cyclotron at the University of California, Berkeley. They first synthesized neptunium-238 (half-life 2.1 days) which subsequently beta-decayed to form a new heavier element with atomic number 94 and atomic weight 238 (half-life 87.7 years). Uranium had been named after the planet Uranus and neptunium after the planet Neptune, and so element 94 was named after Pluto, which at the time was considered to be a planet as well. Wartime secrecy prevented them from announcing the discovery until 1948. Plutonium is the heaviest element to occur in nature as trace quantities arising similarly from the neutron capture of natural uranium-238. Plutonium is much more common on Earth since 1945 as a product of neutron capture and beta decay, where some of the neutrons released by the fission process convert uranium-238 nuclei into plutonium-239.

Both plutonium-239 and plutonium-241 are fissile, meaning that they can sustain a nuclear chain reaction, leading to applications in nuclear weapons and nuclear reactors. Plutonium-240 exhibits a high rate of spontaneous fission, raising the neutron flux of any sample containing it. The presence of plutonium-240 limits a plutonium sample's usability for weapons or its quality as reactor fuel, and the per-centage of plutonium-240 determines its grade (weapons-grade, fuel-grade, or reactor-grade). Plutonium-238 has a half-life of 88 years and emits alpha particles. It is a heat source in radioisotope thermoelectric generators, which are used to power some spacecraft. Plutonium isotopes are expensive and inconvenient to separate, so particular isotopes are usually manufactured in specialized reactors.

Producing plutonium in useful quantities for the first time was a major part of the Manhattan Project during World War II that developed the first atomic bombs. The Fat Man bombs used in the Trinity nuclear test in July 1945, and in the bombing of Nagasaki in August 1945, had plutonium cores. Human radiation experiments studying plutonium were conducted without informed consent, and several criticality accidents, some lethal, occurred after the war. Disposal of plutonium waste from nuclear power plants and dismantled nuclear weapons built during the Cold War is a nuclear-proliferation and environmental concern. Other sources of plutonium in the environment are fallout from numerous above-ground nuclear tests, now banned.

19.1 Characteristics

19.1.1 Physical properties

Plutonium, like most metals, has a bright silvery appearance at first, much like nickel, but it oxidizes very quickly to a dull gray, although yellow and olive green are also reported.[2][3] At room temperature plutonium is in its α (*alpha*) form. This, the most common structural form of the element (allotrope), is about as hard and brittle as gray cast iron unless it is alloyed with other metals to make it soft and ductile. Unlike most metals, it is not a good conductor of heat or electricity. It has a low melting point (640 °C) and an unusually high boiling point (3,228 °C).[2]

Alpha decay, the release of a high-energy helium nucleus, is the most common form of radioactive decay for plutonium.[4] A 5 kg mass of ^{239}Pu contains about 12.5×10^{24} atoms. With a half-life of 24,100 years, about 11.5×10^{12}

of its atoms decay each second by emitting a 5.157 MeV alpha particle. This amounts to 9.68 watts of power. Heat produced by the deceleration of these alpha particles makes it warm to the touch.[5][6]

Resistivity is a measure of how strongly a material opposes the flow of electric current. The resistivity of plutonium at room temperature is very high for a metal, and it gets even higher with lower temperatures, which is unusual for metals.[7] This trend continues down to 100 K, below which resistivity rapidly decreases for fresh samples.[7] Resistivity then begins to increase with time at around 20 K due to radiation damage, with the rate dictated by the isotopic composition of the sample.[7]

Because of self-irradiation, a sample of plutonium fatigues throughout its crystal structure, meaning the ordered arrangement of its atoms becomes disrupted by radiation with time.[8] Self-irradiation can also lead to annealing which counteracts some of the fatigue effects as temperature increases above 100 K.[9]

Unlike most materials, plutonium *increases* in density when it melts, by 2.5%, but the liquid metal exhibits a linear decrease in density with temperature.[7] Near the melting point, the liquid plutonium has also very high viscosity and surface tension as compared to other metals.[8]

19.1.2 Allotropes

Main article: Allotropes of plutonium

Plutonium normally has six allotropes and forms a sev-

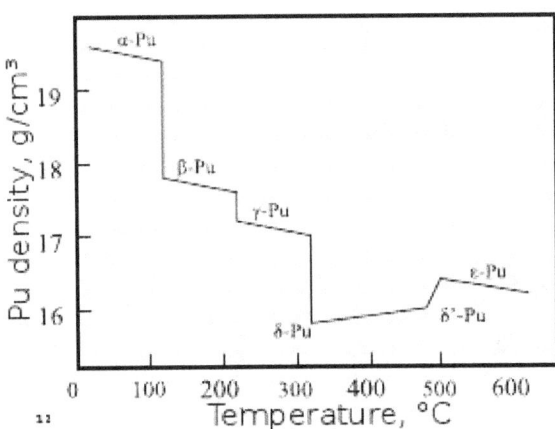

*Plutonium has six allotropes at ambient pressure: **alpha** (α), **beta** (β), **gamma** (γ), **delta** (δ), **delta prime** (δ'), & **epsilon** (ε) [10]*

enth (zeta, ζ) at high temperature within a limited pressure range.[10] These allotropes, which are different structural modifications or forms of an element, have very similar internal energies but significantly varying densities and crystal structures. This makes plutonium very sensitive to

changes in temperature, pressure, or chemistry, and allows for dramatic volume changes following phase transitions from one allotropic form to another.[8] The densities of the different allotropes vary from 16.00 g/cm³ to 19.86 g/cm³.[11]

The presence of these many allotropes makes machining plutonium very difficult, as it changes state very readily. For example, the α form exists at room temperature in unalloyed plutonium. It has machining characteristics similar to cast iron but changes to the plastic and malleable β (*beta*) form at slightly higher temperatures.[12] The reasons for the complicated phase diagram are not entirely understood. The α form has a low-symmetry monoclinic structure, hence its brittleness, strength, compressibility, and poor thermal conductivity.[10]

Plutonium in the δ (*delta*) form normally exists in the 310 °C to 452 °C range but is stable at room temperature when alloyed with a small percentage of gallium, aluminium, or cerium, enhancing workability and allowing it to be welded.[12] The δ form has more typical metallic character, and is roughly as strong and malleable as aluminium.[10] In fission weapons, the explosive shock waves used to compress a plutonium core will also cause a transition from the usual δ phase plutonium to the denser α form, significantly helping to achieve supercriticality.[13] The ε phase, the highest temperature solid allotrope, exhibits anomalously high atomic self-diffusion compared to other elements.[8]

19.1.3 Nuclear fission

A ring of weapons-grade 99.96% pure electrorefined plutonium, enough for one bomb core. The ring weighs 5.3 kg, is ca. 11 cm in diameter and its shape helps with criticality safety.

Plutonium is a radioactive actinide metal whose isotope, plutonium-239, is one of the three primary fissile isotopes (uranium-233 and uranium-235 are the other two); plutonium-241 is also highly fissile. To be considered fissile, an isotope's atomic nucleus must be able to break apart or fission when struck by a slow moving neutron and to release enough additional neutrons to sustain the nuclear chain reaction by splitting further nuclei.*[14]

Pure plutonium-239 may have a multiplication factor (k_{eff}) larger than one, which means that if the metal is present in sufficient quantity and with an appropriate geometry (e.g., a sphere of sufficient size), it can form a critical mass.*[15] During fission, a fraction of the nuclear binding energy, which holds a nucleus together, is released as a large amount of electromagnetic and kinetic energy (much of the latter being quickly converted to thermal energy). Fission of a kilogram of plutonium-239 can produce an explosion equivalent to 21,000 tons of TNT (88,000 GJ). It is this energy that makes plutonium-239 useful in nuclear weapons and reactors.*[5]

The presence of the isotope plutonium-240 in a sample limits its nuclear bomb potential, as plutonium-240 has a relatively high spontaneous fission rate (~440 fissions per second per gram—over 1,000 neutrons per second per gram),*[16] raising the background neutron levels and thus increasing the risk of predetonation.*[17] Plutonium is identified as either weapons-grade, fuel-grade, or reactor-grade based on the percentage of plutonium-240 that it contains. Weapons-grade plutonium contains less than 7% plutonium-240. Fuel-grade plutonium contains from 7% to less than 19%, and power reactor-grade contains 19% or more plutonium-240. Supergrade plutonium, with less than 4% of plutonium-240, is used in U.S. Navy weapons stored in proximity to ship and submarine crews, due to its lower radioactivity.*[18] The isotope plutonium-238 is not fissile but can undergo nuclear fission easily with fast neutrons as well as alpha decay.*[5]

19.1.4 Isotopes and nucleosynthesis

Main article: Isotopes of plutonium

Twenty radioactive isotopes of plutonium have been characterized. The longest-lived are plutonium-244, with a half-life of 80.8 million years, plutonium-242, with a half-life of 373,300 years, and plutonium-239, with a half-life of 24,110 years. All of the remaining radioactive isotopes have half-lives that are less than 7,000 years. This element also has eight metastable states, though all have half-lives less than one second.*[4]

The isotopes of plutonium range in mass number from 228

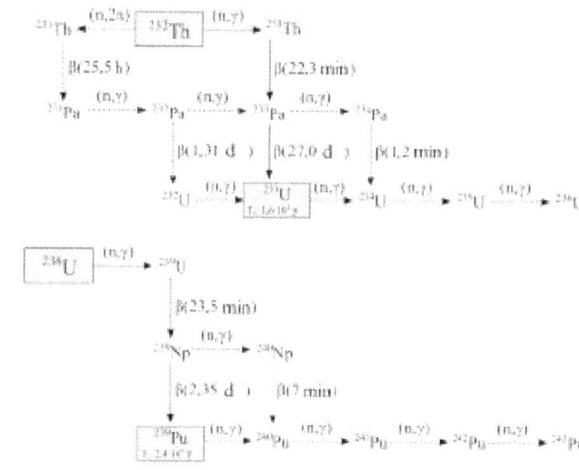

Uranium-plutonium and thorium-uranium chains

to 247. The primary decay modes of isotopes with mass numbers lower than the most stable isotope, plutonium-244, are spontaneous fission and alpha emission, mostly forming uranium (92 protons) and neptunium (93 protons) isotopes as decay products (neglecting the wide range of daughter nuclei created by fission processes). The primary decay mode for isotopes with mass numbers higher than plutonium-244 is beta emission, mostly forming americium (95 protons) isotopes as decay products. Plutonium-241 is the parent isotope of the neptunium decay series, decaying to americium-241 via beta or electron emission.*[4]*[19]

Plutonium-238 and 239 are the most widely synthesized isotopes.*[5] Plutonium-239 is synthesized via the following reaction using uranium (U) and neutrons (n) via beta decay (β^-) with neptunium (Np) as an intermediate:*[20]

Neutrons from the fission of uranium-235 are captured by uranium-238 nuclei to form uranium-239; a beta decay converts a neutron into a proton to form Np-239 (half-life 2.36 days) and another beta decay forms plutonium-239.*[21] Egon Bretscher working on the British Tube Alloys project predicted this reaction theoretically in 1940.*[22]

Plutonium-238 is synthesized by bombarding uranium-238 with deuterons (D, the nuclei of heavy hydrogen) in the following reaction:*[23]

In this process, a deuteron hitting uranium-238 produces two neutrons and neptunium-238, which spontaneously decays by emitting negative beta particles to form plutonium-238.*[24]

19.1.5 Decay heat and fission properties

Plutonium isotopes undergo radioactive decay, which produces decay heat. Different isotopes produce different amounts of heat per mass. The decay heat is usually listed as watt/kilogram, or milliwatt/gram. In larger pieces of plutonium (e.g. a weapon pit) and inadequate heat removal the resulting self-heating may be significant. All isotopes produce weak gamma radiation on decay.

19.1.6 Compounds and chemistry

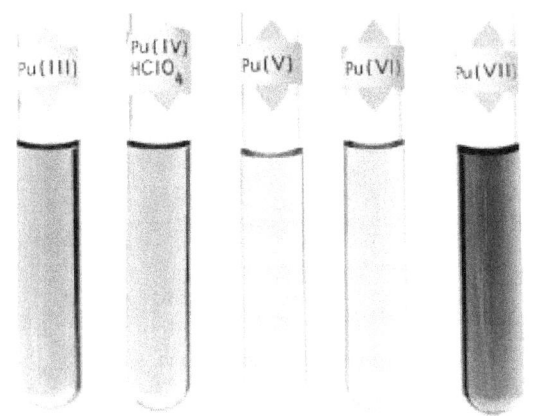

Various oxidation states of plutonium in solution

Plutonium pyrophoricity can cause it to look like a glowing ember under certain conditions.

At room temperature, pure plutonium is silvery in color but gains a tarnish when oxidized.*[26] The element displays four common ionic oxidation states in aqueous solution and one rare one:*[11]

- Pu(III), as Pu*3+ (blue lavender)

- Pu(IV), as Pu*4+ (yellow brown)

- Pu(V), as PuO+
  2 (light pink)*[note 1]

- Pu(VI), as PuO2+
  2 (pink orange)

- Pu(VII), as PuO3−
  5 (green)—the heptavalent ion is rare.

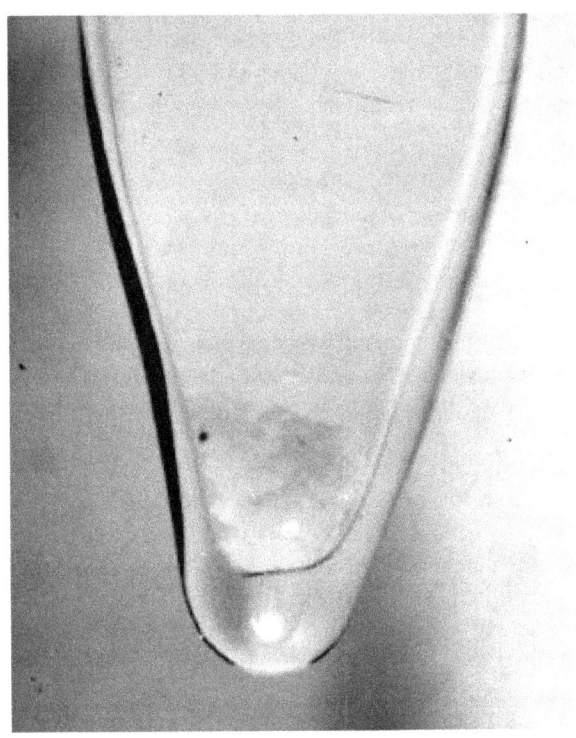

Twenty micrograms of pure plutonium hydroxide

The color shown by plutonium solutions depends on both the oxidation state and the nature of the acid anion.*[28] It is the acid anion that influences the degree of complexing —how atoms connect to a central atom—of the plutonium species.

Metallic plutonium is produced by reacting plutonium tetrafluoride with barium, calcium or lithium at 1200 °C.*[29] It is attacked by acids, oxygen, and steam but not by alkalis and dissolves easily in concentrated hydrochloric, hydroiodic and perchloric acids.*[30] Molten metal must be kept in a vacuum or an inert atmosphere to avoid reaction

with air.*[12] At 135 °C the metal will ignite in air and will explode if placed in carbon tetrachloride.*[31]

Plutonium is a reactive metal. In moist air or moist argon, the metal oxidizes rapidly, producing a mixture of oxides and hydrides.*[2] If the metal is exposed long enough to a limited amount of water vapor, a powdery surface coating of PuO_2 is formed.*[2] Also formed is plutonium hydride but an excess of water vapor forms only PuO_2.*[30]

Plutonium shows enormous, and reversible, reaction rates with pure hydrogen, forming plutonium hydride.*[8] It also reacts readily with oxygen, forming PuO and PuO_2 as well as intermediate oxides; plutonium oxide fills 40% more volume than plutonium metal. The metal reacts with the halogens, giving rise to compounds with the general formula PuX_3 where X can be F, Cl, Br or I and PuF_4 is also seen. The following oxyhalides are observed: PuOCl, PuOBr and PuOI. It will react with carbon to form PuC, nitrogen to form PuN and silicon to form $PuSi_2$.[11][31]

Powders of plutonium, its hydrides and certain oxides like Pu_2O_3 are pyrophoric, meaning they can ignite spontaneously at ambient temperature and are therefore handled in an inert, dry atmosphere of nitrogen or argon. Bulk plutonium ignites only when heated above 400 °C. Pu_2O_3 spontaneously heats up and transforms into PuO_2, which is stable in dry air, but reacts with water vapor when heated.[32]

Crucibles used to contain plutonium need to be able to withstand its strongly reducing properties. Refractory metals such as tantalum and tungsten along with the more stable oxides, borides, carbides, nitrides and silicides can tolerate this. Melting in an electric arc furnace can be used to produce small ingots of the metal without the need for a crucible.[12]

Cerium is used as a chemical simulant of plutonium for development of containment, extraction, and other technologies.[33]

Electronic structure

Plutonium is an element in which the 5f electrons are the transition border between delocalized and localized; it is therefore considered one of the most complex elements.[34] The anomalous behavior of plutonium is caused by its electronic structure. The energy difference between the 6d and 5f subshells is very low. The size of the 5f shell is just enough to allow the electrons to form bonds within the lattice, on the very boundary between localized and bonding behavior. The proximity of energy levels leads to multiple low-energy electron configurations with near equal energy levels. This leads to competing $5f^n7s^2$ and $5f^{n-1}6d^17s^2$ configurations, which causes the complexity of its chemical behavior. The highly directional nature of 5f orbitals is responsible for directional covalent bonds in molecules and complexes of plutonium.[8]

19.1.7 Alloys

Plutonium can form alloys and intermediate compounds with most other metals. Exceptions include lithium, sodium, potassium, rubidium and caesium of the alkali metals; and magnesium, calcium, strontium, and barium of the alkaline earth metals; and europium and ytterbium of the rare earth metals.[30] Partial exceptions include the refractory metals chromium, molybdenum, niobium, tantalum, and tungsten, which are soluble in liquid plutonium, but insoluble or only slightly soluble in solid plutonium.[30] Gallium, aluminium, americium, scandium and cerium can stabilize the δ phase of plutonium for room temperature. Silicon, indium, zinc and zirconium allow formation of metastable δ state when rapidly cooled. High amounts of hafnium, holmium and thallium also allows some retention of the δ phase at room temperature. Neptunium is the only element that can stabilize the α phase at higher temperatures.[8]

Plutonium alloys can be produced by adding a metal to molten plutonium. If the alloying metal is sufficiently reductive, plutonium can be added in the form of oxides or halides. The δ phase plutonium–gallium and plutonium–aluminium alloys are produced by adding plutonium(III) fluoride to molten gallium or aluminium, which has the advantage of avoiding dealing directly with the highly reactive plutonium metal.[35]

- Plutonium–gallium is used for stabilizing the δ phase of plutonium, avoiding the α-phase and α–δ related issues. Its main use is in pits of implosion nuclear weapons.[36]

- **Plutonium–aluminium** is an alternative to the Pu-Ga alloy. It was the original element considered for δ phase stabilization, but its tendency to react with the alpha particles and release neutrons reduces its usability for nuclear weapon pits. Plutonium–aluminium alloy can be also used as a component of nuclear fuel.[37]

- **Plutonium–gallium–cobalt** alloy ($PuCoGa_5$) is an unconventional superconductor, showing superconductivity below 18.5 K, an order of magnitude higher than the highest between heavy fermion systems, and has large critical current.[34][38]

- **Plutonium–zirconium** alloy can be used as nuclear fuel.[39]

- **Plutonium–cerium** and **plutonium–cerium–cobalt** alloys are used as nuclear fuels.[40]

- **Plutonium–uranium**, with about 15–30 mol.% plutonium, can be used as a nuclear fuel for fast breeder reactors. Its pyrophoric nature and high susceptibility to corrosion to the point of self-igniting or disintegrating after exposure to air require alloying with other components. Addition of aluminium, carbon or copper does not improve disintegration rates markedly, zirconium and iron alloys have better corrosion resistance but they disintegrate in several months in air as well. Addition of titanium and/or zirconium significantly increases the melting point of the alloy.[41]

- **Plutonium–uranium–titanium** and **plutonium–uranium–zirconium** were investigated for use as nuclear fuels. The addition of the third element increases corrosion resistance, reduces flammability, and improves ductility, fabricability, strength, and thermal expansion. **Plutonium–uranium–molybdenum** has the best corrosion resistance, forming a protective film of oxides, but titanium and zirconium are preferred for physics reasons.[41]

- **Thorium–uranium–plutonium** was investigated as a nuclear fuel for fast breeder reactors.[41]

19.1.8   Occurrence

Trace amounts of plutonium-238 and plutonium-239 can be found in nature. Small traces of plutonium-239, a few parts per trillion, and its decay products are naturally found in some concentrated ores of uranium,[42] such as the natural nuclear fission reactor in Oklo, Gabon.[43] The ratio of plutonium-239 to uranium at the Cigar Lake Mine uranium deposit ranges from 2.4×10^{-12} to 44×10^{-12}.[44] These trace amounts of ^{239}Pu originate in the following fashion: on rare occasions, ^{238}U undergoes spontaneous fission, and in the process, the nucleus emits one or two free neutrons with some kinetic energy. When one of these neutrons strikes the nucleus of another ^{238}U atom, it is absorbed by the atom, which becomes ^{239}U. With a relatively short half-life, ^{239}U decays to ^{239}Np, which decays into ^{239}Pu.[45][46] Finally, exceedingly small amounts of plutonium-238, attributed to the extremely rare double beta decay of uranium-238, have been found in natural uranium samples.[47]

Due to its relatively long half-life of about 80 million years, it was suggested that plutonium-244 occurs naturally as a primordial nuclide, but early reports of its detection could not be confirmed.[48] However, its long half-life ensured its circulation across the solar system before its extinction,[49] and indeed, ^{244}Pu has not yet been found in matter other than meteorites.[50] The former presence of extinct ^{244}Pu in the early Solar System has been confirmed, since it manifests itself today as an excess of its daughters, either ^{232}Th (from the alpha decay pathway) or xenon isotopes (from its spontaneous fission). The latter are generally more useful, because the chemistries of thorium and plutonium are rather similar (both are predominantly tetravalent) and hence an excess of thorium would not be strong evidence that some of it was formed as a plutonium daughter.[51]

Minute traces of plutonium are usually found in the human body due to the 550 atmospheric and underwater nuclear tests that have been carried out, and to a small number of major nuclear accidents. Most atmospheric and underwa-

ter nuclear testing was stopped by the Limited Test Ban Treaty in 1963, which was signed and ratified by the United States, the United Kingdom, the Soviet Union, and other nations. Continued atmospheric nuclear weapons testing since 1963 by non-treaty nations included those by China (atomic bomb test above the Gobi Desert in 1964, hydrogen bomb test in 1967, and follow-on tests), and France (tests as recently as the 1990s). Because it is deliberately manufactured for nuclear weapons and nuclear reactors, plutonium-239 is the most abundant isotope of plutonium by far.[31]

19.2   History

19.2.1   Discovery

Enrico Fermi and a team of scientists at the University of Rome reported that they had discovered element 94 in 1934.[52] Fermi called the element *hesperium* and mentioned it in his Nobel Lecture in 1938.[53] The sample was actually a mixture of barium, krypton, and other elements, but this was not known at the time.[54] Nuclear fission was discovered in Germany in 1939 by Fritz Strassmann and Otto Hahn. The mechanism of fission was then theoretically explained by Lise Meitner and Otto Frisch.[55]

Glenn T. Seaborg and his team at Berkeley were the first to produce plutonium.

Plutonium (specifically, plutonium-238) was first produced

and isolated on December 14, 1940, and chemically identified on February 23, 1941, by Glenn T. Seaborg, Edwin McMillan, Joseph W. Kennedy, and Arthur Wahl by deuteron bombardment of uranium in the 60-inch (150 cm) cyclotron at the Berkeley Radiation Laboratory at the University of California, Berkeley.[56][57] In the 1940 experiment, neptunium-238 was created directly by the bombardment but decayed by beta emission with a half-life of a little over two days, which indicated the formation of element 94.[31]

A paper documenting the discovery was prepared by the team and sent to the journal *Physical Review* in March 1941,[31] but publication was delayed until a year after the end of World War II due to security concerns.[58] At the Cavendish Laboratory in Cambridge, Egon Bretscher and Norman Feather realized that a slow neutron reactor fuelled with uranium would theoretically produce substantial amounts of plutonium-239 as a by-product. They calculated that element 94 would be fissile, and had the added advantage of being chemically different from uranium, and could easily be separated from it.[22]

McMillan had recently named the first transuranic element neptunium after the planet Neptune, and suggested that element 94, being the next element in the series, be named for what was then considered the next planet, Pluto.[5][note 2] Nicholas Kemmer of the Cambridge team independently proposed the same name, based on the same reasoning as the Berkeley team.[22] Seaborg originally considered the name "plutium", but later thought that it did not sound as good as "plutonium".[60] He chose the letters "Pu" as a joke, which passed without notice into the periodic table.[note 3] Alternative names considered by Seaborg and others were "ultimium" or "extremium" because of the erroneous belief that they had found the last possible element on the periodic table.[62]

19.2.2 Early research

The chemistry of plutonium was found to resemble uranium after a few months of initial study.[31] Early research was continued at the secret Metallurgical Laboratory of the University of Chicago. On August 20, 1942, a trace quantity of this element was isolated and measured for the first time. About 50 micrograms of plutonium-239 combined with uranium and fission products was produced and only about 1 microgram was isolated.[42][63] This procedure enabled chemists to determine the new element's atomic weight.[64][note 4] On December 2, 1942, on a racket court under the west grandstand at the University of Chicago's Stagg Field, researchers headed by Enrico Fermi achieved the first self-sustaining chain reaction in a graphite and uranium pile known as CP-1. Using theoretical in-

The dwarf planet Pluto, after which plutonium is named

formation garnered from the operation of CP-1, DuPont constructed an air-cooled experimental production reactor, known as X-10, and a pilot chemical separation facility at Oak Ridge. The separation facility, using methods developed by Glenn T. Seaborg and a team of researchers at the Met Lab, removed plutonium from uranium irradiated in the X-10 reactor. Information from CP-1 was also useful to Met Lab scientists designing the water-cooled plutonium production reactors for Hanford. Construction at the site began in mid-1943.[65]

In November 1943 some plutonium trifluoride was reduced to create the first sample of plutonium metal: a few micrograms of metallic beads.[42] Enough plutonium was produced to make it the first synthetically made element to be visible with the unaided eye.[66]

The nuclear properties of plutonium-239 were also studied; researchers found that when it is hit by a neutron it breaks apart (fissions) by releasing more neutrons and energy. These neutrons can hit other atoms of plutonium-239 and so on in an exponentially fast chain reaction. This can result in an explosion large enough to destroy a city if enough of the isotope is concentrated to form a critical mass.[31]

During the early stages of research, animals were used to study the effects of radioactive substances on health. These studies began in 1944 at the University of California at Berkeley's Radiation Laboratory and were conducted by Joseph G. Hamilton. Hamilton was looking to answer questions about how plutonium would vary in the body depending on exposure mode (oral ingestion, inhalation, absorption through skin), retention rates, and how plutonium would be fixed in tissues and distributed among the various organs. Hamilton started administering soluble microgram portions of plutonium-239 compounds to rats using different valence states and different methods of introducing the plutonium (oral, intravenous, etc.). Eventually, the lab at Chicago also conducted its own plutonium injection experiments using different animals such as mice, rabbits, fish, and even dogs. The results of the studies at Berkeley and Chicago showed that plutonium's physiological behavior differed significantly from that of radium. The most alarming result was that there was significant deposition of plutonium in the liver and in the "actively metabolizing" portion of bone. Furthermore, the rate of plutonium elimination in the excreta differed between species of animals by as much as a factor of five. Such variation made it extremely difficult to estimate what the rate would be for human beings.*[67]

The Hanford site represents two-thirds of the nation's high-level radioactive waste by volume. Nuclear reactors line the riverbank at the Hanford Site along the Columbia River in January 1960.

19.2.3 Production during the Manhattan Project

During World War II the U.S. government established the Manhattan Project, which was tasked with developing an atomic bomb. The three primary research and production sites of the project were the plutonium production facility at what is now the Hanford Site, the uranium enrichment facilities at Oak Ridge, Tennessee, and the weapons research and design laboratory, now known as Los Alamos National Laboratory.*[68]

The Hanford B Reactor face under construction —the first plutonium-production reactor

The first production reactor that made plutonium-239 was the X-10 Graphite Reactor. It went online in 1943 and was built at a facility in Oak Ridge that later became the Oak Ridge National Laboratory.*[31]*[note 5]

In January 1944, workers laid the foundations for the first

chemical separation building, T Plant located in 200-West. Both the T Plant and its sister facility in 200-West, the U Plant, were completed by October. (U Plant was used only for training during the Manhattan Project.) The separation building in 200-East, B Plant, was completed in February 1945. The second facility planned for 200-East was canceled. Nicknamed Queen Marys by the workers who built them, the separation buildings were awesome canyon-like structures 800 feet long, 65 feet wide, and 80 feet high containing forty process pools. The interior had an eerie quality as operators behind seven feet of concrete shielding manipulated remote control equipment by looking through television monitors and periscopes from an upper gallery. Even with massive concrete lids on the process pools, precautions against radiation exposure were necessary and influenced all aspects of plant design.*[65]

On April 5, 1944, Emilio Segrè at Los Alamos received the first sample of reactor-produced plutonium from Oak Ridge.*[70] Within ten days, he discovered that reactor-bred plutonium had a higher concentration of the isotope plutonium-240 than cyclotron-produced plutonium. Plutonium-240 has a high spontaneous fission rate, raising the overall background neutron level of the plutonium sample.*[71] The original gun-type plutonium weapon, code-named "Thin Man", had to be abandoned as a result—the increased number of spontaneous neutrons meant that nuclear pre-detonation (fizzle) was likely.*[72]

The entire plutonium weapon design effort at Los Alamos was soon changed to the more complicated implosion device, code-named "Fat Man". With an implosion weapon, plutonium is compressed to a high density with explosive lenses —a technically more daunting task than the simple gun-type design, but necessary to use plutonium for weapons purposes. Enriched uranium, by contrast, can be used with either method.[72]

Construction of the Hanford B Reactor, the first industrial-sized nuclear reactor for the purposes of material production, was completed in March 1945. B Reactor produced the fissile material for the plutonium weapons used during World War II.[note 6] B, D and F were the initial reactors built at Hanford, and six additional plutonium-producing reactors were built later at the site.[75]

By the end of January 1945, the highly purified plutonium underwent further concentration in the completed chemical isolation building, where remaining impurities were removed successfully. Los Alamos received its first plutonium from Hanford on February 2. While it was still by no means clear that enough plutonium could be produced for use in bombs by the war's end, Hanford was by early 1945 in operation. Only two years had passed since Col. Franklin Matthias first set up his temporary headquarters on the banks of the Columbia River.[65]

According to Kate Brown, the plutonium production plants at Hanford and Mayak in Russia, over a period of four decades, "both released more than 200 million curies of radioactive isotopes into the surrounding environment — twice the amount expelled in the Chernobyl disaster in each instance".[76] Most of this radioactive contamination over the years were part of normal operations, but unforeseen accidents did occur and plant management kept this secret, as the pollution continued unabated.[76]

In 2004, a safe was discovered during excavations of a burial trench at the Hanford nuclear site. Inside the safe were various items, including a large glass bottle containing a whitish slurry which was subsequently identified as the oldest sample of weapons-grade plutonium known to exist. Isotope analysis by Pacific Northwest National Laboratory indicated that the plutonium in the bottle was manufactured in the X-10 Graphite Reactor at Oak Ridge during 1944.[77][78][79]

19.2.4 Trinity and Fat Man atomic bombs

The first atomic bomb test, codenamed "Trinity" and detonated on July 16, 1945, near Alamogordo, New Mexico, used plutonium as its fissile material.[42] The implosion design of "the gadget", as the Trinity device was codenamed, used conventional explosive lenses to compress a sphere of plutonium into a supercritical mass, which was simultaneously showered with neutrons from the "Urchin", an initiator made of polonium and beryllium (neutron source: (α, n) reaction).[31] Together, these ensured a runaway chain reaction and explosion. The overall weapon weighed over 4 tonnes, although it used just 6.2 kg of plutonium in its core.[80] About 20% of the plutonium used

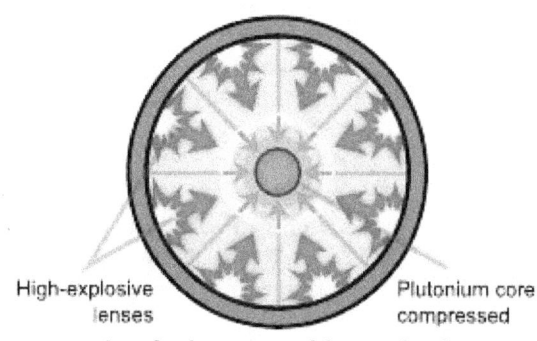

Gun-type assembly method

Conventional chemical explosive   Sub-critical pieces of uranium-235 combined

High-explosive lenses   Plutonium core compressed

Implosion assembly method

Because of the presence of plutonium-240 in reactor-bred plutonium, the implosion design was developed for the "Fat Man" and "Trinity" weapons

in the Trinity weapon underwent fission, resulting in an explosion with an energy equivalent to approximately 20,000 tons of TNT.[81][note 7]

An identical design was used in the "Fat Man" atomic bomb dropped on Nagasaki, Japan, on August 9, 1945, killing 35,000–40,000 people (most of whom were industrial workers) and destroying 68%–80% of war production at Nagasaki.[83] Only after the announcement of the first atomic bombs was the existence and name of plutonium made known to the public by the Manhattan Project's Smyth Report.[84]

19.2.5 Cold War use and waste

See also: Reactor-grade plutonium

Large stockpiles of weapons-grade plutonium were built up by both the Soviet Union and the United States during the Cold War. The U.S. reactors at Hanford and the Savannah River Site in South Carolina produced 103 tonnes,[85] and an estimated 170 tonnes of military-grade plutonium was produced in the USSR.[86][note 8] Each year about 20 tonnes of the element is still produced as a by-product of the nuclear power industry.[11] As much as 1000 tonnes of plutonium may be in storage with more than 200 tonnes of that either inside or extracted from nuclear weapons.[31]

SIPRI estimated the world plutonium stockpile in 2007 as about 500 tonnes, divided equally between weapon and civilian stocks.*[88]

Radioactive contamination at the Rocky Flats Plant primarily resulted from two major plutonium fires in 1957 and 1969. Much lower concentrations of radioactive isotopes were released throughout the operational life of the plant from 1952 to 1992. Prevailing winds from the plant carried airborne contamination south and east, into populated areas northwest of Denver. The contamination of the Denver area by plutonium from the fires and other sources was not publicly reported until the 1970s. According to a 1972 study coauthored by Edward Martell, "In the more densely populated areas of Denver, the Pu contamination level in surface soils is several times fallout", and the plutonium contamination "just east of the Rocky Flats plant ranges up to hundreds of times that from nuclear tests".*[89] As noted by Carl Johnson in Ambio, "Exposures of a large population in the Denver area to plutonium and other radionuclides in the exhaust plumes from the plant date back to 1953."*[90] Weapons production at the Rocky Flats plant was halted after a combined FBI and EPA raid in 1989 and years of protests. The plant has since been shut down, with its buildings demolished and completely removed from the site.*[91]

In the U.S., some plutonium extracted from dismantled nuclear weapons is melted to form glass logs of plutonium oxide that weigh two tonnes.*[31] The glass is made of borosilicates mixed with cadmium and gadolinium.*[note 9] These logs are planned to be encased in stainless steel and stored as much as 4 km (2 mi) underground in bore holes that will be back-filled with concrete.*[31] The U.S. planned to store plutonium in this way at the Yucca Mountain nuclear waste repository, which is about 100 miles (160 km) north-east of Las Vegas, Nevada.*[92]

On March 5, 2009, Energy Secretary Steven Chu told a Senate hearing "the Yucca Mountain site no longer was viewed as an option for storing reactor waste".*[93] Starting in 1999, military-generated nuclear waste is being entombed at the Waste Isolation Pilot Plant in New Mexico.

In a Presidential Memorandum dated January 29, 2010, President Obama established the Blue Ribbon Commission on America's Nuclear Future.*[94] In their final report the Commission put forth recommendations for developing a comprehensive strategy to pursue, including:*[95]

> "Recommendation #1: The United States should undertake an integrated nuclear waste management program that leads to the timely development of one or more permanent deep geological facilities for the safe disposal of spent fuel and high-level nuclear waste".*[95]

19.2.6   Medical experimentation

See also: Human radiation experiments and Albert Stevens

During and after the end of World War II, scientists working on the Manhattan Project and other nuclear weapons research projects conducted studies of the effects of plutonium on laboratory animals and human subjects.*[96] Animal studies found that a few milligrams of plutonium per kilogram of tissue is a lethal dose.*[97]

In the case of human subjects, this involved injecting solutions containing (typically) five micrograms of plutonium into hospital patients thought to be either terminally ill, or to have a life expectancy of less than ten years either due to age or chronic disease condition.*[96] This was reduced to one microgram in July 1945 after animal studies found that the way plutonium distributed itself in bones was more dangerous than radium.*[97] Most of the subjects, Eileen Welsome says, were poor, powerless, and sick.*[98]

From 1945 to 1947, eighteen human test subjects were injected with plutonium without informed consent. The tests were used to create diagnostic tools to determine the uptake of plutonium in the body in order to develop safety standards for working with plutonium.*[96] Ebb Cade was an unwilling participant in medical experiments that involved injection of 4.7 micrograms of Plutonium on 10 April 1945 at Oak Ridge, Tennessee.*[99]*[100] This experiment was under the supervision of Harold Hodge.*[101] Other experiments directed by the United States Atomic Energy Commission and the Manhattan Project continued into the 1970s. *The Plutonium Files* chronicles the lives of the subjects of the secret program by naming each person involved and discussing the ethical and medical research conducted in secret by the scientists and doctors. The episode is now considered to be a serious breach of medical ethics and of the Hippocratic Oath.*[102]

The government covered up most of these radiation mishaps until 1993, when President Bill Clinton ordered a change of policy and federal agencies then made available relevant records. The resulting investigation was undertaken by the president's Advisory Committee on Human Radiation Experiments, and it uncovered much of the material about plutonium research on humans. The committee issued a controversial 1995 report which said that "wrongs were committed" but it did not condemn those who perpetrated them.*[98]

19.3   Applications

The atomic bomb dropped on Nagasaki, Japan in 1945 had a plutonium core.

19.3.1 Explosives

The isotope plutonium-239 is a key fissile component in nuclear weapons, due to its ease of fission and availability. Encasing the bomb's plutonium pit in a tamper (an optional layer of dense material) decreases the amount of plutonium needed to reach critical mass by reflecting escaping neutrons back into the plutonium core. This reduces the amount of plutonium needed to reach criticality from 16 kg to 10 kg, which is a sphere with a diameter of about 10 centimeters (4 in).*[103] This critical mass is about a third of that for uranium-235.*[5]

The Fat Man plutonium bombs used explosive compression of plutonium to obtain significantly higher densities than normal, combined with a central neutron source to begin the reaction and increase efficiency. Thus only 6.2 kg of plutonium was needed for an explosive yield equivalent to 20 kilotons of TNT.*[81]*[104] Hypothetically, as little as 4 kg of plutonium—and maybe even less—could be used to make a single atomic bomb using very sophisticated assembly designs.*[104]

19.3.2 Mixed oxide fuel

Main article: Nuclear reprocessing

Spent nuclear fuel from normal light water reactors contains plutonium, but it is a mixture of plutonium-242, 240, 239 and 238. The mixture is not sufficiently enriched for efficient nuclear weapons, but can be used once as MOX fuel.*[105] Accidental neutron capture causes the amount of plutonium-242 and 240 to grow each time the plutonium is irradiated in a reactor with low-speed "thermal" neutrons, so that after the second cycle, the plutonium can only be consumed by fast neutron reactors. If fast neutron reactors are not available (the normal case), excess plutonium is usually discarded, and forms the longest-lived component of nuclear waste. The desire to consume this plutonium and other transuranic fuels and reduce the radiotoxicity of the waste is the usual reason nuclear engineers give to make fast neutron reactors.*[106]

The most common chemical process, PUREX (*P*lutonium–*UR*anium *EX*traction) reprocesses spent nuclear fuel to extract plutonium and uranium which can be used to form a mixed oxide (MOX) fuel for reuse in nuclear reactors. Weapons-grade plutonium can be added to the fuel mix. MOX fuel is used in light water reactors and consists of 60 kg of plutonium per tonne of fuel; after four years, three-quarters of the plutonium is burned (turned into other elements).*[31] Breeder reactors are specifically designed to create more fissionable material than they consume.*[107]

MOX fuel has been in use since the 1980s, and is widely used in Europe.*[105] In September 2000, the United States and the Russian Federation signed a Plutonium Management and Disposition Agreement by which each agreed to dispose of 34 tonnes of weapons-grade plutonium.*[108] The U.S. Department of Energy plans to dispose of 34 tonnes of weapons-grade plutonium in the United States before the end of 2019 by converting the plutonium to a MOX fuel to be used in commercial nuclear power reactors.*[108]

MOX fuel improves total burnup. A fuel rod is reprocessed after three years of use to remove waste products, which by then account for 3% of the total weight of the rods.*[31] Any uranium or plutonium isotopes produced during those three years are left and the rod goes back into production.*[note 10] The presence of up to 1% gallium per mass in weapons-grade plutonium alloy has the potential to interfere with long-term operation of a light water reactor.*[109]

Plutonium recovered from spent reactor fuel poses little proliferation hazard, because of excessive contamination with non-fissile plutonium-240 and plutonium-242. Separation of the isotopes is not feasible. A dedicated reactor operating on very low burnup (hence minimal exposure of newly formed plutonium-239 to additional neutrons which causes it to be transformed to heavier isotopes of plutonium) is generally required to produce material suitable for use in efficient nuclear weapons. While "weapons-grade" plutonium is defined to contain at least 92% plutonium-239 (of the total plutonium), the United States have managed to detonate an under-20Kt device using plutonium believed to contain only about 85% plutonium-239, so called '"fuel-grade" plutonium.'[110] The "reactor-grade" plutonium produced by a regular LWR burnup cycle typically contains less than 60% Pu-239, with up to 30% parasitic Pu-240/Pu-242, and 10–15% fissile Pu-241.'[110] It is unknown if a device using plutonium obtained from reprocessed civil nuclear waste can be detonated, however such a device could hypothetically fizzle and spread radioactive materials over a large urban area. The IAEA conservatively classifies plutonium of all isotopic vectors as "direct-use" material, that is, "nuclear material that can be used for the manufacture of nuclear explosives components without transmutation or further enrichment".'[110]

19.3.3   Power and heat source

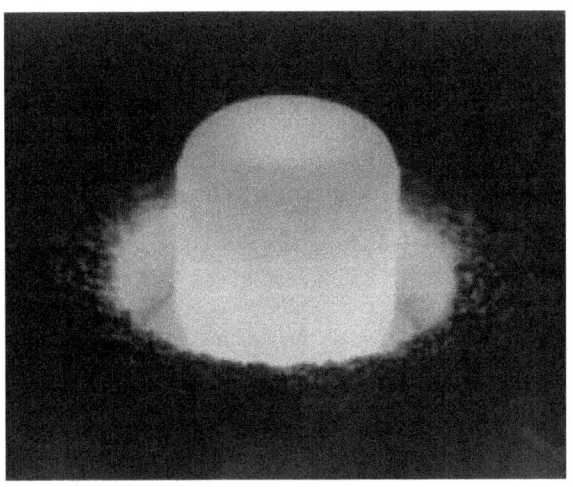

A glowing cylinder of $^{238}PuO_2$

The $^{238}PuO_2$ *radioisotope thermoelectric generator of the* Curiosity *rover*

The isotope plutonium-238 has a half-life of 87.74

years.'[111] It emits a large amount of thermal energy with low levels of both gamma rays/photons and spontaneous neutron rays/particles.'[112] Being an alpha emitter, it combines high energy radiation with low penetration and thereby requires minimal shielding. A sheet of paper can be used to shield against the alpha particles emitted by plutonium-238. One kilogram of the isotope can generate about 570 watts of heat.'[5]'[112]

These characteristics make it well-suited for electrical power generation for devices that must function without direct maintenance for timescales approximating a human lifetime. It is therefore used in radioisotope thermoelectric generators and radioisotope heater units such as those in the Cassini,'[113] Voyager, Galileo and New Horizons'[114] space probes, and the Curiosity Mars rover.'[115]

The twin Voyager spacecraft were launched in 1977, each containing a 500 watt plutonium power source. Over 30 years later, each source is still producing about 300 watts which allows limited operation of each spacecraft.'[116] An earlier version of the same technology powered five Apollo Lunar Surface Experiment Packages, starting with Apollo 12 in 1969.'[31]

Plutonium-238 has also been used successfully to power artificial heart pacemakers, to reduce the risk of repeated surgery.'[117]'[118] It has been largely replaced by lithium-based primary cells, but as of 2003 there were somewhere between 50 and 100 plutonium-powered pacemakers still implanted and functioning in living patients.'[119] Plutonium-238 was studied as a way to provide supplemental heat to scuba diving.'[120] Plutonium-238 mixed with beryllium is used to generate neutrons for research purposes.'[31]

19.4   Precautions

See also: Plutonium in the environment

19.4.1   Toxicity

There are two aspects to the harmful effects of plutonium: the radioactivity and the heavy metal poison effects. Isotopes and compounds of plutonium are radioactive and accumulate in bone marrow. Contamination by plutonium oxide has resulted from nuclear disasters and radioactive incidents, including military nuclear accidents where nuclear weapons have burned.'[121] Studies of the effects of these smaller releases, as well as of the widespread radiation poisoning sickness and death following the atomic bombings of Hiroshima and Nagasaki, have provided considerable information regarding the dangers, symptoms and prognosis of radiation poisoning, which in the case of the Japanese Hibakusha/survivors was largely unrelated to direct plutonium exposure.[122]

During the decay of plutonium, three types of radiation are released—alpha, beta, and gamma. Alpha, beta, and gamma radiation are all forms of ionizing radiation. Either acute or longer-term exposure carries a danger of serious health outcomes including radiation sickness, genetic damage, cancer, and death. The danger increases with the amount of exposure.[31] Alpha radiation can travel only a short distance and cannot travel through the outer, dead layer of human skin. Beta radiation can penetrate human skin, but cannot go all the way through the body. Gamma radiation can go all the way through the body.[123] Even though alpha radiation cannot penetrate the skin, ingested or inhaled plutonium does irradiate internal organs.[31] The skeleton, where plutonium accumulates, and the liver, where it collects and becomes concentrated, are at risk.[30] Plutonium is not absorbed into the body efficiently when ingested; only 0.04% of plutonium oxide is absorbed after ingestion.[31] Plutonium absorbed by the body is excreted very slowly, with a biological half-life of 200 years.[124] Plutonium passes only slowly through cell membranes and intestinal boundaries, so absorption by ingestion and incorporation into bone structure proceeds very slowly.[125][126]

Plutonium is more dangerous when inhaled than when ingested. The risk of lung cancer increases once the total radiation dose equivalent of inhaled plutonium exceeds 400 mSv.[127] The U.S. Department of Energy estimates that the lifetime cancer risk from inhaling 5,000 plutonium particles, each about 3 μm wide, to be 1% over the background U.S. average.[128] Ingestion or inhalation of large amounts may cause acute radiation poisoning and possibly death. However no human being is known to have died because of inhaling or ingesting plutonium, and many people have measurable amounts of plutonium in their bodies.[110]

The "hot particle" theory in which a particle of plutonium dust irradiates a localized spot of lung tissue is not supported by mainstream research—such particles are more mobile than originally thought and toxicity is not measurably increased due to particulate form.[125] When inhaled, plutonium can pass into the bloodstream. Once in the bloodstream, plutonium moves throughout the body and into the bones, liver, or other body organs. Plutonium that reaches body organs generally stays in the body for decades and continues to expose the surrounding tissue to radiation and thus may cause cancer.[129]

A commonly cited quote by Ralph Nader states that a pound of plutonium dust spread into the atmosphere would be enough to kill 8 billion people.[130] However, calculations show that one pound of plutonium could kill no more than 2 million people by inhalation. This makes the toxicity of plutonium roughly equivalent with that of nerve gas.[131] Nader's views were challenged in 1976 by Bernard Cohen, as described in the book *Nuclear Power, Both Sides: The Best Arguments for and Against the Most Controversial Technology*. Cohen's own estimate is that a dose of 200 micrograms would likely be necessary to cause cancer.[132]

Several populations of people who have been exposed to plutonium dust (e.g. people living down-wind of Nevada test sites, Nagasaki survivors, nuclear facility workers, and "terminally ill" patients injected with Pu in 1945–46 to study Pu metabolism) have been carefully followed and analyzed. These studies generally do not show especially high plutonium toxicity or plutonium-induced cancer results, such as Albert Stevens who survived into old age after being injected with plutonium.[125] "There were about 25 workers from Los Alamos National Laboratory who inhaled a considerable amount of plutonium dust during 1940s; according to the hot-particle theory, each of them has a 99.5% chance of being dead from lung cancer by now, but there has not been a single lung cancer among them." [131][133]

Plutonium has a metallic taste.[134]

19.4.2 Criticality potential

A sphere of simulated plutonium surrounded by neutron-reflecting tungsten carbide blocks in a re-enactment of Harry Daghlian's 1945 experiment

Care must be taken to avoid the accumulation of amounts of plutonium which approach critical mass, particularly because plutonium's critical mass is only a third of that of uranium-235.[5] A critical mass of plutonium emits lethal amounts of neutrons and gamma rays.[135] Plutonium in solution is more likely to form a critical mass than the solid form due to moderation by the hydrogen in water.*[11]

Criticality accidents have occurred in the past, some of them with lethal consequences. Careless handling of tungsten carbide bricks around a 6.2 kg plutonium sphere resulted in a fatal dose of radiation at Los Alamos on August 21, 1945, when scientist Harry Daghlian received a dose estimated to be 5.1 sievert (510 rems) and died 25 days later.*[136]*[137] Nine months later, another Los Alamos scientist, Louis Slotin, died from a similar accident involving a beryllium reflector and the same plutonium core (the so-called "demon core") that had previously claimed the life of Daghlian.*[138]

In December 1958, during a process of purifying plutonium at Los Alamos, a critical mass was formed in a mixing vessel, which resulted in the death of a chemical operator named Cecil Kelley. Other nuclear accidents have occurred in the Soviet Union, Japan, the United States, and many other countries.*[139]

19.4.3   Flammability

Metallic plutonium is a fire hazard, especially if the material is finely divided. In a moist environment, plutonium forms hydrides on its surface, which are pyrophoric and may ignite in air at room temperature. Plutonium expands up to 70% in volume as it oxidizes and thus may break its container.*[32] The radioactivity of the burning material is an additional hazard. Magnesium oxide sand is probably the most effective material for extinguishing a plutonium fire. It cools the burning material, acting as a heat sink, and also blocks off oxygen. Special precautions are necessary to store or handle plutonium in any form; generally a dry inert gas atmosphere is required.*[32]*[note 11]

19.5   Transportation

Around 11 metric tons of plutonium may be possessed by Japan alone, with 36 tons pending return after reprocessing in Britain and France. This is probably enough to make 6,000 atomic bombs.*[141]

19.5.1   Land and sea

The usual transportation of plutonium is through the more stable plutonium oxide in a sealed package. A typical transport consists of one truck carrying one protected shipping container, holding a number of packages with a total weight varying from 80 to 200 kg of plutonium oxide. A sea shipment may consist of several containers, each of them hold-

ing a sealed package.*[142] The United States Nuclear Regulatory Commission dictates that it must be solid instead of powder if the contents surpass 0.74 TBq (20 Curie) of radioactive activity.*[143] In a recent example, the Pacific Egret*[144] and Pacific Heron of Pacific Nuclear Transport Ltd. are taking 331 kg (730 lbs) of plutonium to a United States government facility in Savannah River, South Carolina.*[145]*[146]

19.5.2   Air

The U.S. Government air transport regulations permit the transport of plutonium by air, subject to restrictions on other dangerous materials carried on the same flight, packaging requirements, and stowage in the rearmost part of the aircraft.*[147]

In 2012 media revealed that plutonium has been flown out of Norway on commercial passenger airlines—around every other year—including one time in 2011.*[148] Regulations permit an airplane to transport 15 grams of fissionable material.*[148] Such plutonium transportation is without problems, according to a Senior Advisor (seniorrådgiver) at Statens strålevern.*[148]

19.6   Notes

19.6.1   Footnotes

[1] The PuO_2^+ ion is unstable in solution and will disproportionate into Pu^{4+} and PuO_2^{2+}: the Pu^{4+} will then oxidize the remaining PuO_2^+ to PuO_2^{2+}, being reduced in turn to Pu^{3+}. Thus, aqueous solutions of PuO_2^+ tend over time towards a mixture of Pu^{3+} and PuO_2^{2+}. UO_2^+ is unstable for the same reason.*[27]

[2] This was not the first time somebody suggested that an element be named "plutonium". A decade after barium was discovered, a Cambridge University professor suggested it be renamed to "plutonium" because the element was not (as suggested by the Greek root, barys, it was named for) heavy. He reasoned that, since it was produced by the relatively new technique of electrolysis, its name should refer to fire. Thus he suggested it be named for the Roman god of the underworld, Pluto.*[59]

[3] As one article puts it, referring to information Seaborg gave in a talk: "The obvious choice for the symbol would have been Pl, but facetiously, Seaborg suggested Pu, like the words a child would exclaim 'Pee yoo!' when smelling something bad. Seaborg thought that he would receive a

great deal of flak over that suggestion, but the naming committee accepted the symbol without a word." [61]

[4] Room 405 of the George Herbert Jones Laboratory, where the first isolation of plutonium took place, was named a National Historic Landmark in May 1967.

[5] During the Manhattan Project, plutonium was also often referred to as simply "49": the number 4 was for the last digit in 94 (atomic number of plutonium), and 9 was for the last digit in plutonium-239, the weapons-grade fissile isotope used in nuclear bombs. [69]

[6] The American Society of Mechanical Engineers (ASME) established B Reactor as a National Historic Mechanical Engineering Landmark in September 1976. [73] In August 2008, B Reactor was designated a U.S. National Historic Landmark. [74]

[7] The efficiency calculation is based on the fact that 1 kg of plutonium-239 (or uranium-235) fissioning results in an energy release of approximately 17 kt, leading to a rounded estimate of 1.2 kg plutonium actually fissioned to produce the 20 kt yield. [82]

[8] Much of this plutonium was used to make the fissionable cores of a type of thermonuclear weapon employing the Teller–Ulam design. These so-called 'hydrogen bombs' are a variety of nuclear weapon that use a fission bomb to trigger the nuclear fusion of heavy hydrogen isotopes. Their destructive yield is commonly in the millions of tons of TNT equivalent compared with the thousands of tons of TNT equivalent of fission-only devices. [87]

[9] Gadolinium zirconium oxide (Gd
2Zr
2O
7) has been studied because it could hold plutonium for up to 30 million years. [87]

[10] Breakdown of plutonium in a spent nuclear fuel rod: plutonium-239 (~58%), 240 (24%), 241 (11%), 242 (5%), and 238 (2%). [87]

[11] There was a major plutonium-initiated fire at the Rocky Flats Plant near Boulder, Colorado in 1969. [140]

19.6.2 Citations

[1] Magurno & Pearlstein 1981, pp. 835 ff.

[2] "Plutonium, Radioactive". *Wireless Information System for Emergency Responders (WISER)*. Bethesda (MD): U.S. National Library of Medicine, National Institutes of Health. Retrieved November 23, 2008. (public domain text)

[3] "Nitric acid processing". *Actinide Research Quarterly*. Los Alamos (NM): Los Alamos National Laboratory (3rd quarter). 2008. Retrieved February 9, 2010. While plutonium dioxide is normally olive green, samples can be various colors. It is generally believed that the color is a function of

chemical purity, stoichiometry, particle size, and method of preparation, although the color resulting from a given preparation method is not always reproducible.

[4] Sonzogni, Alejandro A. (2008). "Chart of Nuclides". Upton: National Nuclear Data Center, Brookhaven National Laboratory. Retrieved September 13, 2008.

[5] Heiserman 1992, p. 338

[6] Rhodes 1986, pp. 659–660 Leona Marshall: "When you hold a lump of it in your hand, it feels warm, like a live rabbit"

[7] Miner 1968, p. 544

[8] Hecker, Siegfried S. (2000). "Plutonium and its alloys: from atoms to microstructure" (PDF). *Los Alamos Science*. **26**: 290–335. Retrieved February 15, 2009.

[9] Hecker, Siegfried S.; Martz, Joseph C. (2000). "Aging of Plutonium and Its Alloys" (PDF). *Los Alamos Science*. Los Alamos, New Mexico: Los Alamos National Laboratory (26): 242. Retrieved February 15, 2009.

[10] Baker, Richard D.; Hecker, Siegfried S.; Harbur, Delbert R. (1983). "Plutonium: A Wartime Nightmare but a Metallurgist's Dream" (PDF). *Los Alamos Science*. Los Alamos National Laboratory: 148, 150–151. Retrieved February 15, 2009.

[11] Lide 2006, pp. 4–27

[12] Miner 1968, p. 542

[13] "Plutonium Crystal Phase Transitions". GlobalSecurity.org.

[14] "Glossary – Fissile material:". United States Nuclear Regulatory Commission. November 20, 2014. Retrieved February 5, 2015.

[15] Asimov 1988, p. 905

[16] Glasstone, Samuel; Redman, Leslie M. (June 1972). "An Introduction to Nuclear Weapons" (PDF). Atomic Energy Commission Division of Military Applications. p. 12. WASH-1038. Archived from the original (PDF) on August 27, 2009.

[17] Gosling 1999, p. 40

[18] "Plutonium: The First 50 Years" (PDF). U.S. Department of Energy. 1996. DOE/DP-1037. Archived from the original (PDF) on February 18, 2013.

[19] Heiserman 1992, p. 340

[20] Kennedy, J. W.; Seaborg, G. T.; Segrè, E.; Wahl, A. C. (1946). "Properties of Element 94". *Physical Review*. **70** (7–8): 555–556. Bibcode:1946PhRv...70..555K. doi:10.1103/PhysRev.70.555.

[21] Greenwood 1997, p. 1259

[22] Clark 1961, pp. 124–125.

[23] Seaborg, Glenn T.; McMillan, E.; Kennedy, J. W.; Wahl, A. C. (1946). "Radioactive Element 94 from Deuterons on Uranium". *Physical Review*. **69** (7–8): 366. Bibcode:1946PhRv...69..366S. doi:10.1103/PhysRev.69.366.

[24] Bernstein 2007, pp. 76–77.

[25] "Can Reactor Grade Plutonium Produce Nuclear Fission Weapons?". Council for Nuclear Fuel Cycle Institute for Energy Economics. Japan. May 2001.

[26] Heiserman 1992, p. 339

[27] Crooks, William J. (2002). "Nuclear Criticality Safety Engineering Training Module 10 – Criticality Safety in Material Processing Operations, Part 1" (PDF). Retrieved February 15, 2006.

[28] Matlack, George (2002). *A Plutonium Primer: An Introduction to Plutonium Chemistry and its Radioactivity*. Los Alamos National Laboratory. LA-UR-02-6594.

[29] Eagleson 1994, p. 840

[30] Miner 1968, p. 545

[31] Emsley 2001, pp. 324–329

[32] "Primer on Spontaneous Heating and Pyrophoricity – Pyrophoric Metals – Plutonium". Washington (DC): U.S. Department of Energy, Office of Nuclear Safety, Quality Assurance and Environment. 1994. Archived from the original on April 28, 2007.

[33] Crooks, W. J.; et al. (2002). "Low Temperature Reaction of ReillexTM HPQ and Nitric Acid". *Solvent Extraction and Ion Exchange*. **20** (4–5): 543–559. doi:10.1081/SEI-120014371.

[34] Dumé, Belle (November 20, 2002). "Plutonium is also a superconductor". PhysicsWeb.org.

[35] Moody, Hutcheon & Grant 2005, p. 169

[36] Kolman, D. G. & Colletti, L. P. (2009). "The aqueous corrosion behavior of plutonium metal and plutonium–gallium alloys exposed to aqueous nitrate and chloride solutions". *ECS transactions*. Electrochemical Society. **16** (52): 71. ISBN 978-1-56677-751-3.

[37] Hurst & Ward 1956

[38] Curro, N. J. (Spring 2006). "Unconventional superconductivity in PuCoGa5" (PDF). Los Alamos National Laboratory.

[39] McCuaig, Franklin D. "Pu–Zr alloy for high-temperature foil-type fuel" U.S. Patent 4,059,439, Issued on November 22, 1977

[40] Jha 2004, p. 73

[41] Kay 1965, p. 456

[42] Miner 1968, p. 541

[43] "Oklo: Natural Nuclear Reactors". U.S. Department of Energy, Office of Civilian Radioactive Waste Management. 2004. Archived from the original on October 20, 2008. Retrieved November 16, 2008.

[44] Curtis, David; Fabryka-Martin, June; Paul, Dixon; Cramer, Jan (1999). "Nature's uncommon elements: plutonium and technetium". *Geochimica et Cosmochimica Acta*. **63** (2): 275–285. Bibcode:1999GeCoA..63..275C. doi:10.1016/S0016-7037(98)00282-8.

[45] Bernstein 2007, pp. 75–77.

[46] Hoffman, D. C.; Lawrence, F. O.; Mewherter, J. L.; Rourke, F. M. (1971). "Detection of Plutonium-244 in Nature". *Nature*. **234** (5325): 132–134. Bibcode:1971Natur.234..132H. doi:10.1038/234132a0.

[47] Peterson, Ivars (December 7, 1991). "Uranium displays rare type of radioactivity". *Science News*. Wiley-Blackwell. **140** (23): 373. doi:10.2307/3976137. JSTOR 3976137. Retrieved June 29, 2015.

[48] Hoffman, D. C.; Lawrence, F. O.; Mewherter, J. L.; Rourke, F. M. (1971). "Detection of Plutonium-244 in Nature". *Nature*. **234** (5325): 132–134. Bibcode:1971Natur.234..132H. doi:10.1038/234132a0. Nr. 34.

[49] Turner, Grenville; Harrison, T. Mark; Holland, Greg; Mojzsis, Stephen J.; Gilmour, Jamie (2004-01-01). "Extinct ^{244}Pu in Ancient Zircons". *Science*. **306** (5693): 89–91. Bibcode:2004Sci...306...89T. doi:10.1126/science.1101014. JSTOR 3839259.

[50] Hutcheon, I. D.; Price, P. B. (1972-01-01). "Plutonium-244 Fission Tracks: Evidence in a Lunar Rock 3.95 Billion Years Old". *Science*. **176** (4037): 909–911. Bibcode:1972Sci...176..909H. doi:10.1126/science.176.4037.909. JSTOR 1733798.

[51] Kunz, Joachim; Staudacher, Thomas; Allègre, Claude J. (1998-01-01). "Plutonium-Fission Xenon Found in Earth's Mantle". *Science*. **280** (5365): 877–880. Bibcode:1998Sci...280..877K. doi:10.1126/science.280.5365.877. JSTOR 2896480.

[52] Holden, Norman E. (2001). "A Short History of Nuclear Data and Its Evaluation". *51st Meeting of the USDOE Cross Section Evaluation Working Group*. Upton (NY): National Nuclear Data Center, Brookhaven National Laboratory. Retrieved January 3, 2009.

[53] Fermi, Enrico (December 12, 1938). "Artificial radioactivity produced by neutron bombardment: Nobel Lecture" (PDF). Royal Swedish Academy of Sciences.

[54] Darden, Lindley (1998). "The Nature of Scientific Inquiry". College Park: Department of Philosophy, University of Maryland. Retrieved January 3, 2008.

[55] Bernstein 2007, pp. 44–52.

[56] Seaborg, Glenn T. "An Early History of LBNL: Elements 93 and 94". Advanced Computing for Science Department, Lawrence Berkeley National Laboratory. Retrieved September 17, 2008.

[57] Glenn T. Seaborg. "The plutonium story". Lawrence Berkeley Laboratory. University of California. LBL-13492, DE82 004551.

[58] Seaborg & Seaborg 2001, pp. 71–72.

[59] Heiserman 1992, p. 338.

[60] Clark, David L.; Hobart, David E. (2000). "Reflections on the Legacy of a Legend: Glenn T. Seaborg, 1912–1999" (PDF). *Los Alamos Science*. **26**: 56–61, on 57. Retrieved February 15, 2009.

[61] Clark, David L.; Hobart, David E. (2000). "Reflections on the Legacy of a Legend: Glenn T. Seaborg, 1912–1999" (PDF). *Los Alamos Science*. **26**: 56–61, on 57. Retrieved February 15, 2009.

[62] "Frontline interview with Seaborg". *Frontline*. Public Broadcasting Service. 1997. Retrieved December 7, 2008.

[63] Glenn T. Seaborg. "History of MET Lab Section C-I, April 1942 – April 1943". California Univ., Berkeley (USA). Lawrence Berkeley Lab. doi:10.2172/7110621.

[64] "Room 405, George Herbert Jones Laboratory". National Park Service. Retrieved December 14, 2008.

[65] "Periodic Table of Elements". Los Alamos National Laboratory. Retrieved September 15, 2015.

[66] Miner 1968, p. 540

[67] "Plutonium". Atomic Heritage Foundation. Retrieved September 15, 2015.

[68] "Site Selection". *LANL History*. Los Alamos, New Mexico: Los Alamos National Laboratory. Retrieved December 23, 2008.

[69] Hammel, E.F. (2000). "The taming of "49" – Big Science in little time. Recollections of Edward F. Hammel, In: Cooper N.G. Ed. Challenges in Plutonium Science" (PDF). *Los Alamos Science*. **26** (1): 2–9. Retrieved February 15, 2009.

Hecker, S.S. (2000). "Plutonium: an historical overview. In: Challenges in Plutonium Science". *Los Alamos Science*. **26** (1): 1–2. Retrieved February 15, 2009.

[70] Sublette, Carey. "Atomic History Timeline 1942–1944". Washington (DC): Atomic Heritage Foundation. Retrieved December 22, 2008.

[71] Hoddeson et al. 1993, pp. 235–239.

[72] Hoddeson et al. 1993, pp. 240–242.

[73] Wahlen 1989, p. 1.

[74] "Weekly List Actions". National Park Service. August 29, 2008. Retrieved August 30, 2008.

[75] Wahlen 1989, p. iv, 1

[76] Lindley, Robert (2013). "Kate Brown: Nuclear "Plutopias" the Largest Welfare Program in American History". *History News Network*.

[77] Rincon, Paul (March 2, 2009). "BBC NEWS – Science & Environment – US nuclear relic found in bottle". *BBC News*. Retrieved March 2, 2009.

[78] Gebel, Erika (2009). "Old plutonium, new tricks". *Analytical Chemistry*. **81** (5): 1724. doi:10.1021/ac900093b.

[79] Schwantes, Jon M.; Matthew Douglas; Steven E. Bonde; James D. Briggs; et al. (2009). "Nuclear archeology in a bottle: Evidence of pre-Trinity U.S. weapons activities from a waste burial site". *Analytical Chemistry*. **81** (4): 1297–1306. doi:10.1021/ac802286a. PMID 19152306.

[80] Sublette, Carey (July 3, 2007). "8.1.1 The Design of Gadget, Fat Man, and "Joe 1" (RDS-1)". *Nuclear Weapons Frequently Asked Questions, edition 2.18*. The Nuclear Weapon Archive. Retrieved January 4, 2008.

[81] Malik, John (September 1985). "The Yields of the Hiroshima and Nagasaki Explosions" (PDF). Los Alamos. p. Table VI. LA-8819. Retrieved February 15, 2009.

[82] On the figure of 1 kg = 17 kt, see Garwin, Richard (October 4, 2002). "Proliferation of Nuclear Weapons and Materials to State and Non-State Actors: What It Means for the Future of Nuclear Power" (PDF). *University of Michigan Symposium*. Federation of American Scientists. Retrieved January 4, 2009.

[83] Sklar 1984, pp. 22–29.

[84] Bernstein 2007, p. 70.

[85] "Historic American Engineering Record: B Reactor (105-B Building)". Richland: U.S. Department of Energy. 2001. p. 110. DOE/RL-2001-16. Retrieved December 24, 2008.

[86] Cochran, Thomas B. (1997). *Safeguarding nuclear weapons-usable materials in Russia* (PDF). International Forum on Illegal Nuclear Traffic. Washington (DC): Natural Resources Defense Council, Inc. Archived from the original (PDF) on July 5, 2013. Retrieved December 21, 2008.

[87] Emsley 2001.

[88] Stockholm International Peace Research Institute 2007, p. 567.

[89] Poet, SE; Martell, EA (October 1972). "Plutonium-239 and americium-241 contamination in the Denver area.". *Health Physics*. **23** (4): 537–48. doi:10.1097/00004032-197210000-00012. PMID 4634934. Retrieved 12 June 2013.

[90] Johnson, CJ (October 1981). "Cancer Incidence in an area contaminated with radionuclides near a nuclear installation". *Ambio*. **10** (4): 176–182. JSTOR 4312671. Reprinted in "Cancer Incidence in an area contaminated with radionuclides near a nuclear installation.". *Colo Med.* **78**: 385–92. Oct 1981. PMID 7348208.

[91] "Rocky Flats National Wildlife Refuge". U.S. Fish & Wildlife Service. Retrieved 2 July 2013.

[92] Press Secretary (July 23, 2002). "President Signs Yucca Mountain Bill". Washington (DC): Office of the Press Secretary, White House. Archived from the original on March 6, 2008. Retrieved February 9, 2015.

[93] Hebert, H. Josef. 2009. "Nuclear waste won't be going to Nevada's Yucca Mountain, Obama official says." *Chicago Tribune*. March 6, 2009, 4. Accessed 3-6-09.

[94] "About the Commission".

[95] Blue Ribbon Commission on America's Nuclear Future. "Disposal Subcommittee Report to the Full Commission" (PDF).

[96] Moss, William; Eckhardt, Roger (1995). "The Human Plutonium Injection Experiments" (PDF). *Los Alamos Science*. Los Alamos National Laboratory. **23**: 188, 205, 208, 214. Retrieved June 6, 2006.

[97] Voelz, George L. (2000). "Plutonium and Health: How great is the risk?". *Los Alamos Science*. Los Alamos (NM): Los Alamos National Laboratory (26): 78–79.

[98] Longworth, R.C. (November–December 1999). "Injected! Book review: The Plutonium Files: America's Secret Medical Experiments in the Cold War." (PDF). *The Bulletin of the Atomic Scientists*. **55** (6): 58–61.

[99] Moss, William, and Roger Eckhardt. (1995). "The human plutonium injection experiments." Los Alamos Science. 23: 177–233.

[100] Openness, DOE. (June 1998). Human Radiation Experiments: ACHRE Report. Chapter 5: The Manhattan district Experiments; the first injection. Washington, DC. Superintendent of Documents US Government Printing Office.

[101] AEC no. UR-38. 1948 Quarterly Technical Report

[102] Yesley, Michael S. (1995). "'Ethical Harm' and the Plutonium Injection Experiments" (PDF). *Los Alamos Science*. **23**: 280–283. Retrieved February 15, 2009.

[103] Martin 2000, p. 532.

[104] "Nuclear Weapon Design". Federation of American Scientists. 1998. Retrieved December 7, 2008.

[105] "Mixed Oxide (MOX) Fuel". London (UK): World Nuclear Association. 2006. Retrieved December 14, 2008.

[106] Till & Chang 2011, pp. 254–256.

[107] Till & Chang 2011, p. 15.

[108] "Plutonium Storage at the Department of Energy's Savannah River Site: First Annual Report to Congress" (PDF). Defense Nuclear Facilities Safety Board. 2004. pp. A–1. Retrieved February 15, 2009.

[109] Besmann, Theodore M. (2005). "Thermochemical Behavior of Gallium in Weapons-Material-Derived Mixed-Oxide Light Water Reactor (LWR) Fuel". *Journal of the American Ceramic Society*. **81** (12): 3071–3076. doi:10.1111/j.1151-2916.1998.tb02740.x.

[110] "Plutonium". World Nuclear Association. March 2009. Retrieved February 28, 2010.

[111] "Science for the Critical Masses: How Plutonium Changes with Time". Institute for Energy and Environmental Research.

[112] "From heat sources to heart sources: Los Alamos made material for plutonium-powered pumper". *Actinide Research Quarterly*. Los Alamos: Los Alamos National Laboratory (1). 2005. Retrieved February 15, 2009.

[113] "Why the Cassini Mission Cannot Use Solar Arrays" (PDF). NASA/JPL. December 6, 1996. Retrieved March 21, 2014.

[114] St. Fleur, Nicholas. "The Radioactive Heart of the New Horizons Spacecraft to Pluto". New York *Times*. August 7, 2015. The "craft's 125-pound generator [is] called the General Purpose Heat Source-Radioisotope Thermoelectric Generator. [It] was stocked with 24 pounds of plutonium that produced about 240 watts of electricity when it left Earth in 2006, according to Ryan Bechtel, an engineer from the Department of Energy who works on space nuclear power. During the Pluto flyby the battery produced 202 watts, Mr. Bechtel said. The power will continue to decrease as the metal decays, but there is enough of it to command the probe for another 20 years, according to Curt Niebur, a NASA program scientist on the New Horizons mission." Retrieved 2015-08-10.

[115] Mosher, Dave (September 19, 2013). "NASA's Plutonium Problem Could End Deep-Space Exploration". *Wired*. Retrieved February 5, 2015.

[116] "Voyager-Spacecraft Lifetime". Jet Propulsion Laboratory. June 11, 2014. Retrieved February 5, 2015.

[117] Venkateswara Sarma Mallela; V. Ilankumaran & N.Srinivasa Rao (2004). "Trends in Cardiac Pacemaker Batteries". *Indian Pacing Electrophysiol*. **4** (4): 201–212. PMC 1502062. PMID 16943934.

[118] "Plutonium Powered Pacemaker (1974)". Oak Ridge Associated Universities. Retrieved February 6, 2015.

[119] "Plutonium Powered Pacemaker (1974)". Oak Ridge: Orau.org. 2011. Retrieved February 1, 2015.

[120] Bayles, John J.; Taylor, Douglas (1970). "SEALAB III – Diver's Isotopic Swimsuit-Heater System". Port Hueneme: Naval Civil Engineering Lab. AD0708680.

[121] "Toxicological Profile for Plutonium" (PDF). U.S. Department of Health and Human Services, Agency for Toxic Substances and Disease Registry (ATSDR). November 2010. Retrieved February 9, 2015.

[122] Little, M P (June 2009). "Cancer and non-cancer effects in Japanese atomic bomb survivors". *J Radiol Prot*. **29** (2A): A43–59. Bibcode:2009JRP....29...43L. doi:10.1088/0952-4746/29/2A/S04. PMID 19454804.

[123] "Plutonium, CAS ID #: 7440-07-5". Centers for Disease Control and Prevention (CDC) Agency for Toxic Substances and Disease Registry. Retrieved February 5, 2015.

[124] "Radiological control technical training" (PDF). U.S. Department of Energy. Archived from the original (PDF) on June 30, 2007. Retrieved December 14, 2008.

[125] Cohen, Bernard L. "The Myth of Plutonium Toxicity". Archived from the original on August 26, 2011.

[126] Cohen, Bernard L. (May 1977). "Hazards from Plutonium Toxicity". *The Radiation Safety Journal: Health Physics*. **32** (5): 359–379. doi:10.1097/00004032-197705000-00003.

[127] Brown, Shannon C.; Margaret F. Schonbeck; David McClure; et al. (July 2004). "Lung cancer and internal lung doses among plutonium workers at the Rocky Flats Plant: a case-control study". *American Journal of Epidemiology*. Oxford Journals. **160** (2): 163–172. doi:10.1093/aje/kwh192. PMID 15234938. Retrieved February 15, 2009.

[128] "ANL human health fact sheet—plutonium" (PDF). Argonne National Laboratory. 2001. Archived from the original (PDF) on February 16, 2013. Retrieved June 16, 2007.

[129] "Radiation Protection, Plutonium: What does plutonium do once it gets into the body?". U.S. Environmental Protection Agency. Retrieved March 15, 2011.

[130] "Did Ralph Nader say that a pound of plutonium could cause 8 billion cancers?". Retrieved January 3, 2013.

[131] Bernard L. Cohen. "The Nuclear Energy Option, Chapter 13, Plutonium and Bombs". Retrieved March 28, 2011. (Online version of Cohen's book *The Nuclear Energy Option* (Plenum Press, 1990) ISBN 0-306-43567-5).

[132] Kaku & Trainer 1983, p. 77.

[133] Voelz, G. L. (1975). "What We Have Learned About Plutonium from Human Data". *The Radiation Safety Journal Health Physics*; 29.

[134] Welsome 2000, p. 17.

[135] Miner 1968, p. 546

[136] Roark, Kevin N. (2000). "Criticality accidents report issued". Los Alamos (NM): Los Alamos National Laboratory. Archived from the original on October 8, 2008. Retrieved November 16, 2008.

[137] Hunner 2004, p. 85.

[138] "Raemer Schreiber". *Staff Biographies*. Los Alamos: Los Alamos National Laboratory. Archived from the original on January 3, 2013. Retrieved November 16, 2008.

[139] McLaughlin, Monahan & Pruvost 2000, p. 17.

[140] Albright, David; O'Neill, Kevin (1999). "The Lessons of Nuclear Secrecy at Rocky Flats". *ISIS Issue Brief*. Institute for Science and International Security (ISIS). Archived from the original on July 8, 2008. Retrieved December 7, 2008.

[141] 2 British ships arrive in Japan to carry plutonium to US

[142] "Transport of Radioactive Materials". World Nuclear Association. Retrieved February 6, 2015.

[143] "§ 71.63 Special requirement for plutonium shipments". United States Nuclear Regulatory Commission. Retrieved February 6, 2015.

[144] "Pacific Egret". Retrieved 22 March 2016.

[145] Yamaguchi, Mari. "Two British ships arrive in Japan to carry plutonium to US". Retrieved 22 March 2016.

[146] "Two British ships arrive in Japan to transport plutonium for storage in U.S." Retrieved 22 March 2016.

[147] "Part 175.704 Plutonium shipments". *Code of Federal Regulations 49 —Transportation*. Retrieved August 1, 2012.

[148] Av Ida Søraunet Wangberg og Anne Kari Hinna. "Klassekampen : Flyr plutonium med rutefly". Klassekampen.no. Retrieved August 13, 2012.

19.7 References

- Asimov, Isaac (1988). "Nuclear Reactors". *Understanding Physics*. New York: Barnes & Noble Publishing. ISBN 0-88029-251-2.

- Bernstein, Jeremy (2007). *Plutonium: a History of the World's most Dangerous Element*. Washington, D.C.: Joseph Henry Press. ISBN 978-0-309-10296-4. OCLC 76481517.

- Clark, Ronald (1961). *The Birth of the Bomb: The Untold Story of Britain's Part in the Weapon That Changed the World*. London: Phoenix House. OCLC 824335.

- Eagleson, Mary (1994). *Concise Encyclopedia Chemistry*. Berlin: Walter de Gruyter. ISBN 978-3-11-011451-5.

- Emsley, John (2001). "Plutonium". *Nature's Building Blocks: An A–Z Guide to the Elements*. Oxford (UK): Oxford University Press. ISBN 0-19-850340-7.

- Gosling, F.G. (1999). *The Manhattan Project: Making the Atomic Bomb* (PDF). Oak Ridge: United States Department of Energy. ISBN 0-7881-7880-6. DOE/MA-0001-01/99. Retrieved February 15, 2009.

- Greenwood, N. N.; Earnshaw, A. (1997). *Chemistry of the Elements* (2nd ed.). Oxford (UK): Butterworth-Heinemann. ISBN 0-7506-3365-4.

- Heiserman, David L. (1992). "Element 94: Plutonium". *Exploring Chemical Elements and their Compounds*. New York (NY): TAB Books. pp. 337–340. ISBN 0-8306-3018-X.

- Hoddeson, Lillian; Henriksen, Paul W.; Meade, Roger A.; Westfall, Catherine L. (1993). *Critical Assembly: A Technical History of Los Alamos During the Oppenheimer Years, 1943–1945*. New York: Cambridge University Press. ISBN 0-521-44132-3. OCLC 26764320.

- Hunner, Jon (2004). *Inventing Los Alamos*. ISBN 978-0-8061-3891-6.

- Hurst, D. G.; Ward, A. G. (1956). *Canadian Research Reactors* (PDF). Ottawa: Atomic Energy of Canada Limited. OCLC 719819357. Retrieved February 6, 2015.

- Jha, D.K. (2004). *Nuclear Energy*. Discovery Publishing House. ISBN 81-7141-884-8.

- Kaku, Michio; Trainer, Jennifer (1983). *Nuclear Power, Both Sides: The Best Arguments for and Against the Most Controversial Technology*. W. W. Norton & Company. Retrieved December 8, 2013.

- Kay, A. E. (1965). *plutonium 1965*. Taylor & Francis.

- Lide, David R., ed. (2006). *Handbook of Chemistry and Physics* (87th ed.). Boca Raton: CRC Press, Taylor & Francis Group. ISBN 0-8493-0487-3.

- Magurno, B.A.; Pearlstein, S., eds. (1981). *Proceedings of the conference on nuclear data evaluation methods and procedures. BNL-NCS 51363.* (PDF). **II**. Upton: Brookhaven National Laboratory. Retrieved August 6, 2014.

- Martin, James E. (2000). *Physics for Radiation Protection*. Wiley-Interscience. ISBN 0-471-35373-6.

- McLaughlin, Thomas P.; Monahan, Shean P.; Pruvost, Norman L. (2000). *A Review of Criticality Accidents* (PDF). Los Alamos: Los Alamos National Laboratory. LA-13638. Retrieved February 6, 2015.

- Miner, William N.; Schonfeld, Fred W. (1968). "Plutonium". In Clifford A. Hampel. *The Encyclopedia of the Chemical Elements*. New York (NY): Reinhold Book Corporation. pp. 540–546. LCCN 68029938.

- Moody, Kenton James; Hutcheon, Ian D.; Grant, Patrick M. (2005). *Nuclear forensic analysis*. CRC Press. ISBN 0-8493-1513-1.

- Rhodes, Richard (1986). *The Making of the Atomic Bomb*. New York: Simon & Schuster. ISBN 0-671-65719-4.

- Seaborg, G. T.; Seaborg, E. (2001). *Adventures in the Atomic Age: From Watts to Washington*. Farrar, Straus and Giroux. ISBN 0-374-29991-9.

- Sklar, Morty (1984). *Nuke-Rebuke: Writers & Artists Against Nuclear Energy & Weapons*. The Contemporary anthology series. The Spirit That Moves Us Press.

- Stockholm International Peace Research Institute (2007). *SIPRI Yearbook 2007: Armaments, Disarmament, and International Security*. Oxford University Press. ISBN 978-0-19-923021-1. ISSN 0953-0282.

- Till, C.E.; Chang, Y.I. (2011). *Plentiful Energy: The Story of the Integral Fast Reactor, the Complex History of a Simple Reactor Technology, with Emphasis on Its Scientific Basis for Non-specialists*. Charles E. Till and Yoon Il Chang. ISBN 978-1-4663-8460-6.

- Wahlen, R.K. (1989). *History of 100-B Area* (PDF). Richland, Washington: Westinghouse Hanford Company. WHC-EP-0273. Archived from the original (PDF) on March 27, 2009. Retrieved February 15, 2009.

- Welsome, Eileen (2000). *The Plutonium Files: America's Secret Medical Experiments in the Cold War*. New York: Random House. ISBN 0-385-31954-1.

19.8 External links

- "Alsos Digital Library for Nuclear Issues - Plutonium". Washington and Lee University.

- Sutcliffe, W.G.; et al. (1995). "A Perspective on the Dangers of Plutonium". Lawrence Livermore National Laboratory. Archived from the original on September 29, 2006.

- Johnson, C.M.; Davis, Z.S. (1997). "Nuclear Weapons: Disposal Options for Surplus Weapons-Usable Plutonium". *CRS Report for Congress # 97-564 ENR*. Retrieved February 15, 2009.

- "Physical, Nuclear, and Chemical, Properties of Plutonium". IEER. 2005. Retrieved February 15, 2009.

- Bhadeshia, H. "Plutonium crystallography".

- Samuels, D. (2005). "End of the Plutonium Age". *Discover Magazine*. **26** (11).

- Pike, J.; Sherman, R. (2000). "Plutonium production". Federation of American Scientists. Retrieved February 15, 2009.

- "Plutonium Manufacture and Fabrication". Nuclearweaponarchive.org.

- Ong, C. (1999). "World Plutonium Inventories". Nuclear Files.org. Archived from the original on August 5, 2014. Retrieved February 15, 2009.

- "Challenges in Plutonium Science". *Los Alamos Science*. I & II (26). 2000. Retrieved February 15, 2009.

- "Plutonium". Royal Society of Chemistry. Retrieved February 6, 2015.

- "Plutonium". The Periodic Table of Videos. University of Nottingham. Retrieved February 6, 2015.

Chapter 20

Plutonium-238

Plutonium-238 (also known as **Pu-238** or ^{238}**Pu**) is a radioactive isotope of plutonium that has a half-life of 87.7 years.

Plutonium-238 is a very powerful alpha emitter. This makes the plutonium-238 isotope suitable for usage in radioisotope thermoelectric generators (RTGs) and radioisotope heater units – one gram of plutonium-238 generates approximately 0.5 W of thermal power.

20.1 History

20.1.1 Initial Production

Plutonium-238 was the first isotope of plutonium to be discovered. It was synthesized by Glenn Seaborg and associates in December, 1940 by bombarding uranium-238 with deuterons, creating neptunium-238, which then decays to form plutonium-238. Plutonium-238 decays to uranium-234 and then further along the radium series to lead-206. Plutonium-238 was produced by irradiating Neptunium-237 (half life 2.144M years), which is a by-product of the production of Plutonium-239 weapons-grade material. As produced by Savannah River in their weapons reactor, shut down in 1988, Plutonium-238 was mixed with about 16% Plutonium-239. [1]

The first application was its use in a weapons component made at Mound for the Weapons Design Agency Lawrence Livermore Laboratory (LLL). Mound was chosen for this work because of its experience in producing the Polonium-210 fueled Urchin initiator and its work with several heavy elements in a Reactor Fuels program. Two Mound scientists spent 1959 at LLL in joint development while the Special Metallurgical Building was constructed at Mound to house the project. Meanwhile the first sample of Plutonium-238 came to Mound in 1959. [2]

The weapons project was planned for about 1 Kg./year of Pu-238 over a 3-year period. But the Pu-238 component could not be produced to the specifications despite a 2 year

effort beginning at Mound in mid-1961. A maximum effort was undertaken with 3 shifts a day, 6 days a week and ramp-up of Savannah River's Pu-238 production over a 3 year period to about 20 Kg./year. A loosening of the specifications resulted in productivity of about 3%, and production finally began in 1964.

20.1.2 Use as Radioisotope Thermoelectric Generator

Beginning on January 1, 1957, Mound RTG inventors Jordan & Birden were working on an Army Signal Corps contract (R-65-8- 998 11-SC-03-91) to conduct research on radioactive materials and thermocouples suitable for the direct conversion of heat to electrical energy using Polonium-210 as the heat source.

Capt. R. T. Carpenter had chosen Pu-238 as the fuel for the first RTG to be launched into space as auxiliary power for the Transit IV Navy navigational satellite, June 29, 1961. As of January 21, 1963, the decision had yet to be made as to what isotope would be used to fuel the large RTGs for NASA programs. [3]

Then early in 1964 Mound scientists developed a different method of fabricating the weapon component that resulted in a production efficiency of around 98%. This made available the excess Savannah River Pu-238 production for Space Electric Power use just in time to meet the needs of the SNAP-27 RTG on the moon, the Pioneer spacecraft, the Viking Mars Landers, more Transit Navy navigation satellites (precursor to today's GPS) and Voyager spacecraft, all of the Pu-238 heat sources for which were fabricated at Mound.

See the revised table from "RTG: A Source of Power: A History of the Radioisotopic Thermoelectric Generators Fueled at Mound" [4]

The Radioisotope Heater Units were the unsung heroes of space exploration beginning with the Apollo Radioisotope Heaters (ALRH) warming the Seismic Experiment placed on the Moon in the first landing (Apollo 11) through their use on Moon & Mars rovers to the 120 LWRHU's heating the experiments on the Galileo spacecraft.

With Plutonium-238 becoming available for non-military uses, numerous applications were proposed and tested, including the Cardiac Pacemaker program that began on June 1, 1966, in conjunction with NUMEC. *[5] When it was recognized that the heat source would not remain intact through cremation, the program was cancelled because 100% assurance could not be guaranteed that a cremation event would not occur.

An addition to the Special Metallurgical building weapon component production facility was completed at the end of 1964 for Pu-238 heat source fuel fabrication. A temporary fuel production facility was also installed in the Research Building in 1969 for Transit fuel fabrication. With completion of the weapons component project, the Special Metallurgical Building, nicknamed "Snake Mountain" because of the difficulties encountered in handling large quantities of Pu-238, ceased operations on June 30, 1968, with Pu-238 operations taken over by the new Plutonium Processing Building, especially designed and constructed for handling large quantities of Pu-238. *[6] Plutonium-238 is given the highest relative hazard number (152) of all 256 radionuclides evaluated by Karl Z. Morgan, et al, in 1963 *[7]

20.2   Production

Reactor-grade plutonium from spent nuclear fuel contains various isotopes of plutonium. Pu-238 makes up only one or two percent, but it may be responsible for much of the short-term decay heat because of its short half-life relative to other plutonium isotopes. Reactor-grade plutonium is not useful for producing Pu-238 for RTGs because difficult isotopic separation would be needed.

Pure plutonium-238 is prepared by neutron irradiation of neptunium-237, one of the minor actinides that can be recovered from spent nuclear fuel during reprocessing, or by the neutron irradiation of americium in a reactor. *[8] In both cases, the targets are subjected to a chemical treatment, including dissolution in nitric acid to extract the plutonium-238. A 100 kg sample of light water reactor fuel that has been irradiated for three years contains only about 700 grams of neptunium-237, and the neptunium must be extracted selectively. Significant amounts of pure Pu-238 could also be produced in a thorium fuel cycle. *[9]

20.2.1   United States supply

The United States stopped producing bulk Pu-238 in 1988; *[10] since 1993, all of the Pu-238 used in American spacecraft has been purchased from Russia. In total, 16.5 kilograms (36 lb) has been purchased but Russia is no longer producing Pu-238 and their own supply is reportedly running low. *[11] *[12]

The United States Pu-238 inventory supports both NASA (civil space) and other national security applications. *[13] The Department of Energy maintains separate inventory accounts for the two categories. As of March 2015, a total of 35 kilograms (77 pounds) of Pu-238 was available for civil space uses. *[13] Out of the inventory, 1 kilogram (2.2 lb) remains in good enough condition to meet NASA specifications for power delivery; it is this pool of Pu-238 that will be used in a multi-mission radioisotope thermoelectric generator (MMRTG) for the 2020 Mars Rover mission and two additional MMRTGs for a notional 2024 NASA mission. *[13] 21 kilograms (46 lb) will remain after that, with approximately 4 kilograms (8.8 lb) just barely meeting the NASA specification. *[13] This 21 kilograms (46 lb) can be brought up to NASA specifications if it is blended with a smaller amount of newly produced Pu-238 having a higher energy density. *[13]

To restart production a sustained year-to-year funding would maintain the infrastructure and knowledge base in order to avoid significant recapture costs. *[13] Approximately $50 million per year, formerly funded by the Department of Energy (DoE), was transitioned to a full cost recovery model as part of the FY 2014 federal budget. *[13] NASA has also provided additional funding to refurbish critical equipment at Los Alamos National Laboratory (LANL). *[13] DoE manages the operation of its nuclear facilities in order to ensure nuclear safety/security, to meet mission needs, and to work with other DoE programs. *[13] A project to re-establish Pu-238 production capability has a total estimated cost range of $85-$125 million over 9 years, but actual project costs are likely to increase since available funding has not supported the planned pace, thus drawing out the schedule. *[13] After production has been restarted it is predicted that it would take at least five years to get enough for a single spacecraft mission. *[14]

The Advanced Test Reactor at the Idaho National Laboratory and the High Flux Isotope Reactor at the Oak Ridge National Laboratory were both seen as potential producers. *[12]

In February 2013, it was reported that a small amount of Pu-238 was successfully produced by Oak Ridge's High Flux Isotope Reactor – this was the first time the United States had produced Pu-238 since production ended in the late 1980s. *[15] On December 22, 2015, the Oak Ridge National Laboratory reported that its researchers had successfully produced 50 grams (1.8 ounces) of Pu-238.[16][17] After an analysis of this sample, production of 300 to 400 grams (11 to 14 oz) of the material per year is planned to begin and then, through automation and scale-up processes, production will increase to an average of 1.5 kilograms (3.3 lb) per year.[16]

20.3   Applications

The main application of Pu-238 is as the heat source in radioisotope thermoelectric generators (RTGs). The RTG was invented in 1954 by Mound scientists Ken Jordan and John Birden. They were inducted into the National Inventors Hall of Fame in 2013. http://invent.org/inductee-detail/?IID=473 They immediately produced a working prototype using a ^{210}polonium heat source, and on January 1, 1957 entered into an Army Signal Corps contract (R-65-8- 998 11-SC-03-91) to conduct research on radioactive materials and thermocouples suitable for the direct conversion of heat to electrical energy using Polonium-210 as the heat source.

RTG technology was first developed by Los Alamos National Laboratory during the 1960s and 1970s to provide radioisotope thermoelectric generator power for cardiac pacemakers. Of the 250 plutonium-powered pacemakers Medtronic manufactured, twenty-two were still in service more than twenty-five years later, a feat that no battery-powered pacemaker could achieve.[18]

This same RTG power technology has been used in spacecraft such as Voyager 1 and 2, Cassini–Huygens and New Horizons, and in other devices, such as the Mars Science Laboratory, for long-term nuclear power generation.[19]

20.4   See also

- Atomic battery
- Plutonium-239
- Polonium-210

20.5   References

[1] "MLM-CF-67-1-71 Plutonium 238 Oxide Shipment No. 33" (PDF). 1966-12-30.

[2] "Little Known Pu Stories" (PDF).

[3] G. R. Grove to D. L. Scot (1963-01-21). "Trip Report" (PDF).

[4] Carol Craig. "RTG: A Source of Power; A History of the Radioisotopic Thermoelectric Generators Fueled at Mound (MLM-MU-82-72-0006)" (PDF).

[5] (PDF)   https://dl.dropboxusercontent.com/u/77675434/Heat%20Source%20Datasheets/CARDIAC%20PACEMAKER.pdf. Missing or empty |title= (help)

[6] "Final Safety Analysis Report, January 15, 1975 (MLM-ENG-105)".

[7] Karl Z. Morgan; et al. (1964-03-01). "Health Physics Journal, Vol. 10, No. 3 - Relative Hazard of the Various Radioactive Materials".

[8] "Process for producing ultra-pure ... - Google Patents". Google.com. Retrieved 2011-09-19.

[9] http://www.thoriumenergyalliance.com/downloads/plutonium-238.pdf

[10] Steven D. Howe; Douglas Crawford; Jorge Navarro; Terry Ring. "Economical Production of Pu - 238: Feasibility Study" (PDF). Center for Space Nuclear Research. Retrieved 2013-03-19.

[11] "Commonly Asked Questions About Radioisotope Power Systems" (PDF). Idaho National Laboratory. July 2005. Archived from the original (PDF) on September 28, 2011. Retrieved 2011-10-24.

[12] "Plutonium-238 Production Project" (PDF). Department of Energy. 5 February 2011. Archived from the original (PDF) on February 3, 2012. Retrieved 2 July 2012.

[13] Caponiti, Alice. "Space and Defense Power Systems Program Information Briefing" (PDF). *Lunar and Planetary Institute*. NASA. Retrieved 24 March 2015.

[14] "Plutonium Shortage Could Stall Space Exploration". NPR. Retrieved 2011-09-19.

[15] Clark, Stephen (20 March 2013). "U.S. laboratory produces first plutonium in 25 years". Spaceflightnow. Retrieved 21 March 2013.

[16] Walli, Ron (22 December 2015). "ORNL achieves milestone with plutonium-238 sample". Oak Ridge National Laboratory. Retrieved 22 December 2015.

[17] Harvey, Chelsea (30 December 2015). "This is the fuel NASA needs to make it to the edge of the solar system - and beyond". *The Washington Post*. Retrieved 4 January 2016.

[18] Kathy DeLucas; Jim Foxx; Robert Nance (January–March 2005). "From heat sources to heart sources: Los Alamos made material for plutonium-powered pumper". *Actinide Research Quarterly*. Los Alamos National Laboratory. Retrieved 2015-07-09.

[19] Alexandra Witze. Nuclear power: Desperately seeking plutonium. NASA has 35 kg of ^{238}Pu to power its deep-space missions - but that will not get it very far., *Nature*. 25 Nov 2014

20.6 External links

- Story of Seaborg's discovery of Pu-238, especially pages 34-35.

- NLM Hazardous Substances Databank – Plutonium, Radioactive

Chapter 21

Americium-241

Americium-241 (^{241}**Am**) is an isotope of americium. Like all isotopes of americium, it is radioactive. ^{241}Am is the most common isotope of americium. It is the most prevalent isotope of americium in nuclear waste. Americium-241 has a half-life of 432.2 years. It is commonly found in ionization type smoke detectors. It is a potential fuel for long-lifetime radioisotope thermoelectric generators (RTGs). Its common parent nuclides are β^+– from ^{241}Pu, EC from ^{241}Cm and α from ^{245}Bk. ^{241}Am is fissile and the critical mass of a bare sphere is 57.6-75.6 kilograms and a sphere diameter of 19–21 centimeters.[1] Americium-241 has a specific activity of 3.43 Ci/g (Curies per gram or 117.29 Gigabequerels (GBq) per gram).[2] It is commonly found in the form of americium-241 dioxide (^{241}AmO$_2$). This isotope also has one meta state; ˙241mAm, with an exitation energy of 2.2 MeV, and a half-life of 1.23 μs. Its presence in plutonium is determined by the original concentration of plutonium-241 and the sample age. Because of the low penetration of alpha radiation, americium-241 only poses a health risk when ingested or inhaled. Older samples of plutonium containing plutonium-241 contain a buildup of ^{241}Am. A chemical removal of americium-241 from reworked plutonium (e.g. during reworking of plutonium pits) may be required in some cases.

21.1 Nucleosynthesis

Americium-241 has been produced in small quantities in nuclear reactors for decades, and many kilograms of ^{241}Am have been accumulated by now.[3] Nevertheless, since it was first offered for sale in 1962, its price, about 1,500 USD per gram of ^{241}Am, remains almost unchanged owing to the very complex separation procedure.[4]

Americium-241 is not synthesized directly from uranium – the most common reactor material – but from the plutonium isotope ^{239}Pu. The latter needs to be produced first, according to the following nuclear process:

$$^{238}_{92}\text{U} \xrightarrow{(n,\gamma)} {}^{239}_{92}\text{U} \xrightarrow[23.5\text{ min}]{\beta^-} {}^{239}_{93}\text{Np} \xrightarrow[2.3565\text{ d}]{\beta^-} {}^{239}_{94}\text{Pu}$$

The capture of two neutrons by ^{239}Pu (a so-called (n,γ) reaction), followed by a β-decay, results in ^{241}Am:

$$^{239}_{94}\text{Pu} \xrightarrow{2\,(n,\gamma)} {}^{241}_{94}\text{Pu} \xrightarrow[14.35\text{ yr}]{\beta^-} {}^{241}_{95}\text{Am}$$

The plutonium present in spent nuclear fuel contains about 12% of ^{241}Pu. Because it converts to ^{241}Am, ^{241}Pu can be extracted and may be used to generate further ^{241}Am.[4] However, this process is rather slow: half of the original amount of ^{241}Pu decays to ^{241}Am after about 14 years, and the ^{241}Am amount reaches a maximum after 70 years.[5]

The obtained ^{241}Am can be used for generating heavier americium isotopes by further neutron capture inside a nuclear reactor. In a light water reactor (LWR), 79% of ^{241}Am converts to ^{242}Am and 10% to its nuclear isomer ˙242mAm:[6]

$$79\%: {}^{241}_{95}\text{Am} \xrightarrow{(n,\gamma)} {}^{242}_{95}\text{Am}$$

21.2 Decay

Main article: Radioactive decay

Americium-241 decays mainly via alpha decay, with a weak gamma ray byproduct. The α-decay is shown as follows:

$$^{241}_{95}\text{Am} \xrightarrow{432.2\text{y}} {}^{237}_{93}\text{Np} + {}^{4}_{2}\alpha^{2+} + \gamma\ 59.5409\text{ keV}$$

The α-decay energies are 5.486 MeV for 85% of the time (the one which is widely accepted for standard α-decay energy), 5.443 MeV for 13% of the time, and 5.388 MeV for the remaining 2%.[7] The γ-ray energy is 59.5409 keV for the most part, with little amounts of other energies such as 13.9 keV, 17.8 keV and 26.4 keV.[8]

The second most common type of decay for americium-241 is cluster decay, with a branching ratio of less than $7.4 \times 10^{*} - 16$. Also shown as follows:

$$^{241}_{95}\text{Am} \longrightarrow ^{207}_{81}\text{Tl} + ^{34}_{14}\text{Si}$$

The least common (rarest) type of decay that americium-241 undergoes is spontaneous fission, with a branching ratio of $4 \times 10^{*} - 12$ and happening 1.2 times a second per gram of ^{241}Am. It is written as such (the asterisk denotes an excited nucleus):

$$^{241}_{95}\text{Am} \longrightarrow ^{241}_{95}\text{Am}^{*} \longrightarrow 3^{1}_{0}\text{n} + \text{fission products} + \text{energy}$$

21.3 Applications

21.3.1 Ionization-type smoke detector

Main article: Smoke detector

Americium-241 is the only synthetic isotope to have found its way into the household, where the most common type of smoke detector (the ionization-type) uses $^{241}\text{AmO}_2$ (americium-241 dioxide) as its source of ionizing radiation.[9] This isotope is preferred over ^{226}Ra as ^{241}Am emits 5 times more alpha particles and also emits relatively little harmful gamma radiation. With its half-life of 432.2 years, the americium in a smoke detector decreases and includes about 4.4% neptunium after 19 years, and about 7.4% after 32 years. The amount of americium in a typical new smoke detector is 0.29 microgram (about one-third the weight of a grain of sand) with an activity of 1 microcurie/37 kilobequerels (1.0 μCi/37 kBq). Some old industrial smoke detectors (notably from the Pyrotronics Corporation) can contain up to 80 μCi. The amount of ^{241}Am declines slowly as it decays into neptunium-237 a different transuranic element with a much longer half-life (about 2.14 million years). The radiated alpha-particles pass through an ionization chamber, an air-filled space between two electrodes, which allows a small, constant electric current to pass between the capacitor plates due to the radiation ionizing the air space between. Any smoke that enters the chamber blocks/absorbs some of the alpha particles from freely passing through and reduces the ionization and therefore causes a drop in the current. The alarm's circuitry detects this drop in the current and as a result, triggers the piezoelectric buzzer to sound. Compared to the alternative optical smoke detector, the ionization smoke detector is cheaper and can detect particles which are too small to produce significant light scattering. However, it is more prone to false alarms.[10][11][12][13]

21.3.2 Radionuclide

As ^{241}Am has a roughly similar half-life to ^{238}Pu (432.2 years vs. 87 years), it has been proposed as an active isotope of radioisotope thermoelectric generators, for use in spacecraft.[14][15] Even though americium-241 produces less heat and electricity than plutonium-238 (the power yield is 114.7 mW/g for ^{241}Am vs. 390 mW/g for ^{238}Pu).[14] and although its radiation poses a bigger threat to humans owing to gamma and neutron emission, the European Space Agency is still considering to use americium-241 for its space probes, as a result of the global shortage of plutonium-238.[16]

21.3.3 Neutron source

Oxides of ^{241}Am pressed with beryllium can be very efficient neutron sources, since it emits alpha particle during its radioactive decay:

$$^{241}_{95}\text{Am} \xrightarrow{432.2y} ^{237}_{93}\text{Np} + ^{4}_{2}\alpha^{2+} + \gamma \ 59.5 \text{ keV}$$

Here americium acts as the alpha source, and beryllium produces neutrons owing to its large cross-section for the (α,n) nuclear reaction:

$$^{9}_{4}\text{Be} + ^{4}_{2}\alpha^{2+} \longrightarrow ^{12}_{6}\text{C} + ^{1}_{0}\text{n} + \gamma$$

The most widespread use of $^{241}\text{AmBe}$ neutron sources is a neutron probe – a device used to measure the quantity of water present in soil, as well as moisture/density for quality control in highway construction. ^{241}Am neutron sources are also used in well logging applications, as well as in neutron radiography, tomography and other radiochemical investigations.

21.3.4 Production of other elements

Americium-241 is sometimes used as a starting material for the production of other transuranic elements and transactinides – for example, neutron bombardment of ^{241}Am yields ^{242}Am:

$$^{241}_{95}\text{Am} \xrightarrow{(n,\gamma)} ^{242}_{95}\text{Am}$$

From there, 82.7% of ^{242}Am decays to ^{242}Cm and 17.3% to ^{242}Pu:

$$82.7\% \to ^{241}_{95}\text{Am} \xrightarrow{(n,\gamma)} ^{242}_{95}\text{Am} \xrightarrow[16.02 \text{ h}]{\beta^{-}} ^{242}_{96}\text{Cm}$$

$$17.3\% \to ^{241}_{95}\text{Am} \xrightarrow{(n,\gamma)} ^{242}_{95}\text{Am} \xrightarrow[16.02 \text{ h}]{\beta^{+}} ^{242}_{94}\text{Pu}$$

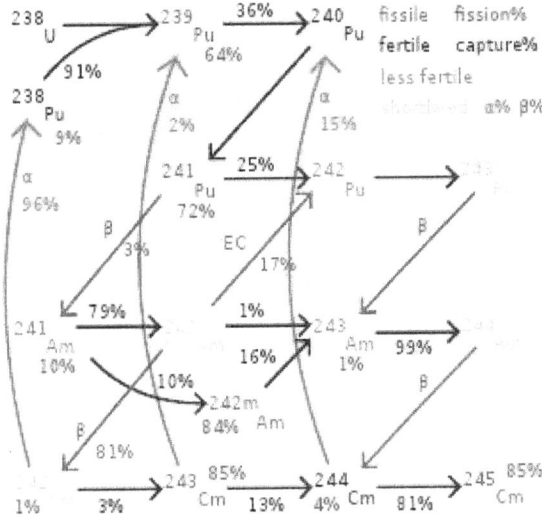

Chart displaying actinides and their decays and transmutations.

In the nuclear reactor, ^{242}Am is also up-converted by neutron capture to ^{243}Am and ^{244}Am, which transforms by β-decay to ^{244}Cm:

$$^{242}_{95}\text{Am} \xrightarrow{(n,\gamma)} \, ^{243}_{95}\text{Am} \xrightarrow{(n,\gamma)} \, ^{244}_{95}\text{Am} \xrightarrow[10.1\,h]{\beta^-} \, ^{244}_{96}\text{Cm}$$

Irradiation of ^{241}Am by ^{12}C or ^{22}Ne ions yields the isotopes ^{253}Es (einsteinium) or ^{263}Db (dubnium), respectively.[17] Furthermore, the element berkelium (^{243}Bk isotope) had been first intentionally produced and identified by bombarding ^{241}Am with alpha particles, in 1949, by the same Berkeley group, using the same 60-inch cyclotron that had been used for many previous experiments. Similarly, nobelium was produced at the Joint Institute for Nuclear Research, Dubna, Russia, in 1965 in several reactions, one of which included irradiation of ^{243}Am with ^{15}N ions. Besides, one of the synthesis reactions for lawrencium, discovered by scientists at Berkeley and Dubna, included bombardment of ^{243}Am with ^{18}O.[18]

21.3.5   Spectrometer

Americium-241 has been used as a portable source of both gamma rays and alpha particles for a number of medical and industrial uses. The 59.5409 keV gamma ray emissions from ^{241}Am in such sources can be used for indirect analysis of materials in radiography and X-ray fluorescence spectroscopy, as well as for quality control in fixed nuclear density gauges and nuclear densometers. For example, this isotope has been employed to gauge glass thickness to help create flat glass.[3] Americium-241 is also

suitable for calibration of gamma-ray spectrometers in the low-energy range, since its spectrum consists of nearly a single peak and negligible Compton continuum (at least three orders of magnitude lower intensity).[19]

21.3.6   Medicine

Americium-241 gamma rays has been used to provide passive diagnosis of thyroid function. This medical application is now obsolete. Americium-241's gamma rays can provide reasonable quality radiographs, with a 10-minute exposure time. ^{241}Am radiographs have only been taken for experimentally and are never used due to the long exposure time which is carcinogenic (due to gamma exposure), as well as the fact that like before, it takes a long time.[20]

21.4   Hazards

Americium-241 is a form of americium therefore having the same general hazards. Americium and its isotopes are both extremely toxic and radioactive. Although α-particles can be stopped by a sheet of paper, there are serious health concerns for ingestion of α-emitters. Americium and its isotopes are also very chemically toxic as well, in the form of heavy-metal toxicity. As little as 0.03 μCi (1,110 Bq) is the maximum permissible body burden for ^{241}Am.[21]

Americium-241 is an α-emitter with a weak γ-ray byproduct. Safely handling americium-241 requires knowing and following proper safety precautions, as without them it would be extremely dangerous. Its specific gamma dose constant is 3.14 x 10*−1 mR/hr/mCi or 8.48 x10*−5 mSv/hr/MBq at 1 meter.[22]

If consumed, americium-241 is excreted within a few days and only 0.05% is absorbed in the blood. From there, roughly 45% of it goes to the liver and 45% to the bones, and the remaining 10% is excreted. The uptake to the liver depends on the individual and increases with age. In the bones, americium is first deposited over cortical and trabecular surfaces and slowly redistributes over the bone with time. The biological half-life of ^{241}Am is 50 years in the bones and 20 years in the liver, whereas in the gonads (testicles and ovaries) it remains permanently; in all these organs, americium promotes formation of cancer cells as a result of its radioactivity.[23]

Americium-241 often enters landfills from discarded smoke detectors. The rules associated with the disposal of smoke detectors are relaxed in most jurisdictions. In the U.S., the "Radioactive Boy Scout" David Hahn was able to concentrate americium-241 from smoke detectors after managing to buy a hundred of them at remainder prices and also stealing a few."[24]"[25]"[26]"[27] There have been a few cases of exposure to americium-241, the worst case being that of Harold McCluskey, who at the age of 64 was exposed to 500 times the occupational standard for americium-241 as a result of an explosion in his lab. McCluskey died at the age of 75, not as a result of exposure, but of a heart disease which he had before the accident."[28]"[29]

21.5 References

[1] http://typhoon.tokai-sc.jaea.go.jp/icnc2003/Proceeding/paper/6.5_022.pdf%5B%5D Dias et al.

[2] "The Preparation, Properties and Uses of Americium-241, Alpha, Gamma and Neutron Sources" (PDF). Oak Ridge National Laboratory.

[3] Greenwood, p. 1262

[4] Smoke detectors and americium. World Nuclear Association, January 2009. Retrieved 28 November 2010

[5] BREDL Southern Anti-Plutonium Campaign, Blue Ridge Environmental Defense League, Retrieved 28 November 2010

[6] Sasahara, A.; et al. (2004). "Neutron and Gamma Ray Source Evaluation of LWR High Burn-up UO_2 and MOX Spent Fuels". Journal of Nuclear Science and Technology. 41 (4): 448–456. doi:10.3327/jnst.41.448. article/200410/000020041004A0333355.php Abstract

[7] "AMERICIUM-241".

[8] "GAMMA RAY SPECTRUM OF AM-241 IN A BACK SCATTERING GEOMETRY USING A HIGH PURITY GERMANIUM DETECTOR" (PDF).

[9] "Smoke Detectors and Americium". Nuclear Issues Briefing Paper, 35. May 2002, archived from the original on 2008-03-03, retrieved 2015-08-26

[10] Residential Smoke Alarm Performance, Thomas Cleary. Building and Fire Research Laboratory, National Institute of Standards and Technology; UL Smoke and Fire Dynamics Seminar. November 2007

[11] Bukowski, R. W. et al. (2007) Performance of Home Smoke Alarms Analysis of the Response of Several Available Technologies in Residential Fire Settings. NIST Technical Note 1455-1

[12] "Smoke detectors and americium-241 fact sheet" (PDF). Canadian Nuclear Society. Retrieved 31 August 2009.

[13] Gerberding, Julie Louise (2004). "Toxicological Profile For Americium" (PDF: 2.1 MB). United States Department of Health and Human Services/Agency for Toxic Substances and Disease Registry. Archived (PDF) from the original on 6 September 2009. Retrieved 29 August 2009.

[14] Basic elements of static RTGs

[15] G.L. Kulcinski, NEEP 602 Course Notes (Spring 2000), Nuclear Power in Space, University of Wisconsin Fusion Technology Institute (see last page)

[16] Space agencies tackle waning plutonium stockpiles, Spaceflight now. 9 July 2010

[17] Binder, Harry H. (1999). Lexikon der chemischen Elemente: das Periodensystem in Fakten, Zahlen und Daten : mit 96 Abbildungen und vielen tabellarischen Zusammenstellungen. ISBN 978-3-7776-0736-8.

[18] Greenwood, p. 1252

[19] Nuclear Data Viewer 2.4, NNDC

[20] "Americium-241 Uses" (PDF).

[21] "Americium Am".

[22] "AMERICIUM-241 [241Am]".

[23] Frisch, Franz Crystal Clear, 100 x energy, Bibliographisches Institut AG, Mannheim 1977, ISBN 3-411-01704-X, p. 184

[24] Ken Silverstein. The Radioactive Boy Scout: When a teenager attempts to build a breeder reactor. Harper's Magazine, November 1998

[25] "'Radioactive Boy Scout' Charged in Smoke Detector Theft". Fox News. 4 August 2007. Archived from the original on 8 December 2007. Retrieved 28 November 2007.

[26] "Man dubbed 'Radioactive Boy Scout' pleads guilty". Detroit Free Press. Associated Press. 27 August 2007. Archived from the original on 29 September 2007. Retrieved 27 August 2007.

[27] "'Radioactive Boy Scout' Sentenced to 90 Days for Stealing Smoke Detectors". Fox News. 4 October 2007. Archived from the original on 13 November 2007. Retrieved 28 November 2007.

[28] Cary, Annette (25 April 2008). "Doctor remembers Hanford's 'Atomic Man'". Tri-City Herald. Archived from the original on 10 February 2010. Retrieved 17 June 2008.

[29] AP wire (3 June 2005). "Hanford nuclear workers enter site of worst contamination accident". Archived from the original on 13 June 2005. Retrieved 17 June 2007.

Chapter 22

Isotopes of californium

Californium (Cf) is an artificial element, and thus a standard atomic mass cannot be given. Like all artificial elements, it has no stable isotopes. The first isotope to be synthesized was ^{245}Cf in 1950. There are 20 known radioisotopes ranging from ^{237}Cf to ^{256}Cf and one nuclear isomer, *249mCf. The longest-lived isotope is ^{251}Cf with a half-life of 900 years.

22.1 Californium-252

Californium-252 (Cf-252) undergoes spontaneous fission and is used in small sized neutron sources. Fission neutrons have an energy range of 0 to 13 MeV with a mean value of 2.3 MeV and a most probable value of 1 MeV.*[6]

22.1.1 Uses

This isotope produces high neutron emissions and can be used for a number of applications in industries such as nuclear energy, medicine, and petrochemical exploration.

Nuclear Reactors

The neutron sources produced from Cf-252 are most notably used in the start-up of nuclear reactors. Once a reactor is filled with nuclear fuel, the stable neutron emissions from the source material initiates the chain reaction known as fission.

Military & Defense

The portable isotopic neutron spectroscopy (PINS) used by United States Armed Forces, the National Guard, Homeland Security, and U.S. Customs and Border Protection, employs the use of Cf-252 sources to detect hazardous contents found inside artillery projectiles, mortar projectiles, rockets, bombs, land mines, and improvised explosive devices (IED).*[7]*[8]

Oil & Petroleum

In the oil industry, Cf-252 neutron sources are used to find layers of petroleum and water in a well. Instrumentation is lowered into the well which bombards the formation with high energy neutrons to determine porosity, permeability, and hydrocarbon presence along the length of the borehole.*[9]

Medicine

Californium-252 has also been used in the treatment of serious forms of cancer. In patients suffering from certain types of brain and cervical cancer, Cf-252 can be used as a more cost-effective substitute for radium.*[10]

22.2 Table

[1] Abbreviations:
  EC: Electron capture
  SF: Spontaneous fission

[2] Lightest nuclide known to undergo spontaneous fission as the main decay mode

[3] High neutron cross-section, tends to absorb neutrons

[4] Most common isotope

[5] High neutron emitter, average 3.7 neutrons per fission

22.2.1 Notes

- Values marked # are not purely derived from experimental data, but at least partly from systematic trends. Spins with weak assignment arguments are enclosed in parentheses.

- Uncertainties are given in concise form in parentheses after the corresponding last digits. Uncertainty values denote one standard deviation, except isotopic composition and standard atomic mass from IUPAC, which use expanded uncertainties.

22.3 References

[1] Plus radium (element 88). While actually a sub-actinide, it immediately precedes actinium (89) and follows a three-element gap of instability after polonium (84) where no isotopes have half-lives of at least four years (the longest-lived isotope in the gap is radon-222 with a half life of less than four *days*). Radium's longest lived isotope, at 1,600 years, thus merits the element's inclusion here.

[2] Specifically from thermal neutron fission of U-235, e.g. in a typical nuclear reactor.

[3] Milsted, J.; Friedman, A. M.; Stevens, C. M. (1965). "The alpha half-life of berkelium-247; a new long-lived isomer of berkelium-248". *Nuclear Physics*. **71** (2): 299. doi:10.1016/0029-5582(65)90719-4.
"The isotopic analyses disclosed a species of mass 248 in constant abundance in three samples analysed over a period of about 10 months. This was ascribed to an isomer of Bk^{248} with a half-life greater than 9 y. No growth of Cf^{248} was detected, and a lower limit for the β^- half-life can be set at about 10^4 y. No alpha activity attributable to the new isomer has been detected; the alpha half-life is probably greater than 300 y."

[4] This is the heaviest isotope with a half-life of at least four years before the "Sea of Instability".

[5] Excluding those "classically stable" isotopes with half-lives significantly in excess of ^{232}Th; e.g., while ^{113m}Cd has a half-life of only fourteen years, that of ^{113}Cd is nearly eight quadrillion years.

[6] Dicello, J. F.; Gross, W.; Kraljevic, U. (1972). "Radiation Quality of Californium-252". *Physics in Medicine and Biology*. **17** (3): 345. Bibcode:1972PMB....17..345D. doi:10.1088/0031-9155/17/3/301.

[7] "Portable Isotopic Neutron Spectroscopy (PINS) for the Military". *Frontier Technology Corp*. Retrieved 2016-02-24.

[8] Martin, R. C.; Knauer, J. B.; Balo, P. A. (2000-11-01). "Production, distribution and applications of californium-252 neutron sources". *Applied Radiation and Isotopes: Including Data, Instrumentation and Methods for Use in Agriculture, Industry and Medicine*. **53** (4-5): 785–792. doi:10.1016/s0969-8043(00)00214-1. ISSN 0969-8043. PMID 11003521.

[9] "Californium-252 & Antimony-Beryllium Sources". *Frontier Technology Corp*. Retrieved 2016-02-24.

[10] Maruyama, Y.; van Nagell, J. R.; Yoneda, J.; Donaldson, E.; Hanson, M.; Martin, A.; Wilson, L. C.; Coffey, C. W.; Feola, J. (1984-10-01). "Five-year cure of cervical cancer treated using californium-252 neutron brachytherapy". *American Journal of Clinical Oncology*. **7** (5): 487–493. doi:10.1097/00000421-198410000-00018. ISSN 0277-3732. PMID 6391143.

[11] "Universal Nuclide Chart". nucleonica. (registration required (help)).

- Isotope masses from:

  - G. Audi; A. H. Wapstra; C. Thibault; J. Blachot; O. Bersillon (2003). "The NUBASE evaluation of nuclear and decay properties" (PDF). *Nuclear Physics A*. **729**: 3–128. Bibcode:2003NuPhA.729....3A. doi:10.1016/j.nuclphysa.2003.11.001.

- Isotopic compositions and standard atomic masses from:

  - J. R. de Laeter; J. K. Böhlke; P. De Bièvre; H. Hidaka; H. S. Peiser; K. J. R. Rosman; P. D. P. Taylor (2003). "Atomic weights of the elements. Review 2000 (IUPAC Technical Report)". *Pure and Applied Chemistry*. **75** (6): 683–800. doi:10.1351/pac200375060683.

  - M. E. Wieser (2006). "Atomic weights of the elements 2005 (IUPAC Technical Report)". *Pure and Applied Chemistry*. **78** (11): 2051–2066. doi:10.1351/pac200678112051. Lay summary.

- Half-life, spin, and isomer data selected from the following sources. See editing notes on this article's talk page.

  - G. Audi; A. H. Wapstra; C. Thibault; J. Blachot; O. Bersillon (2003). "The NUBASE evaluation of nuclear and decay properties" (PDF). *Nuclear Physics A*. **729**: 3–128. Bibcode:2003NuPhA.729....3A. doi:10.1016/j.nuclphysa.2003.11.001.

  - National Nuclear Data Center. "NuDat 2.1 database". Brookhaven National Laboratory. Retrieved September 2005. Check date values in: |access-date= (help)

  - N. E. Holden (2004). "Table of the Isotopes". In D. R. Lide. *CRC Handbook of Chemistry and Physics* (85th ed.). CRC Press. Section 11. ISBN 978-0-8493-0485-9.

- Other

Chapter 23

Caesium-137

For the band, see Cesium 137 (band).

Caesium-137 (137
55Cs
, Cs-137), **cesium-137**, or **radiocaesium**, is a radioactive isotope of caesium which is formed as one of the more common fission products by the nuclear fission of uranium-235 and other fissionable isotopes in nuclear reactors and nuclear weapons. It is among the most problematic of the short-to-medium-lifetime fission products because it easily moves and spreads in nature due to the high water solubility of caesium's most common chemical compounds, which are salts.

23.1 Decay

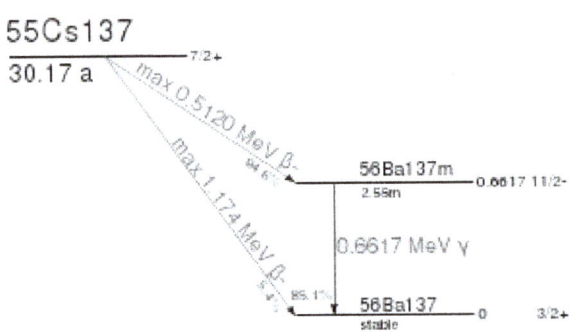

Cs-137 Decay Scheme

Caesium-137 has a half-life of about 30.17 years.[1] About 95 percent decays by beta emission to a metastable nuclear isomer of barium: barium-137m (* 137mBa, Ba-137m). The remainder directly populates the ground state of barium-137, which is stable. Ba-137m has a half-life of about 153 seconds, and is responsible for all of the emissions of gamma rays in samples of caesium-137. One gram of caesium-137 has an activity of 3.215 terabecquerel (TBq).[3]

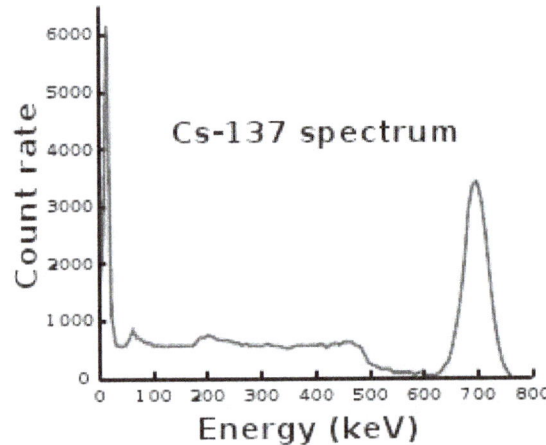

Cs-137 γ-spectrum

The main photon peak of Ba-137m is 662 keV.[4]

23.2 Uses

Caesium-137 has a number of practical uses. In small amounts, it is used to calibrate radiation-detection equipment.[5] In medicine, it is used in radiation therapy.[5] In industry, it is used in flow meters, thickness gauges,[5] moisture-density gauges (for density readings, with americium-241/beryllium providing the moisture reading),[6] and in gamma ray well logging devices.[6]

Caesium-137 is not widely used for industrial radiography because it is quite chemically reactive, and hence difficult to handle. The salts of caesium are also soluble in water, and this complicates the safe handling of caesium. Cobalt-60,
60
27Co
, is preferred for radiography, since it is chemically a rather nonreactive metal and produces higher energy gamma-ray photons.

As a man-made isotope, caesium-137 has been used to date

wine and detect counterfeits[7] and as a relative-dating material for assessing the age of sedimentation occurring after 1954.[8]

23.3 Health risk of radioactive caesium

Caesium-137 reacts with water, producing a water-soluble compound (caesium hydroxide). The biological behavior of caesium is similar to that of potassium and rubidium. After entering the body, caesium gets more or less uniformly distributed throughout the body, with the highest concentrations in soft tissue.[9]:114 The biological half-life of caesium is rather short, at about 70 days.[10] A 1972 experiment showed that when dogs are subjected to a whole body burden of 3800 μCi/kg (140 MBq/kg, or approximately 44 μg/kg) of caesium-137 (and 950 to 1400 rads), they die within 33 days, while animals with half of that burden all survived for a year.[11]

Accidental ingestion of caesium-137 can be treated with Prussian blue, which binds to it chemically and reduces the biological half-life to 30 days.[12]

23.4 Radioactive caesium in the environment

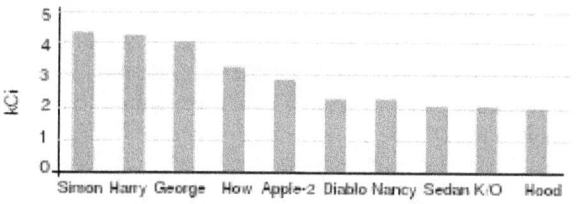

The ten highest deposits of caesium-137 from U.S. nuclear testing at the Nevada Test Site. Test explosions "Simon" and "Harry" were both from Operation Upshot–Knothole in 1953, while the test explosions "George" and "How" were from Operation Tumbler–Snapper in 1952

Caesium-134 and caesium-137 were released into the environment during nearly all nuclear weapon tests and some nuclear accidents, most notably the Chernobyl disaster and the Fukushima Daiichi disaster.

As of 2005 and for the next few hundred years, caesium-137 is the principal source of radiation in the zone of alienation around the Chernobyl nuclear power plant. Together with caesium-134, iodine-131, and strontium-90, caesium-137 was among the isotopes distributed by the reactor explosion that constitute the greatest risk to health. The

mean contamination of caesium-137 in Germany following the Chernobyl disaster was 2000 to 4000 Bq/m². This corresponds to a contamination of 1 mg/km² of caesium-137, totaling about 500 grams deposited over all of Germany. In Scandinavia, some reindeer and sheep exceeded the Norwegian legal limit (3000 Bq/kg) 26 years after Chernobyl.[13] As of 2016 the Chernobyl caesium-137 has decayed by half, but can have been locally concentrated by much larger factors.

In April 2011, elevated levels of caesium-137 were also being found in the environment after the Fukushima Daiichi nuclear disasters in Japan. In July 2011, meat from 11 cows shipped to Tokyo from Fukushima Prefecture was found to have 1,530 to 3,200 becquerels per kilogram of Cs-137, considerably exceeding the Japanese legal limit of 500 becquerels per kilogram at that time.[14] In March 2013, the Japanese utility that owns the tsunami-damaged nuclear power plant said that it had detected a record 740,000 becquerels per kilogram of radioactive caesium in a fish caught close to the plant. That is 7,400 times the government limit for safe human consumption.[15]

Caesium-137 is reported to be the major health concern in Fukushima. The government is under pressure to clean up radioactivity from Fukushima from as much land as possible so that some of the 110,000 people can return. A number of techniques are being considered that will be able to strip out 80% to 95% of the caesium from contaminated soil and other materials efficiently and without destroying the organic material in the soil. These include hydrothermal blasting. The caesium precipitated with ferric ferricyanide (Prussian blue) would be the only waste requiring special burial sites.[16] The aim is to get annual exposure from the contaminated environment down to 1 millisievert (mSv) above background. The most contaminated area where radiation doses are greater than 50 mSv/year must remain off limits, but some areas that are currently less than 5 mSv/year may be decontaminated, allowing 22,000 residents to return.

Caesium-137 in the environment is substantially anthropogenic (human-made). Unlike most other radioisotopes, caesium-137 is not produced from the same element's nonradioactive isotopes but as a byproduct of the nuclear fission of much heavier elements;[17] until the building of the first artificial nuclear reactor, the Chicago Pile-1, in late 1942, caesium-137 had not occurred on Earth in significant amounts for about 1.7 billion years. By observing the characteristic gamma rays emitted by this isotope, one can determine whether the contents of a given sealed container were made before or after the first atomic bomb explosion (Trinity test, 16 July 1945), which spread some of it into the atmosphere, quickly distributing trace amounts of it around the globe. This procedure has been used by researchers to check the authenticity of certain rare wines, most notably the purported "Jefferson bottles".[18] The short life of Cs137 across the Earth's entire surface also means one can date soils and sediments.

23.5 Incidents and accidents

23.5.1 1987 Goiânia, Goiás, Brazil

Caesium-137 gamma sources have been involved in several radiological accidents and incidents. Perhaps the best-known case is the Goiânia accident of 1987, in which an improperly disposed of radiation therapy system from an abandoned clinic in the city of Goiânia, Brazil, was scavenged from a junkyard, and the glowing caesium salt sold to curious, uneducated buyers. This led to four deaths and several serious injuries from radiation exposure.[19] Caesium gamma-ray sources that have been encased in metallic housings can be mixed in with scrap metal on its way to smelters, resulting in production of steel contaminated with radioactivity.[20]

23.5.2 1989 Kramatorsk, Donetsk, Ukraine

The Kramatorsk radiological accident happened in 1989 when a small capsule containing highly radioactive caesium-137 was found inside the concrete wall of an apartment building in Kramatorsk, Ukrainian SSR. It is believed that the capsule, originally a part of a measurement device, was lost in the late 1970s and ended up mixed with gravel used to construct the building in 1980. Over 9 years, two families lived in the apartment. By the time the capsule was discovered, 6 residents of the building had died from leukemia and 17 more had received varying doses of radiation.

23.5.3 1998, Los Barrios, Cádiz, Spain

In the Acerinox accident of 1998, the Spanish recycling company Acerinox accidentally melted down a mass of radioactive caesium-137 that came from a gamma-ray generator.[21]

23.5.4 2009 Tongchuan, Shaanxi, China

In 2009, a Chinese cement company (in Tongchuan, Shaanxi Province) was demolishing an old, unused cement plant and did not follow standards for handling radioactive materials. This caused some caesium-137 from a measuring instrument to be included with eight truckloads of scrap metal on its way to a steel mill, where the radioactive caesium was melted down into the steel.[22]

23.5.5 March 2015, University of Tromsø, Norway

In March 2015, the Norwegian University of Tromsø lost 8 radioactive samples including samples of caesium-137, Am-241 and strontium-90. The samples were moved out of a secure location to be used for education. When the samples were supposed to be returned the university was unable to find them. As of 4 November 2015 the samples are still missing.[23]

23.5.6 3 March 2016 Helsinki, Uusimaa, Finland

On 3 and 4 March 2016, unusually high levels of caesium-137 were detected in the air in Helsinki, Finland. According to STUK, the country's nuclear regulator, measurements showed 4,000 µBq/m^3 —about 1,000 times the usual level.[24] An investigation by the agency traced the source to a building from which STUK and a radioactive waste treatment company operate.[25]

23.6 See also

- Commonly used gamma-emitting isotopes

23.7 References

[1] National Institute of Standards and Technology. "Radionuclide Half-Life Measurements". Retrieved 7 November 2011.

[2] The Lund/LBNL Nuclear Data Search. "Nuclide Table". Archived from the original on 22 May 2015. Retrieved 14 March 2009.

[3] "NIST Nuclide Half-Life Measurements". NIST. Retrieved 13 March 2011.

[4] Delacroix, D.; Guerre, J. P.; Leblanc, P.; Hickman, C. (2002). Radionuclide and Radiation Protection Handbook. Nuclear Technology Publishing. ISBN 1870965876.

[5] "CDC Radiation Emergencies | Radioisotope Brief: Cesium-137 (Cs-137)". CDC. Retrieved 5 November 2013.

[6] "Cesium | Radiation Protection | US EPA". EPA. 3 June 2012. Archived from the original on 6 September 2015. Retrieved 4 March 2015.

[7] "How Atomic Particles Helped Solve A Wine Fraud Mystery". NPR. 3 June 2014. Retrieved 4 March 2015.

[8] Williams, H. F. L. (1995). "Assessing the impact of weir construction on recent sedimentation using cesium-137". *Environmental Geology*. **26** (3): 166–171. doi:10.1007/BF00768738. ISSN 0943-0105.

[9] Delacroix, D.; Guerre, J. P.; Leblanc, P.; Hickman, C. (2002). *Radionuclide and Radiation Protection Data Handbook 2002* (2nd ed.). Nuclear Technology Publishing. ISBN 1-870965-87-6.

[10] R. Nave. "Biological Half-life". *Hyperphysics*.

[11] H.C. Redman; et al. (1972). "Toxicity of 137-CsCl in the Beagle. Early Biological Effects". *Radiation Research*. **50** (3): 629–648. doi:10.2307/3573559. JSTOR 3573559. PMID 5030090.

[12] "CDC Radiation Emergencies | Facts About Prussian Blue". CDC. Retrieved 5 November 2013.

[13] Michael Sandelson; Lyndsey Smith (21 May 2012). "Higher radiation in Jotunheimen than first believed". The Foreigner. Retrieved 21 May 2012.

[14] "High levels of caesium in Fukushima beef". Independent Online. 9 July 2011.

[15] "Fish Near Fukushima Reportedly Contains High Cesium Level". Huffington Post. 17 March 2013.

[16] Dennis Normile. "Cooling a Hot Zone," Science, 339 (1 March 2013) pp. 1028-1029.

[17] Takeshi Okumura (21 October 2003). "The material flow of radioactive cesium-137 in the U.S. 2000" (PDF). *http://www.epa.gov/*. US Environmental Protection Agency. External link in |work= (help)

[18] "News Analysis: Christie's Is Counterfeit Crusader's Biggest Target Yet | Collecting News | Collecting". Wine Spectator. Retrieved 5 November 2013.

[19] *The Radiological Accident in Goiânia*. IAEA. 1988.

[20] "Radioactive Scrap Metal". *NuclearPolicy.com*. Nuclear Free Local Authorities. October 2000.

[21] J.M. LaForge (1999). "Radioactive Caesium Spill Cooks Europe". *Earth Island Journal*. Earth Island Institute. **14** (1).

[22] "Chinese 'find' radioactive ball". BBC News. 27 March 2009.

[23] "UiT har mistet radioaktivt stoff – kan ha blitt kastet". iTromsø. 4 November 2015.

[24] "High level of radioactive cesium detected in Helsinki air - Xinhua | English.news.cn". *news.xinhuanet.com*. Retrieved 2016-03-10.

[25] "Cesium 137 now traced back to the property's garage and parts of its basement premises - Tiedote-en - STUK". *www.stuk.fi*. Retrieved 2016-03-10.

23.8 Bibliography

- Rolf A. Olsen (1994). *Transfer of Radiocaesium from Soil to Plants and Fungi in Seminatural Ecosystems*; Nordic Radioecology—The Transfer of Radionuclides through Nordic Ecosystems to Man; Studies in Environmental Science; Volume 62, pages 265–286 (abstract)

23.9 External links

- NLM Hazardous Substances Databank – Cesium, Radioactive
- Cesium-137 dirty bombs by Theodore Liolios

Chapter 24

Cobalt-60

This article is about the nuclide cobalt-60. For other uses, see Cobalt-60 (disambiguation).

Cobalt-60, 60

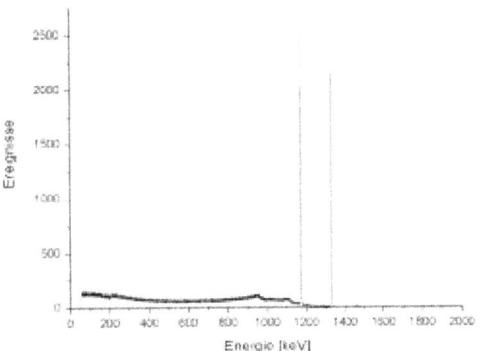

γ-ray spectrum of cobalt-60

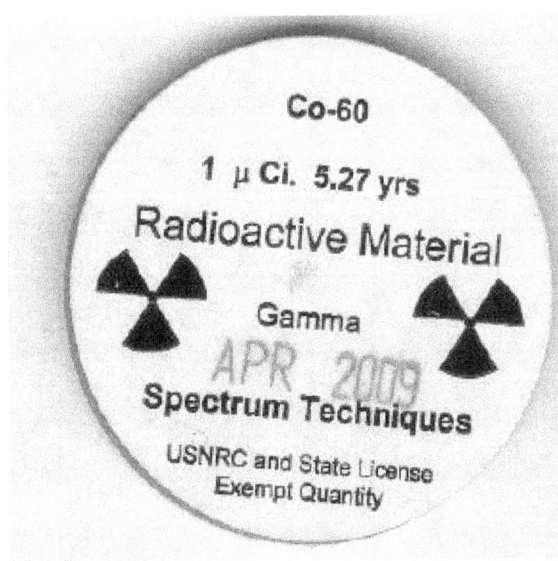

A container containing cobalt-60.

Co

, is a synthetic radioactive isotope of cobalt with a half-life of 5.2714 years. It is produced artificially in nuclear reactors. Deliberate industrial production depends on neutron activation of bulk samples of the monoisotopic and mononuclidic cobalt isotope 59

Co

.[3] Measurable quantities are also produced as a by-product of typical nuclear power plant operation and may be detected externally when leaks occur. In the latter case (in the absence of added cobalt) the incidentally produced 60

Co

is largely the result of multiple stages of neutron activation of iron isotopes in the reactor's steel structures[4] via the creation of 59

Co

precursor. The simplest case of the latter would result from the activation of 58

Fe

. 60

Co

decays by beta decay to the stable isotope nickel-60 (60

Ni

). The activated nickel nucleus emits two gamma rays with energies of 1.17 and 1.33 MeV, hence the overall nuclear equation of the reaction is 59

27Co

+ n → 60

27Co

→ 60

28Ni

+ e⁻ +

ν

e + gamma rays.

24.1 Activity

Corresponding to its half-life the radioactive activity of one gram of 60

Co

180

is 44 TBq (about 1100 curies). The *absorbed dose constant* is related to the decay energy and time. For 60
Co
it is equal to 0.35 mSv/(GBq h) at one meter from the source. This allows calculation of the equivalent dose, which depends on distance and activity.

Example: a 60
Co
source with an activity of 2.8 GBq, which is equivalent to 60 μg of pure 60
Co
. generates a dose of 1 mSv at one meter distance within one hour. The swallowing of 60
Co
reduces the distance to a few millimeters, and the same dose is achieved within seconds.

Test sources, such as those used for school experiments, have an activity of <100 kBq . Devices for nondestructive material testing use sources with activities of 1 TBq and more.

The high γ-energies result in a significant mass difference between 60
Ni
and 60
Co
of 0.003 u. This amounts to nearly 20 watts per gram, nearly 30 times larger than that of 238
Pu
.

24.2   Decay

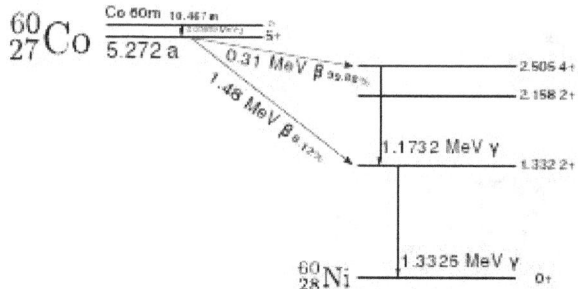

The decay scheme of 60
Co
and 60m
Co
.

The diagram shows a (simplified) decay scheme of 60
Co
and 60m
Co

. The main β-decay transitions are shown. The probability for population of the middle energy level of 2.1 MeV by β-decay is 0.0022%, with a maximum energy of 665.26 keV. Energy transfers between the three levels generate six different gamma-ray frequencies.[5] In the diagram the two important ones are marked. Internal conversion energies are well below the main energy levels.

60m
Co
is a nuclear isomer of 60
Co
with a half-life of 10.467 minutes. It decays by internal transition to 60
Co
, emitting 58.6 keV gamma rays, or with a low probability (0.22%) by β-decay into 60
Ni
.[6]

24.3   Applications

Security screening of cars at the Super Bowl using 60
Co
gamma-ray scanner

Prototype irradiator for food irradiation to prevent spoilage, 1984. The 60
Co
is in the central pipes

The main advantage of 60
Co
is that it is a high intensity gamma-ray emitter with a relatively long half-life, 5.27 years. compared to other gamma
ray sources of similar intensity. The β-decay energy is low
and easily shielded; however, the gamma-ray emission lines
have energies around 1.3 MeV, and are highly penetrating.
The main uses for 60
Co
are:

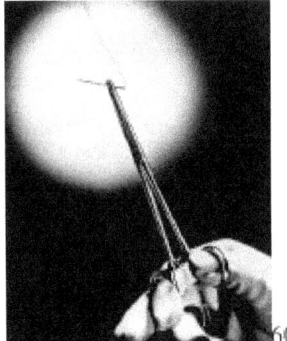

60
Co
needle implanted in tumors for radiotherapy. around 1955.

- As a tracer for cobalt in chemical reactions

- Sterilization of medical equipment.*[7]

- Radiation source for medical radiotherapy.*[8] Cobalt
  therapy, using beams of gamma rays from 60
  Co
  teletherapy machines to treat cancer, has been widely
  used since the 1950s.

- Radiation source for industrial radiography.*[8]

- Radiation source for leveling devices and thickness
  gauges.*[8]

- Radiation source for pest insect sterilization.*[9]

- As a radiation source for food irradiation and blood
  irradiation.*[7]

- As a radiation source for laboratory mutagenesis use.

- As a radiation source for stereoregular polymerization
  reactions, for example in preparation of syndiotactic
  polymethyl methacrylate.

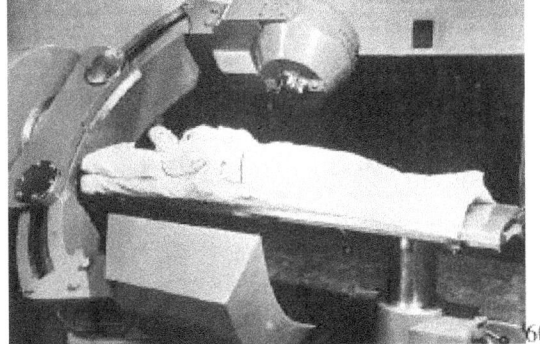

60
Co
teletherapy machine for cancer radiotherapy, early 1950s.

Brookhaven
plant mutation experiment using 60
Co
source in the pipe, center.

Cobalt has been discussed as a "salting" element to add to
nuclear weapons, to produce a cobalt bomb, an extremely
"dirty" weapon which would contaminate large areas with
60
Co
nuclear fallout, rendering them uninhabitable. In one hypothetical design, the tamper of the weapon would be made
of 59
Co
. When the bomb exploded, the excess neutrons from the
nuclear fission would irradiate the cobalt and transmute it
into 60
Co
. No nation is known to have done any serious development
of this type of weapon.

60
Co
source for sterilizing screwflies in the 1959 Screwworm
Eradication Program.

24.4   Occurrence

There is no natural 60
Co
in existence; thus, synthetic 60
Co
is created by bombarding a 59
Co
target with a slow neutron source. Californium-252, moderated through water, can be used for this purpose, as can the neutron flux in a nuclear reactor. The CANDU reactors can be used to activate 59
Co
. by substituting the stainless steel control rods with cobalt rods.[10] In the United States, it is now being produced in a BWR at Hope Creek Nuclear Generating Station. The cobalt targets are substituted here for a small number of fuel assemblies.[11]

59
Co
+ n → 60
Co

24.5   Safety

After entering a living mammal (such as a human being), some of the 60
Co
is excreted in feces. The remainder is taken up by tissues, mainly the liver, kidneys, and bones, where the prolonged exposure to gamma radiation can cause cancer. Over time, the absorbed cobalt is eliminated in urine.[8]

24.5.1   Steel contamination

Cobalt is an element of steel alloys. Uncontrolled disposal of 60
Co
in scrap metal is responsible for the radioactivity found in several iron-based products.[12][13]

In August 2012, Petco recalled several models of steel pet food bowls after US Customs and Border Protection determined that they were emitting low levels of radiation. The source of the radiation was determined to be 60
Co
that had contaminated the steel.[14]

In May 2013 a batch of metal-studded belts sold by online retailer Asos were confiscated and held in a US radioactive storage facility after testing positive for cobalt-60.[15]

24.5.2   Incidents involving medical radiation sources

In 2000, a disused radiotherapy head containing a 60
Co
source was stored at an unsecured location in Bangkok, Thailand and then accidentally sold to scrap collectors. Unaware of the dangers, a junkyard employee dismantled the head and extracted the source, which remained unprotected for a period of days at the junkyard. Ten people, including the scrap collectors and workers at the junkyard, were exposed to high levels of radiation and became ill. Three of the junkyard workers subsequently died as a result of their exposure, which was estimated to be over 6 Gy. Afterward, the source was safely recovered by the Thai authorities.[16]

In December 2013, a truck carrying a disused 111 TBq ^{60}Co teletherapy source from a hospital in Tijuana to a radioactive waste storage center was hijacked at a gas station near Mexico City.[17][18][19] The truck was recovered shortly after, but it was discovered that the thieves had removed the source from its shielding. It was found abandoned and intact in a field close by.[19][20] Despite early reports with lurid headlines asserting that the thieves were "likely doomed".[21] the radiation sickness was mild enough that the suspects were quickly released to police custody.[22] and no one is known to have died from the incident.[23]

24.6   Parity

Main article: Wu experiment

In 1957 Chien-Shiung Wu et al. discovered the β-decay process violated parity, implying nature has a handedness.[24]

In the Wu experiment her group aligned radioactive 60
Co
nuclei by cooling the source to low temperatures in a magnetic field. Wu's observation was that more β-rays were emitted in the opposite direction to the nuclear spin. This asymmetry violates parity conservation.

24.7 Suppliers

Argentina, Canada and Russia are the largest suppliers of cobalt-60 in the world.

24.8 See also

- Cobalt bomb
- Harold E. Johns

24.9 References

[1] National Institute of Standards and Technology. "Radionuclide Half-Life Measurements". Retrieved 2011-11-07.

[2] Korea Atomic Energy Research Institute. "Nuclide Table". Retrieved 2009-03-14.

[3] Malkoske, G. R. *Cobalt-60 production in CANDU power reactors*

[4] US EPA [http://www.epa.gov/radiation/radionuclides/cobalt.html#wheredoes Radiation Protection: Cobalt

[5] "Table of Isotopes decay data". Retrieved April 16, 2012.

[6] "Table of Isotopes decay data". Retrieved April 16, 2012.

[7] *Gamma Irradiators For Radiation Processing* (PDF). IAEA. 2005.

[8] "Cobalt | Radiation Protection | US EPA". EPA. Retrieved April 16, 2012.

[9] Croatia fruit farmers fight flies

[10] Isotope Production: Dual Use Power Plants

[11] PSEG Nuclear's Hope Creek reactor back on line, begins production of Cobalt-60

[12] radioactive contamination of steel

[13] "Lessons Learned The Hard Way". *IAEA Bulletin 47-2*. International Atomic Energy Agency. Archived from the original on 18 July 2010. Retrieved 16 April 2010.

[14] "Petco Recalls Some Stainless Steel Pet Bowls Due to Cobalt-60 Contamination". Retrieved 21 August 2012.

[15] "Asos Belts Seized Over Radioactive Studs". Sky News. 2013-05-28. Retrieved 2013-12-05.

[16] *The Radiological Accident in Samut Prakarn* (PDF). IAEA. 2002. Retrieved 2012-04-14.

[17] "Truck with dangerous radioactive materials hijacked in Mexico - IAEA —RT News". RT. 2013-12-04. Retrieved 2013-12-05.

[18] "Mexico Informs IAEA of Theft of Dangerous Radioactive Source". IAEA. Retrieved 2013-12-05.

[19] "Mexico Says Stolen Radioactive Source Found in Field". IAEA. 2013-12-05. Retrieved 2013-12-05.

[20] Will Grant (2013-12-05). "BBC News - Mexico radioactive material found, thieves' lives 'in danger'". BBC. Retrieved 2013-12-05.

[21] Gabriela Martinez, and Joshua Partlow (6 December 2013). "Thieves who stole lethal radioactive cobalt-60 in Mexico likely doomed". *Los Angeles Daily News*. Retrieved 12 March 2015.

[22] M. Alex Johnson (6 December 2013). "Six released from Mexican hospital but detained in theft of cobalt-60". *NBC News*. Retrieved 12 March 2015.

[23] Mary Cuddehe (13 November 2014). "What Happens When A Truck Carrying Radioactive Material Gets Robbed In Mexico". *BuzzFeed*. Retrieved 12 March 2015.

[24] Wu, C. S.; Ambler, E; Hayward, R. W.; Hoppes, D. D.; Hudson, R. P. (1957). "Experimental Test of Parity Conservation in Beta Decay". *Physical Review*. 105 (4): 1413–1415. Bibcode:1957PhRv..105.1413W. doi:10.1103/PhysRev.105.1413.

24.10 External links

- Cobalt-60, Centers for Disease Control and Prevention.
- NLM Hazardous Substances Databank – Cobalt, Radioactive
- Beta decay of Cobalt-60, HyperPhysics, Georgia State University.
- Dr. Henry Kelly. Cobalt-60 as a Dirty Bomb, Federation of American Scientists. March 6, 2002.

Chapter 25

Isotopes of iridium

There are two natural **isotopes of iridium** (**Ir**), and 34 radioisotopes, the most stable radioisotope being ^{192}Ir with a half-life of 73.83 days, and many nuclear isomers, the most stable of which is *192m2Ir with a half-life of 241 years. All other isomers have half-lives under a year, most under a day.

Relative atomic mass: 192.217(3).

25.1   Iridium-192

Iridium-192 (symbol ^{192}Ir) is a radioactive isotope of iridium, with a half-life of 73.83 days.*[1] It decays by emitting beta (β) particles and gamma (γ) radiation. About 96% of ^{192}Ir decays occur via emission of β and γ radiation, leading to ^{192}Pt. Some of the β particles are captured by other ^{192}Ir nuclei, which are then converted to ^{192}Os. Electron capture is responsible for the remaining 4% of ^{192}Ir decays.*[2]

Iridium-192 is also a strong gamma ray emitter. There are seven principal energy packets produced during its disintegration process ranging from just over 0.2 to about 0.6 MeV. Iridium-192 is commonly used as a gamma ray source in industrial radiography to locate flaws in metal components.*[3] It is also used in radiotherapy as a radiation source, in particular in brachytherapy.

Iridium-192 has accounted for the majority of cases tracked by the U.S. Nuclear Regulatory Commission in which radioactive materials have gone missing in quantities large enough to make a dirty bomb.*[4]

25.2   Table

[1] Abbreviations:
   EC: Electron capture
   IT: Isomeric transition

[2] Bold for stable isotopes, bold italics for nearly-stable isotopes (half-life longer than the age of the universe)

[3] Believed to undergo α decay to ^{187}*Re*

[4] Believed to undergo α decay to ^{189}Re

25.2.1   Notes

- Values marked # are not purely derived from experimental data, but at least partly from systematic trends. Spins with weak assignment arguments are enclosed in parentheses.

- Uncertainties are given in concise form in parentheses after the corresponding last digits. Uncertainty values denote one standard deviation, except isotopic composition and standard atomic mass from IUPAC, which use expanded uncertainties.

25.3   References

[1] "Radioisotope Brief: Iridium-192 (Ir-192)". Retrieved 20 March 2012.

[2] Braggerly, L. L. (1956). "The radioactive decay of Iridium-192 (Pd.D. Thesis)" (PDF). Pasadena, Calif.: California Institute of Technology: 1, 2, 7.

[3] Charles Hellier (2003). *Handbook of Nondestructive Evaluation*. McGraw-Hill. p. 6.20. ISBN 0-07-028121-1.

[4] Steve Coll (March 12, 2007). "The Unthinkable". *The New Yorker*. Retrieved 2007-03-09.

[5] "Universal Nuclide Chart". nucleonica. (registration required (help)).

- Isotope masses from:

  - G. Audi; A. H. Wapstra; C. Thibault; J. Blachot; O. Bersillon (2003). "The NUBASE evaluation of nuclear and decay properties" (PDF). *Nuclear Physics A*. **729**: 3–128. Bibcode:2003NuPhA.729....3A. doi:10.1016/j.nuclphysa.2003.11.001.

- Isotopic compositions and standard atomic masses from:

  - J. R. de Laeter; J. K. Böhlke; P. De Bièvre; H. Hidaka; H. S. Peiser; K. J. R. Rosman; P. D. P. Taylor (2003). "Atomic weights of the elements. Review 2000 (IUPAC Technical Report)". *Pure and Applied Chemistry*. **75** (6): 683–800. doi:10.1351/pac200375060683.

  - M. E. Wieser (2006). "Atomic weights of the elements 2005 (IUPAC Technical Report)". *Pure and Applied Chemistry*. **78** (11): 2051–2066. doi:10.1351/pac200678112051. Lay summary.

- Half-life, spin, and isomer data selected from the following sources. See editing notes on this article's talk page.

  - G. Audi; A. H. Wapstra; C. Thibault; J. Blachot; O. Bersillon (2003). "The NUBASE evaluation of nuclear and decay properties" (PDF). *Nuclear Physics A*. **729**: 3–128. Bibcode:2003NuPhA.729....3A. doi:10.1016/j.nuclphysa.2003.11.001.

  - National Nuclear Data Center. "NuDat 2.1 database". Brookhaven National Laboratory. Retrieved September 2005. Check date values in: |access-date= (help)

  - N. E. Holden (2004). "Table of the Isotopes". In D. R. Lide. *CRC Handbook of Chemistry and Physics* (85th ed.). CRC Press. Section 11. ISBN 978-0-8493-0485-9.

25.4   External links

- NLM Hazardous Substances Databank – Iridium, Radioactive (referring to iridium-192)

Chapter 26

Isotopes of polonium

Polonium (**Po**) has 33 isotopes, all of which are radioactive, with between 186 and 227 nucleons. ^{210}Po with a half-life of 138.376 days has the longest half-life of naturally occurring polonium. ^{209}Po with a half-life of 125 years has the longest half-life of all isotopes of polonium. ^{209}Po and ^{208}Po (half-life 2.9 years) can be made through the alpha, proton, or deuteron bombardment of lead or bismuth in a cyclotron.

26.1   Polonium-210

^{210}Po is an alpha emitter that has a half-life of 138.376 days; it decays directly to stable ^{206}Pb. A milligram of ^{210}Po emits as many alpha particles per second as 5 grams of ^{226}Ra.[1] A few curies (1 curie equals 37 gigabecquerels) of ^{210}Po emit a blue glow caused by excitation of surrounding air. A single gram of ^{210}Po generates 140 watts of power.[2] Because it emits many alpha particles, which are stopped within a very short distance in dense media and release their energy, ^{210}Po has been used as a lightweight heat source to power thermoelectric cells in artificial satellites; for instance, a ^{210}Po heat source was also in each of the Lunokhod rovers deployed on the surface of the Moon, to keep their internal components warm during the lunar nights.[3] Some anti-static brushes, used for neutralizing static electricity on materials like photographic film, contain a few microcuries of ^{210}Po as a source of charged particles.[4] ^{210}Po is also used in initiators for atomic bombs through the (α,n) reaction with beryllium.

The majority of the time ^{210}Po decays by emission of an alpha particle only, not by emission of an alpha particle and a gamma ray. About one in 100,000 decays results in the emission of a gamma ray.[5] This low gamma ray production rate makes it more difficult to find and identify this isotope. Rather than gamma ray spectroscopy, alpha spectroscopy is the best method of measuring this isotope.

^{210}Po occurs in minute amounts in nature, where it is an intermediate isotope in the uranium series (also known as the uranium series) decay chain. It is generated via beta decay from ^{210}Bi.

^{210}Po is extremely toxic, with one microgram being enough to kill the average adult (250,000 times more toxic than hydrogen cyanide by weight).[6] ^{210}Po was used to kill Russian dissident and ex-FSB officer Alexander V. Litvinenko in 2006,[7] and was suspected as a possible cause of Yasser Arafat's death, following exhumation and analysis of his corpse in 2012–2013.[8]

26.2   Table

[1] Abbreviations:
   EC: Electron capture
   IT: Isomeric transition

[2] Bold for stable isotopes, bold italics for nearly stable isotopes (half-life longer than the age of the universe)

[3] Intermediate decay product of Uranium-238

[4] Intermediate decay product of Uranium-235

[5] Intermediate decay product of Thorium-232

26.2.1   Notes

- Values marked # are not purely derived from experimental data, but at least partly from systematic trends. Spins with weak assignment arguments are enclosed in parentheses.

- Uncertainties are given in concise form in parentheses after the corresponding last digits. Uncertainty values denote one standard deviation, except isotopic composition and standard atomic mass from IUPAC, which use expanded uncertainties.

- Half-life abbreviations are y=year, d=day, min=minute, s=second, ms=millisecond, μs=microsecond, ns=nanosecond.

- A superscripted *m* (or *m2*, etc.) refers to an isomer of that particular isotope.

26.3   References

[1] C. R. Hammond. "The Elements" (PDF). Fermi National Accelerator Laboratory. p. 4-22.

[2] "Polonium" (PDF). Argonne National Laboratory. Archived from the original (PDF) on 2012-03-10.

[3] Andrew Wilson. *Solar System Log*. (London: Jane's Publishing Company Ltd. 1987). p. 64.

[4] "Staticmaster Alpha Ionizing Brush". Company 7.

[5] 210PO A DECAY Archived February 24. 2015. at the Wayback Machine.

[6] Sublette, Carey. "Polonium Poisoning".

[7] Cowell, Alan (November 24, 2006). "Radiation Poisoning Killed Ex-Russian Spy". *The New York Times*.

[8] "Arafat's death: what is Polonium-210?". *Al Jazeera*. July 10, 2012.

[9] J. R. de Laeter; J. K. Böhlke; P. De Bièvre; H. Hidaka; H. S. Peiser; K. J. R. Rosman; P. D. P. Taylor (2003). "Atomic weights of the elements. Review 2000 (IUPAC Technical Report)". *Pure and Applied Chemistry*. **75** (6): 683–800. doi:10.1351/pac200375060683.

[10] M. E. Wieser (2006). "Atomic weights of the elements 2005 (IUPAC Technical Report)". *Pure and Applied Chemistry*. **78** (11): 2051–2066. doi:10.1351/pac200678112051. Lay summary.

[11] G. Audi; A. H. Wapstra; C. Thibault; J. Blachot; O. Bersillon (2003). "The NUBASE evaluation of nuclear and decay properties" (PDF). *Nuclear Physics A*. **729**: 3–128. Bibcode:2003NuPhA.729....3A. doi:10.1016/j.nuclphysa.2003.11.001.

[12] National Nuclear Data Center. "NuDat 2.1 database". Brookhaven National Laboratory. Retrieved September 2005. Check date values in: |access-date= (help)

[13] N. E. Holden (2004). "Table of the Isotopes". In D. R. Lide. *CRC Handbook of Chemistry and Physics* (85th ed.). CRC Press. p. **11**–50. ISBN 978-0-8493-0485-9.

[14] "Universal Nuclide Chart". nucleonica. (registration required (help)).

[15] Boutin, Chad. "Polonium's Most Stable Isotope Gets Revised Half-Life Measurement". *nist.gov*. NIST Tech Beat. Retrieved 9 September 2014.

Chapter 27

Isotopes of radium

Radium (Ra) has no stable or nearly stable isotopes, and thus a standard atomic mass cannot be given. The longest lived, and most common, isotope of radium is ^{226}Ra with a half-life of 1,600 years. ^{226}Ra occurs in the decay chain of ^{238}U (often referred to as the radium series.) Radium has 33 known isotopes from ^{202}Ra to ^{234}Ra.

27.1 Table

[1] Abbreviations:
    CD: Cluster decay
    EC: Electron capture
    IT: Isomeric transition

[2] Bold for stable isotopes

[3] Intermediate decay product of ^{238}U

[4] Used for treating bone cancer

[5] Intermediate decay product of ^{235}U

[6] Intermediate decay product of ^{232}Th

[7] Source of element's name

[8] Intermediate decay product of ^{238}U

27.1.1 Notes

- Values marked # are not purely derived from experimental data, but at least partly from systematic trends. Spins with weak assignment arguments are enclosed in parentheses.

- Uncertainties are given in concise form in parentheses after the corresponding last digits. Uncertainty values denote one standard deviation, except isotopic composition and standard atomic mass from IUPAC, which use expanded uncertainties.

27.2 References

[1] Plus radium (element 88). While actually a sub-actinide, it immediately precedes actinium (89) and follows a three-element gap of instability after polonium (84) where no isotopes have half-lives of at least four years (the longest-lived isotope in the gap is radon-222 with a half life of less than four *days*). Radium's longest lived isotope, at 1,600 years, thus merits the element's inclusion here.

[2] Specifically from thermal neutron fission of U-235, e.g. in a typical nuclear reactor.

[3] Milsted, J.; Friedman, A. M.; Stevens, C. M. (1965). "The alpha half-life of berkelium-247; a new long-lived isomer of berkelium-248". *Nuclear Physics*. **71** (2): 299. doi:10.1016/0029-5582(65)90719-4.
    "The isotopic analyses disclosed a species of mass 248 in constant abundance in three samples analysed over a period of about 10 months. This was ascribed to an isomer of Bk^{248} with a half-life greater than 9 y. No growth of Cf^{248} was detected, and a lower limit for the β^- half-life can be set at about 10^4 y. No alpha activity attributable to the new isomer has been detected; the alpha half-life is probably greater than 300 y."

[4] This is the heaviest isotope with a half-life of at least four years before the "Sea of Instability".

[5] Excluding those "classically stable" isotopes with half-lives significantly in excess of ^{232}Th; e.g., while ^{113m}Cd has a half-life of only fourteen years, that of ^{113}Cd is nearly eight quadrillion years.

[6] "Universal Nuclide Chart". nucleonica. (registration required (help)).

- Isotope masses from:

  - G. Audi; A. H. Wapstra; C. Thibault; J. Blachot; O. Bersillon (2003). "The NUBASE evaluation of nuclear and decay properties" (PDF). *Nuclear Physics A*. **729**: 3–128. Bibcode:2003NuPhA.729....3A. doi:10.1016/j.nuclphysa.2003.11.001.

- Isotopic compositions and standard atomic masses from:

  - J. R. de Laeter; J. K. Böhlke; P. De Bièvre; H. Hidaka; H. S. Peiser; K. J. R. Rosman; P. D. P. Taylo (2003). "Atomic weights of the elements. Review 2000 (IUPAC Technical Report)". *Pure and Applied Chemistry*. **75** (6): 683–800. doi:10.1351/pac200375060683.

  - M. E. Wieser (2006). "Atomic weights of the elements 2005 (IUPAC Technical Report)". *Pure and Applied Chemistry*. **78** (11): 2051–2066. doi:10.1351/pac200678112051. Lay summary.

- Half-life, spin, and isomer data selected from the following sources. See editing notes on this article's talk page.

  - G. Audi; A. H. Wapstra; C. Thibault; J. Blachot; O. Bersillon (2003). "The NUBASE evaluation of nuclear and decay properties" (PDF). *Nuclear Physics A*. **729**: 3–128. Bibcode:2003NuPhA.729....3A. doi:10.1016/j.nuclphysa.2003.11.001.

  - National Nuclear Data Center. "NuDat 2.1 database". Brookhaven National Laboratory. Retrieved September 2005. Check date values in: |access-date= (help)

  - N. E. Holden (2004). "Table of the Isotopes". In D. R. Lide. *CRC Handbook of Chemistry and Physics* (85th ed.). CRC Press. Section 11. ISBN 978-0-8493-0485-9.

Chapter 28

Strontium-90

This article is about the chemical isotope. For the band, see Strontium 90 (band).

**Strontium-90 (90
Sr**

) is a radioactive isotope of strontium produced by nuclear fission, with a half-life of 28.8 years. It undergoes β^- – decay into yttrium-90, with a decay energy of 0.546 MeV.[1] Strontium-90 has applications in medicine and industry and is an isotope of concern in fallout from nuclear weapons and nuclear accidents.[2]

28.1   Radioactivity

Naturally occurring strontium is nonradioactive and nontoxic at levels normally found in the environment, but ^{90}Sr is a radiation hazard.[3] ^{90}Sr undergoes β^- – decay with a half-life of 28.79 years and a decay energy of 0.546 MeV distributed to an electron, an anti-neutrino, and the yttrium isotope ^{90}Y, which in turn undergoes β^- – decay with half-life of 64 hours and decay energy 2.28 MeV distributed to an electron, an anti-neutrino, and ^{90}Zr (zirconium), which is stable.[4] Note that ^{90}Sr/Y is almost a pure beta particle source; the gamma photon emission from the decay of ^{90}Y is so infrequent that it can normally be ignored.

^{90}Sr has a specific activity of 5.21 TBq/g.[5]

28.2   Fission product

^{90}Sr is a product of nuclear fission. It is present in significant amount in spent nuclear fuel and in radioactive waste from nuclear reactors and in nuclear fallout from nuclear tests. For thermal neutron fission as in today's nuclear power plants, the fission product yield from U-235 is 5.7%, from U-233 6.6%, but from Pu-239 only 2.0%.[6]

28.3   Biological effects

28.3.1   Biological activity

Strontium-90 is a "bone seeker" that exhibits biochemical behavior similar to calcium, the next lighter group 2 element.[3][7] After entering the organism, most often by ingestion with contaminated food or water, about 70–80% of the dose gets excreted.[2] Virtually all remaining strontium-90 is deposited in bones and bone marrow, with the remaining 1% remaining in blood and soft tissues.[2] Its presence in bones can cause bone cancer, cancer of nearby tissues, and leukemia. Exposure to ^{90}Sr can be tested by a bioassay, most commonly by urinalysis.[3]

The biological half-life of strontium-90 in humans has variously been reported as from 14 to 600 days,[8][9] 1000 days,[10] 18 years,[11] 30 years[12] and, at an upper limit, 49 years.[13] The wide ranging published biological half life figures are explained by strontium's complex metabolism within the body. However, by averaging all excretion paths, the overall biological half life is estimated to be about 18 years.[14]

The elimination rate of strontium-90 is strongly affected by age and sex, due to differences in bone metabolism.[15]

Together with the caesium isotopes ^{134}Cs, ^{137}Cs, and iodine isotope ^{131}I it was among the most important isotopes regarding health impacts after the Chernobyl disaster. As strontium has an affinity to the calcium-sensing receptor of parathyroid cells that is similar to that of calcium, the increased risk of liquidators of the Chernobyl power plant to suffer from primary hyperparathyroidism could be explained by binding of strontium-90.[16]

28.4   Uses

28.4.1 Radioisotope Thermoelectric Generators (RTGs)

The radioactive decay of strontium-90 generates a significant amount of heat, 0.536 W/g in the form of pure strontium metal or approximately 0.256 W/g as strontium titanate[17] and is cheaper than the alternative ^{238}Pu. It is used as a heat source in many Russian/Soviet radioisotope thermoelectric generators, usually in the form of strontium titanate.[18] It was also used in the US "Sentinel" series of RTGs.[19]

28.4.2 Industrial applications

^{90}Sr finds use in industry as a radioactive source for thickness gauges.[2]

28.4.3 Medical applications

^{90}Sr finds extensive use in medicine as a radioactive source for superficial radiotherapy of some cancers. Controlled amounts of ^{90}Sr and ^{89}Sr can be used in treatment of bone cancer. It is also used as a radioactive tracer in medicine and agriculture.[2]

28.5 ^{90}Sr contamination in the environment

Strontium-90 is not quite as likely as caesium-137 to be released as a part of a nuclear reactor accident because it is much less volatile, but is probably the most dangerous component of the radioactive fallout from a nuclear weapon.[20]

A study of hundreds of thousands of deciduous teeth, collected by Dr. Louise Reiss and her colleagues as part of the Baby Tooth Survey, found a large increase in ^{90}Sr levels in through the 1950s and early 1960s. The study's final results showed that children born in St. Louis, Missouri in 1963 had levels of ^{90}Sr in their deciduous teeth that was 50 times higher than that found in children born in 1950, before the advent of large-scale atomic testing. Commentators on the study said that the fallout was likely to cause increased cases of diseases in those who absorb strontium-90 into their bones.[21]

An article with the study's initial findings was circulated to U.S. President John F. Kennedy in 1961, and helped convince him to sign the Partial Nuclear Test Ban Treaty with the United Kingdom and Soviet Union, ending the above-ground nuclear weapons testing that placed the greatest amounts of nuclear fallout into the atmosphere.[22]

The Chernobyl disaster released roughly 10 PBq, or about 5% of the core inventory, of strontium-90 into the environment.[23] The Fukushima Daiichi disaster released 0.1-1 PBq of strontium-90 in the form of contaminated cooling water into the Pacific Ocean.[24]

28.6 External links

- NLM Hazardous Substances Databank – Strontium, Radioactive

28.7 References

[1] "Table of Isotopes decay data". Lund University. Retrieved 2014-10-13.

[2] "Strontium | Radiation Protection | US EPA". EPA. 24 April 2012. Retrieved 18 June 2012.

[3] TOXICOLOGICAL PROFILE FOR STRONTIUM (PDF). Agency for Toxic Substances and Disease Registry. April 2004. retrieved 2014-10-13

[4] Decay data from National Nuclear Data Center at the Brookhaven National Laboratory in the US.

[5] Delacroix, D.; Guerre, J. P.; Leblanc, P.; Hickman, C. (2002). Radionuclide and Radiation Protection Data Handbook 2002 (2nd ed.). Nuclear Technology Publishing. ISBN 1-870965-87-6.

[6] "Livechart - Table of Nuclides - Nuclear structure and decay data". IAEA. Retrieved 2014-10-13.

[7] "NRC: Glossary -- Bone seeker". US Nuclear Regulatory Commission. 7 May 2014. Retrieved 2014-10-13.

[8] Tiller, B. L. (2001). "4.5 Fish and Wildlife Surveillance", Hanford Site 2001 Environmental Report (PDF), DOE, retrieved 2014-01-14

[9] Driver, C.J. (1994). Ecotoxicity Literature Review of Selected Hanford Site Contaminants (PDF), DOE, doi:10.2172/10136486, retrieved 2014-01-14

[10] "Freshwater Ecology and Human Influence". Area IV Envirothon. Retrieved 2014-01-14.

[11] "Radioisotopes That May Impact Food Resources" (PDF). Epidemiology, Health and Social Services. State of Alaska. Retrieved 2014-01-14.

[12] "Human Health Fact Sheet: Strontium" (PDF). Argonne National Laboratory. October 2001. Retrieved 2014-01-14.

[13] "Biological Half-life". HyperPhysics. Retrieved 2014-01-14.

[14] Glasstone, Samuel; Dolan, Philip J. (1977). "XII: Biological Effects". *The effects of Nuclear Weapons* (PDF). p. 605. Retrieved 2014-01-14.

[15] Shagina, N B; Bougrov, N G; Degteva, M O; Kozheurov, V P; Tolstykh, E I (2006). "An application of in vivo whole body counting technique for studying strontium metabolism and internal dose reconstruction for the Techa River population". *Journal of Physics: Conference Series*. **41**: 433–440. doi:10.1088/1742-6596/41/1/048. ISSN 1742-6588.

[16] Boehm BO, Rosinger S, Belyi D, Dietrich JW (August 2011). "The Parathyroid as a Target for Radiation Damage". *New England Journal of Medicine*. **365** (7): 676–678. doi:10.1056/NEJMc1104982. PMID 21848480. Retrieved 19 August 2011.

[17] Harris, Dale; Epstein, Joseph (1968). *Properties of Selected Radioisotopes* (PDF). NASA

[18] Standring, WJF; Selnæs, ØG; Sneve, M; Finne, IE; Hosseini, A; Amundsen, I; Strand, P (2005). *Assessment of environmental, health and safety consequences of decommissioning radioisotope thermal generators (RTGs) in Northwest Russia* (PDF) (StrålevernRapport 2005:4). Østerås: Norwegian Radiation Protection Authority

[19] "Power Sources for Remote Arctic Applications" (PDF). Washington, DC: U.S. Congress, Office of Technology Assessment. June 1994. OTA-BP-ETI-129.

[20] "Nuclear Fission Fragments". HyperPhysics. Retrieved 18 June 2012.

[21] Schneir, Walter (April 25, 1959). "Strontium-90 in U.S. Children". *The Nation*. **188** (17): 355–357.

[22] Hevesi, Dennis. "Dr. Louise Reiss, Who Helped Ban Atomic Testing, Dies at 90". *The New York Times*. January 10, 2011. Accessed January 10, 2011.

[23] "II: The release, dispersion and deposition of radionuclides", *Chernobyl: Assessment of Radiological and Health Impacts* (PDF), NEA, 2002

[24] Povinec, P. P.; Aoyama, M.; Biddulph, D.; et al. (2013). "Cesium, iodine and tritium in NW Pacific waters – a comparison of the Fukushima impact with global fallout". *Biogeosciences*. **10** (8): 5481–5496. doi:10.5194/bg-10-5481-2013. ISSN 1726-4189.

Chapter 29

Goiânia accident

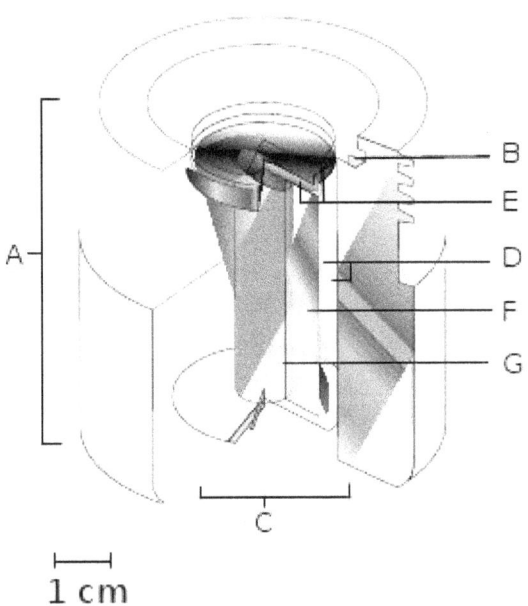

1 cm

A teletherapy radiation capsule composed of the following:
A.) an international standard source holder (usually lead),
B.) a retaining ring, and
C.) a teletherapy "source" composed of
D.) two nested stainless steel canisters welded to
E.) two stainless steel lids surrounding
F.) a protective internal shield (usually uranium metal or a tungsten alloy) and
G.) a cylinder of radioactive source material, often but not always cobalt-60. In the Goiânia incident it was caesium-137. The diameter of the "source" is 30 mm.

The **Goiânia accident** was a radioactive contamination accident that occurred on September 13, 1987, at Goiânia, in the Brazilian state of Goiás, after an old radiotherapy source was stolen from an abandoned hospital site in the city. It was subsequently handled by many people, resulting in four deaths. About 112,000 people were examined for radioactive contamination and 249 were found to have significant levels of radioactive material in or on their bodies.[1][2]

In the cleanup operation, topsoil had to be removed from several sites, and several houses were demolished. All the objects from within those houses were removed and examined. *Time* magazine has identified the accident as one of the world's "worst nuclear disasters" and the International Atomic Energy Agency called it "one of the world's worst radiological incidents".[3][4]

29.1   Description of the source

The radiation source in the Goiânia accident was a small capsule containing about 93 grams (3.3 oz) of highly radioactive caesium chloride (a caesium salt made with a radioisotope, caesium-137) encased in a shielding canister made of lead and steel. The source was positioned in a container of the wheel type, where the wheel turns inside the casing to move the source between the storage and irradiation positions.[1]

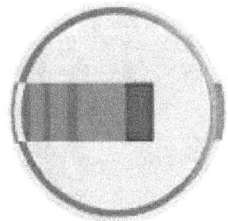

A wheel type radiotherapy device which has a long collimator to focus the radiation into a narrow beam. The caesium chloride radioactive source is the blue rectangle, and gamma rays are represented by the beam emerging from the aperture.

The activity of the source was 74 terabecquerels (TBq) in 1971. The International Atomic Energy Agency (IAEA) describes the container – 51 millimeters (2 inches) in diameter and 48 mm (1.8 inches) long – as an "international standard capsule". The specific activity of the active solid was about 814 TBq kg*−1 of caesium-137 (half life of 30 years). The dose rate at one meter from the source was 4.56 grays per hour (456 rad·h*−1). While the serial number of the device was unknown, thus hindering definitive identification, the device was thought to have been made in the United States at Oak Ridge National Laboratory and was used as a radiation source for radiation therapy at the Goiânia hospital.[1]

The IAEA states that the source contained 50.9 TBq (1.380 Ci) when it was taken and that about 44 TBq (1200 Ci, 87%) of contamination had been recovered during the cleanup operation. This means that 7 TBq (190 Ci) remained in the environment; it will have decayed to about 3.5 TBq (95 Ci) by 2016.

29.2 Events

29.2.1 Hospital abandonment

Culture and Convention Center, built on the site where the IGR had been located

The Instituto Goiano de Radioterapia (IGR), a private radiotherapy institute in Goiânia,[1] was just 1 km (0.6 mi) northwest of Praça Cívica, the administrative center of the city. It moved to its new premises in 1985, leaving behind a caesium-137-based teletherapy unit that had been purchased in 1977.[5] The fate of the abandoned site was disputed in court between IGR and the Society of Saint Vincent de Paul, then owner of the premises.[6] On September 11, 1986, the Court of Goiás stated it had knowledge of the abandoned radioactive material in the building.[6]

Four months before the theft, on May 4, 1987, Saura Taniguti, then director of Ipasgo, the institute of insurance for civil servants, used police force to prevent one of the owners of IGR, Carlos Figueiredo Bezerril, from removing the objects that were left behind.[6] Figueiredo then warned the president of Ipasgo, Lício Teixeira Borges, that he should take responsibility "for what would happen with the caesium bomb".[6]

The court posted a security guard to protect the hazardous abandoned equipment.[7] Meanwhile, the owners of IGR wrote several letters to the National Nuclear Energy Commission, warning them about the danger of keeping a teletherapy unit at an abandoned site, but they could not remove the equipment by themselves once a court order prevented them from doing so.[6][7]

29.2.2 Theft of the source

On September 13, 1987, taking advantage of the absence of the guard,[7] Roberto dos Santos Alves and Wagner Mota Pereira illegally entered the partially demolished facility. They partially disassembled the teletherapy unit, and placed the source assembly – which they thought might have some scrap value – in a wheelbarrow, taking it to Alves's home.[1] There, they began dismantling the equipment. That same evening, they both began to vomit. Nevertheless, they continued in their efforts. The following day, Pereira began to experience diarrhea and dizziness and his left hand began to swell. He soon developed a burn on this hand in the same size and shape as the aperture – he eventually underwent partial amputation of several fingers.[8]

On September 15, Pereira visited a local clinic where his symptoms were diagnosed as the result of something he had eaten, and he was told to return home and rest.[1] Alves, however, continued with his efforts to dismantle the equipment. In the course of this effort, he eventually freed the caesium capsule from its protective rotating head. His prolonged exposure to the radioactive material led to his right forearm becoming ulcerated, requiring amputation.[9]

29.2.3 Source is partially broken

On September 16, Alves succeeded in puncturing the capsule's aperture window with a screwdriver, allowing him to see a deep blue light coming from the tiny opening he had created.[1] He inserted the screwdriver and successfully scooped out some of the glowing substance. Thinking it was perhaps a type of gunpowder, he tried to light it, but the powder would not ignite. The exact mechanism by which the light was generated was not known at the time the IAEA report was written, though it was thought to be either ionized air glow, fluorescence or Cherenkov radiation associated with the absorption of moisture by the source; similar blue light was observed in 1988 at Oak Ridge National

Laboratory during the disencapsulation of a ^{137}Cs source.

29.2.4   Source is sold and dismantled

On September 18, Alves sold the items to a nearby scrapyard. That night, Devair Alves Ferreira (the owner of the scrapyard) noticed the blue glow from the punctured capsule. Thinking the capsule's contents were valuable or even supernatural, he immediately brought it into his house. Over the next three days, he invited friends and family to view the strange glowing substance.

On September 21 at the scrapyard, one of Ferreira's friends (given as EF1 in the IAEA report) succeeded in freeing several rice-sized grains of the glowing material from the capsule using a screwdriver. Alves Ferreira began to share some of them with various friends and family members. That same day, his wife, 37-year-old Gabriela Maria Ferreira, began to fall ill. On September 25, 1987, Devair Alves Ferreira sold the scrap metal to a second scrapyard.

29.2.5   Ivo and his daughter

The day before the sale to the second scrapyard, on September 24, Ivo, Devair's brother, successfully scraped some additional dust out of the source and took it to his house a short distance away. There he spread some of it on the concrete floor. His six-year-old daughter, Leide das Neves Ferreira, later ate a sandwich while sitting on this floor. She was also fascinated by the blue glow of the powder, applying it to her body and showing it off to her mother. Dust from the powder fell on the sandwich she was consuming; she eventually absorbed 1.0 GBq and received a total dose of 6.0 Gy, more than a fatal dose even with treatment.[10]

29.2.6   Gabriela Maria Ferreira notifies authorities

Gabriela Maria Ferreira had been the first to notice that many people around her had become severely ill at the same time.[11]

On September 28, 1987 – 15 days after the item was found —she reclaimed the materials from the rival scrapyard and transported them to a hospital. Because the remains of the source were kept in a plastic bag, the level of contamination at the hospital was low.

29.2.7   Source's radioactivity is detected

In the morning of September 29, 1987 a visiting medical physicist [12] used a scintillation counter to confirm the presence of radioactivity and persuaded the authorities to take immediate action. The city, state, and national governments were all aware of the incident by the end of the day.

29.3   Health outcomes

News of the radiation incident was broadcast on local, national, and international media. Within days, nearly 130,000 people swarmed local hospitals concerned that they might have been exposed.[2] Of those, 250 were indeed found to be contaminated—some with radioactive residue still on their skin—through the use of Geiger counters.[2] Eventually, 20 people showed signs of radiation sickness and required treatment.[2]

29.3.1   Fatalities

Ages in years are given, with dosages listed in grays (Gy).

- **Leide das Neves Ferreira**, age 6 (6.0 Gy), was the daughter of Ivo Ferreira. When an international team arrived to treat her, she was discovered confined to an isolated room in the hospital because the hospital staff were afraid to go near her. She gradually experienced swelling in the upper body, hair loss, kidney and lung damage, and internal bleeding. She died on October 23, 1987, of "septicemia and generalized infection" at the Marcilio Dias Navy Hospital, in Rio de Janeiro.[13] She was buried in a common cemetery in Goiânia, in a special fiberglass coffin lined with lead to prevent the spread of radiation. Despite these measures, news of her impending burial caused a riot of more than 2,000 people in the cemetery on the day of her burial, all fearing that her corpse would poison the surrounding land. Rioters tried to prevent her burial by using stones and bricks to block the cemetery roadway.[14] She was buried despite this interference.

- **Gabriela Maria Ferreira**, aged 37 (5.7 Gy), wife of scrapyard owner Devair Ferreira, became sick about three days after coming into contact with the substance. Her condition worsened, and she developed internal bleeding, especially in the limbs, eyes, and digestive tract, and hair loss. She suffered mental confusion, diarrhea, and acute renal insufficiency before also dying on October 23, 1987, of "septicemia and generalized infection",[13][15] about a month after exposure.

- **Israel Baptista dos Santos**, aged 22 (4.5 Gy), was an employee of Devair Ferreira who worked on the radioactive source primarily to extract the lead. He developed serious respiratory and lymphatic complications, was eventually admitted to hospital, and died six days later on October 27, 1987.

- **Admilson Alves de Souza**, aged 18 (5.3 Gy), was also an employee of Devair Ferreira who worked on the radioactive source. He developed lung damage, internal bleeding, and heart damage, and died October 18, 1987.

Devair Ferreira himself survived despite receiving 7 Gy of radiation. He died in 1994 of cirrhosis aggravated by depression and binge drinking.[16]

29.3.2 Other individuals

The outcomes for the 46 most contaminated people are shown in the bar chart below. Several people survived high doses of radiation. This is thought in some cases to be because the dose was fractionated. Given time, the body's repair mechanisms will reverse cell damage caused by radiation. If the dose is spread over a long time period, these mechanisms can mitigate the effects of radiation poisoning.

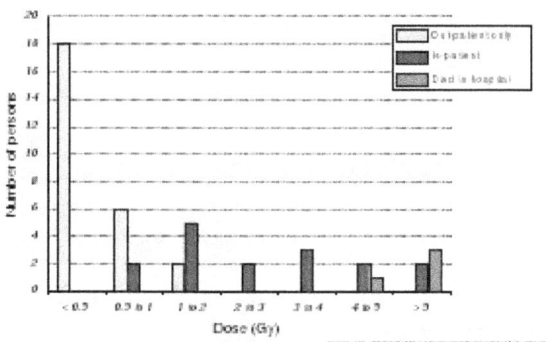

This is a barchart showing the outcome for the 46 most contaminated people for whom a dose estimate has been made. The people are divided into seven groups according to dose.

29.3.3 Other affected people

Afterwards, about 112,000 people were examined for radioactive contamination; 249 were found to have significant levels of radioactive material in or on their body.[1] Of this group, 129 people had internal contamination. The majority of the internally contaminated people only suffered small doses (< 50 mSv, less than a 1 in 400 risk of getting cancer as a result).

A thousand people were identified as having suffered a dose which was greater than one year of background radiation; it is thought that 97% of these people had a dose of between 10 and 200 mSv (between 1 in 2,000 and 1 in 100 risk of developing cancer as a result).

In 2007, the Oswaldo Cruz Foundation determined that the rate of caesium-137 related diseases are the same in Goiânia accident survivors as they are in the population at large. Nevertheless, compensation is still distributed to survivors, who suffer radiation-related prejudices in everyday life.[17]

29.4 Legal matters

In light of the deaths caused, the three doctors who had owned and run IGR were charged with criminal negligence. Because the accidents occurred before the promulgation of the Federal Constitution of 1988 and because the substance was acquired by the clinic and not by the individual owners, the court could not declare the owners of IGR liable. One of the medical doctors owning IGR and the clinic's physicist were ordered to pay R$100,000 for the derelict condition of the building. The two thieves were not included as defendants in the public civil suit.

In 2000, CNEN, the National Nuclear Energy Commission, was ordered by the 8th Federal Court of Goiás to pay compensation of R$ 1.3 million (near US$750,000) and to guarantee medical and psychological treatment for the direct and indirect victims of the accident and their descendants down to the third generation.[18]

29.5 Cleanup

29.5.1 Objects and places

Topsoil had to be removed from several sites, and several houses were demolished. All the objects from within those houses were removed and examined. Those that were found to be free of radioactivity were wrapped in plastic bags, while those that were contaminated were either decontaminated or disposed of as waste. In industry, the choice between decontaminating or disposing objects is based on only the economic value of the object and the ease of decontamination. In this case, the IAEA recognized that to reduce the psychological impact of the event, greater effort should have been taken to clean up items of personal value, such as jewelry and photographs. It is not clear from the IAEA report to what degree this was practiced.

29.5.2   Means and methods

After the houses were emptied, vacuum cleaners were used to remove dust, and plumbing was examined for radioactivity. Painted surfaces could be scraped, while floors were treated with acid and Prussian blue mixtures. Roofs were vacuumed and hosed, but two houses had to have their roofs removed. The waste from the cleanup was moved out of the city to a remote place for storage.

Potassium alum dissolved in hydrochloric acid was used on clay, concrete, soil, and roofs. Caesium has a high affinity for many clays.

Organic solvents, followed by potassium alum dissolved in hydrochloric acid, were used to treat waxed/greased floors and tables. Sodium hydroxide solutions, also followed by dissolved potassium alum, were used to treat synthetic floors, machines and typewriters.

Prussian blue was used to internally decontaminate many people, although by the time it was applied, much of the radioactive material had already migrated from the bloodstream to the muscle tissue, greatly hampering its effectiveness. Urine from victims was treated with ion exchange resin to compact the waste for ease of storage.

29.5.3   Recovery considerations

The cleanup operation was much harder for this event than it could have been because the source was opened and the active material was water soluble. A sealed source need only be picked up, placed in a lead container, and transported to the radioactive waste storage. In the recovery of lost sources, the IAEA recommends careful planning and using a crane or other device to place shielding (such as a pallet of bricks or a concrete block) near the source to protect recovery workers.

29.5.4   Contamination locations

The Goiânia accident spread significant radioactive contamination throughout the Aeroporto, Central, and Ferroviários districts. Even after the cleanup, 7 TBq of radioactivity remained unaccounted for.

Some of the key contamination sites:

- **Goiânia's Instituto Goiano de Radioterapia (IGR)** (16°40′29″S 49°15′51″W / 16.6746°S 49.2641°W)[1] suffered no actual exposure or breach of radioactive contents, but the site is noteworthy as the source of deadly, unsecured material. The IGR clinic no longer exists, having been re-

placed around 2000 with the modernized Centro de Convenções de Goiânia (Goiânia Convention Center).

- **Roberto dos Santos' house** (16°40′07″S 49°15′48″W / 16.66848°S 49.26341°W)[1] on Rua 57. The radioactive source was here for about six days, and it was partially broken into.

- **Devair Ferreira's scrapyard** (16°40′02″S 49°15′59″W / 16.66713°S 49.26652°W),[1] on Rua 15A ("Junkyard I") in the Aeroporto section of the city, had possession of the items for 7 days. The caesium container was entirely dismantled, spreading significant contamination. Extreme radiation levels of up to 1.5 Sv·h^{-1} were found by investigators in the middle of the scrapyard.

- **Ivo Ferreira's house** (16°39′50″S 49°16′09″W / 16.66401°S 49.26911°W)[1] ("Junkyard II"), at 1F Rua 6. Some of the contamination was spread about the house, causing the fatalities of Leide das Neves Ferreira and Gabriela Maria Ferreira. The adjacent junkyard scavenged the remainder of parts from the IGR facility. The premises were heavily contaminated, with radiation dose rates up to 2 Sv·h^{-1}.

- **Junkyard III** (16°40′09″S 49°16′48″W / 16.66915°S 49.28003°W).[1] This junkyard had possession of the items for 3 days until they were sent away.

- **Vigilância Sanitária** (16°40′30″S 49°16′23″W / 16.675°S 49.273°W).[1] Here, the substance was quarantined, and an official cleanup response began.

Other contamination was also found in or on:[19]

- 50,000 rolls of toilet paper
- Three buses
- 42 houses
- 14 cars
- five pigs

29.6   Legacy

29.6.1   Disposal of the capsule

The original teletherapy capsule was seized by the Brazilian military as soon as it was discovered, and since then the empty capsule has been on display at the *Escola de Instrução Especializada* ("School of Specialized Instruction") in Rio de Janeiro as a memento to those who participated in the cleanup of the contaminated area.

29.6.2 Research

In 1991, a group of researchers collected blood samples from highly exposed survivors of the incident. Subsequent analysis resulted in the publication of numerous scientific articles.[20][21][22][23]

29.6.3 Film

A 1990 film *Césio 137 – O Pesadelo de Goiânia* ("Caesium-137 – The Nightmare of Goiânia"), a dramatisation of the incident, was made by Roberto Pires.[24] It won several awards at the 1990 Festival de Brasília.[25]

29.6.4 Foundation

The state government of Goiás established the Fundação Leide das Neves Ferreira in February 1988, both to study the extent of contamination of the population as a result of the incident and to render aid to those affected.[26]

29.7 See also

- List of civilian radiation accidents

29.8 References

[1] *The Radiological accident in Goiânia* (PDF). Vienna: International Atomic Energy Agency. 1988. ISBN 92-0-129088-8.

[2] Foderaro, Lisa (July 8, 2010). "Columbia Scientists Prepare for a Threat: A Dirty Bomb". *The New York Times*.

[3] The Worst Nuclear Disasters

[4] Yukiya Amano (March 26, 2012). "Time to better secure radioactive materials". *Washington Post*.

[5] Puig, Diva E. "Apuntes sobre energía nuclear. Lo que los abogados deben saber sobre la tecnología nuclear" (in Spanish). Noticias Jurídicas. Retrieved May 2, 2014. |chapter= ignored (help)

[6] (Portuguese) Godinho, Iúri. "Os médicos e o acidente radioativo". *Jornal Opção*. February 8, 2004. Archived September 18, 2009, at the Wayback Machine.

[7] (Portuguese) Borges, Weber. "O jornalista que foi vítima do césio". *Jornal Opção*. May 27, 2007.

[8] Planeta Diário: July 2010

[9] Aint No Way to Go: All That Glitters

[10] "Brazil Deadly Glitter". *Time*. October 19, 1987.

[11] "2 Die of Radiation Poisoning in Brazil". *Los Angeles Times*. October 24, 1987.

[12] "País está preparado para atuar em acidente radioativo" [Country is prepared to act in radioactive incident] (in Portuguese). Ministry of Science, Technology and Innovation (MCTI). 13 September 2012. Retrieved 10 November 2013. Note: person named only as "WF" in the IAEA report.

[13] "Vida Verde" (in Portuguese). 1987. p. 15.

[14] "Memorial Césio 137" (in Portuguese). Greenpeace.

[15] Malheiros, Tania (1996). *Histórias secretas do Brasil nuclear* (in Portuguese). Rio de Janeiro: WVA. p. 122. ISBN 9788585644086.

[16] Irene, Mirelle (13 September 2012). "Goiânia, 25 anos depois: 'perguntam até se brilhamos', diz vítima". *Terra*. Retrieved 5 December 2013.

[17] UOL. Vítimas do césio 137 voltam a receber remédios e pedem assistência médica para todos. September 25, 2012

[18] "Case Law and Administrative Decisions, Judgement of the Federal Court in the Public Civil Action concerning the Goiânia Accident" (PDF). OECD. 2000. Archived from the original on 2013-12-06. ()

[19] Steinhauser, Friedrich (November 2007). "Countering Radiological Terrorism: Consequences of the Radiation Exposure Incident in Goiania (Brazil)". *Volume 29 NATO Science for Peace and Security Series: Human and Societal Dynamics*: 7.

[20] Da Cruz, AD; Curry, J; Curado, MP; Glickman, BW (1996). "Monitoring hprt mutant frequency over time in T-lymphocytes of people accidentally exposed to high doses of ionizing radiation". *Environmental and molecular mutagenesis*. **27** (3): 165–75. doi:10.1002/(SICI)1098-2280(1996)27:3<165::AID-EM1>3.0.CO;2-E. PMID 8625952.

[21] Saddi, V; Curry, J; Nohturfft, A; Kusser, W; Glickman, BW (1996). "Increased hprt mutant frequencies in Brazilian children accidentally exposed to ionizing radiation". *Environmental and molecular mutagenesis*. **28** (3): 267–75. doi:10.1002/(SICI)1098-2280(1996)28:3<267::AID-EM11>3.0.CO;2-D. PMID 8908186.

[22] Da Cruz, AD; Volpe, JP; Saddi, V; Curry, J; Curadoc, MP; Glickman, BW (1997). "Radiation risk estimation in human populations: lessons from the radiological accident in Brazil". *Mutation research*. **373** (2): 207–14. doi:10.1016/S0027-5107(96)00199-6. PMID 9042402.

[23] Skandalis, A; Da Cruz, AD; Curry, J; Nohturfft, A; Curado, MP; Glickman, BW (1997). "Molecular analysis of T-lymphocyte HPRT– mutations in individuals exposed to ionizing radiation in Goiânia, Brazil". *Environmental and molecular mutagenesis.* **29** (2): 107–16. doi:10.1002/(SICI)1098-2280(1997)29:2<107::AID-EM1>3.0.CO;2-B. PMID 9118962.

[24] *Césio 137 - O Pesadelo de Goiânia* at the Internet Movie Database

[25] UraniumFilmFestival.org: Roberto Pires

[26] Camargo Da Silva, T. (1997). Leibing, Annette, ed. *Biomedical Discourses and Health Care Experiences: The Goiâna Radiological Disaster. The Medical Anthropologies in Brazil. Curare* Sonderband. **12**. Berlin: Verlag für Wissenschaft Und Bildung. pp. 72–73. ISBN 9783861355687.

29.9   External links

- Detailed Report from the International Atomic Energy Agency, Vienna, 1988

- Similar accidents over the world (short overview)

- The Goiânia Radiation Incident

- Radioactive waste sold as scrap in India

- Q&A: Health effects of radiation exposure, *BBC News*, 21 July 2011.

Chapter 30

Biological warfare

For the use of biological agents by terrorists, see bioterrorism.

"Germ Warfare" redirects here. For the *M*A*S*H* episode, see Germ Warfare (M*A*S*H).

Biological warfare (BW)—also known as **germ warfare** —is the use of biological toxins or infectious agents such as bacteria, viruses, and fungi with the intent to kill or incapacitate humans, animals or plants as an act of war. Biological weapons (often termed "bio-weapons", "biological threat agents", or "bio-agents") are living organisms or replicating entities (viruses, which are not universally considered "alive") that reproduce or replicate within their host victims. Entomological (insect) warfare is also considered a type of biological weapon. This type of warfare is distinct from nuclear warfare and chemical warfare, which together with biological warfare make up NBC, the military acronym for nuclear, biological, and chemical warfare using weapons of mass destruction (WMDs). None of these are conventional weapons, which are deployed primarily for their explosive, kinetic, or incendiary potential.

Biological weapons may be employed in various ways to gain a strategic or tactical advantage over the enemy, either by threats or by actual deployments. Like some of the chemical weapons, biological weapons may also be useful as area denial weapons. These agents may be lethal or non-lethal, and may be targeted against a single individual, a group of people, or even an entire population. They may be developed, acquired, stockpiled or deployed by nation states or by non-national groups. In the latter case, or if a nation-state uses it clandestinely, it may also be considered bioterrorism.[1]

There is an overlap between biological warfare and chemical warfare, as the use of toxins produced by living organisms is considered under the provisions of both the Biological Weapons Convention and the Chemical Weapons Convention. Toxins and psychochemical weapons are often referred to as *midspectrum agents*. Unlike bioweapons, these midspectrum agents do not reproduce in

their host and are typically characterized by shorter incubation periods.[2]

30.1 Overview

Offensive biological warfare, including mass production, stockpiling and use of biological weapons, was outlawed by the 1972 Biological Weapons Convention (BWC). The rationale behind this treaty, which has been ratified or acceded to by 170 countries as of April 2013,[3] is to prevent a biological attack which could conceivably result in large numbers of civilian casualties and cause severe disruption to economic and societal infrastructure. Many countries, including signatories of the BWC, currently pursue research into the defense or protection against BW, which is not prohibited by the BWC.

A nation or group that can pose a credible threat of mass casualty has the ability to alter the terms on which other nations or groups interact with it. Biological weapons allow for the potential to create a level of destruction and loss of life far in excess of nuclear, chemical or conventional weapons, relative to their mass and cost of development and storage. Therefore, biological agents may be useful as strategic deterrents in addition to their utility as offensive weapons on the battlefield.[4][5]

As a tactical weapon for military use, a significant problem with a BW attack is that it would take days to be effective, and therefore might not immediately stop an opposing force. Some biological agents (smallpox, pneumonic plague) have the capability of person-to-person transmission via aerosolized respiratory droplets. This feature can be undesirable, as the agent(s) may be transmitted by this mechanism to unintended populations, including neutral or even friendly forces. While containment of BW is less of a concern for certain criminal or terrorist organizations, it remains a significant concern for the military and civilian populations of virtually all nations.

30.2   History

Main article: History of biological warfare

Rudimentary forms of biological warfare have been practiced since antiquity.[6] During the 6th century BC, the Assyrians poisoned enemy wells with a fungus that would render the enemy delirious. In 1346, the bodies of Mongol warriors of the Golden Horde who had died of plague were thrown over the walls of the besieged Crimean city of Kaffa. Specialists disagree over whether this operation may have been responsible for the spread of the Black Death into Europe.[7][8][9][10]

Historians have long debated inconclusively whether the British Army used smallpox against Native Americans during the Siege of Fort Pitt in 1763.[11] It has been claimed that the British Marines used smallpox in New South Wales in 1789.[12]

By 1900 the germ theory and advances in bacteriology brought a new level of sophistication to the techniques for possible use of bio-agents in war. Biological sabotage—in the form of anthrax and glanders—was undertaken on behalf of the Imperial German government during World War I (1914–1918), with indifferent results.[13] The Geneva Protocol of 1925 prohibited the use of chemical weapons and biological weapons.

With the onset of World War II, the Ministry of Supply in the United Kingdom established a BW program at Porton Down, headed by the microbiologist Paul Fildes. The research was championed by Winston Churchill and soon tularemia, anthrax, brucellosis, and botulism toxins had been effectively weaponized. In particular, Gruinard Island in Scotland, during a series of extensive tests was contaminated with anthrax for the next 56 years. Although the UK never offensively used the biological weapons it developed on its own, its program was the first to successfully weaponize a variety of deadly pathogens and bring them into industrial production.[14] Other nations, notably France and Japan had begun their own biological weapons programs.[15]

When the USA entered the war, Allied resources were pooled at the request of the British and the U.S. established a large research program and industrial complex at Fort Detrick, Maryland in 1942 under the direction of George W. Merck.[16] The biological and chemical weapons developed during that period were tested at the Dugway Proving Grounds in Utah. Soon there were facilities for the mass production of anthrax spores, brucellosis, and botulism toxins, although the war was over before these weapons could be of much operational use.[17]

The most notorious program of the period was run by

Shiro Ishii, commander of Unit 731, which performed live human vivisections and other biological experimentation.

the secret Imperial Japanese Army Unit 731 during the war, based at Pingfan in Manchuria and commanded by Lieutenant General Shirō Ishii. This unit did research on BW, conducted often fatal human experiments on prisoners, and produced biological weapons for combat use.[18] Although the Japanese effort lacked the technological sophistication of the American or British programs, it far outstripped them in its widespread application and indiscriminate brutality. Biological weapons were used against both Chinese soldiers and civilians in several military campaigns.[19] In 1940, the Japanese Army Air Force bombed Ningbo with ceramic bombs full of fleas carrying the bubonic plague.[20] Many of these operations were ineffective due to inefficient delivery systems,[18] although up to 400,000 people may have died.[21] During the Zhejiang-Jiangxi Campaign in 1942, around 1,700 Japanese troops died out of a total 10,000 Japanese soldiers who fell ill with disease when their own biological weapons attack rebounded on their own forces.[22][23]

During the final months of World War II, Japan planned to use plague as a biological weapon against U.S. civilians in San Diego, California, during Operation Cherry Blossoms at Night. The plan was set to launch on 22 September 1945.

but it was not executed because of Japan's surrender on 15 August 1945.[24][25][26][27]

In Britain, the 1950s saw the weaponization of plague, brucellosis, tularemia and later equine encephalomyelitis and vaccinia viruses, but the programme was unilaterally cancelled in 1956. The United States Army Biological Warfare Laboratories weaponized anthrax, tularemia, brucellosis, Q-fever and others.

In 1969, the UK and the Warsaw Pact, separately, introduced proposals to the UN to ban biological weapons, and US President Richard Nixon terminated production of biological weapons, allowing only scientific research for defensive measures. The Biological and Toxin Weapons Convention was signed by the US, UK, USSR and other nations, as a ban on "development, production and stockpiling of microbes or their poisonous products except in amounts necessary for protective and peaceful research" in 1972. However, the Soviet Union continued research and production of massive offensive biological weapons in a program called Biopreparat, despite having signed the convention.[28] By 2011, 165 countries had signed the treaty and none are proven—though nine are still suspected[29]—to possess offensive BW programs.[29]

The international biological hazard symbol

30.3 Modern BW operations

30.3.1 Offensive

It has been argued that rational state actors would never use biological weapons offensively. The argument is that biological weapons cannot be controlled: the weapon could backfire and harm the army on the offensive, perhaps having even worse effects than on the target. An agent like smallpox or other airborne viruses would almost certainly spread worldwide and ultimately infect the user's home country. However, this argument does not necessarily apply to bacteria. For example, anthrax can easily be controlled and even created in a garden shed; the FBI suspects it can be done for as little as $2,500 using readily available laboratory equipment.[30] Also, using microbial methods, bacteria can be suitably modified to be effective in only a narrow environmental range, the range of the target that distinctly differs from the army on the offensive. Thus only the target might be affected adversely. The weapon may be further used to bog down an advancing army making them more vulnerable to counterattack by the defending force.

Anti-personnel

Ideal characteristics of a biological agent to be used as a weapon against humans are high infectivity, high virulence,

non-availability of vaccines, and availability of an effective and efficient delivery system. Stability of the weaponized agent (ability of the agent to retain its infectivity and virulence after a prolonged period of storage) may also be desirable, particularly for military applications, and the ease of creating one is often considered. Control of the spread of the agent may be another desired characteristic.

The primary difficulty is not the production of the biological agent, as many biological agents used in weapons can often be manufactured relatively quickly, cheaply and easily. Rather, it is the weaponization, storage and delivery in an effective vehicle to a vulnerable target that pose significant problems.

For example, *Bacillus anthracis* is considered an effective agent for several reasons. First, it forms hardy spores, perfect for dispersal aerosols. Second, this organism is not considered transmissible from person to person, and thus rarely if ever causes secondary infections. A pulmonary anthrax infection starts with ordinary influenza-like symptoms and progresses to a lethal hemorrhagic mediastinitis within 3–7 days, with a fatality rate that is 90% or higher in untreated patients.[31] Finally, friendly personnel can be protected with suitable antibiotics.

A large-scale attack using anthrax would require the creation of aerosol particles of 1.5 to 5 μm: larger particles would not reach the lower respiratory tract, while smaller particles would be exhaled back out into the atmosphere. At this size, conductive powders tend to aggregate because of electrostatic charges, hindering dispersion. So the material must be treated to insulate and neutralize the charges. The weaponized agent must be resistant to degradation by rain and ultraviolet radiation from sunlight, while retaining the ability to efficiently infect the human lung. There are other technological difficulties as well, chiefly relating to storage of the weaponized agent.

Agents considered for weaponization, or known to be weaponized, include bacteria such as *Bacillus anthracis*, *Brucella spp.*, *Burkholderia mallei*, *Burkholderia pseudomallei*, *Chlamydophila psittaci*, *Coxiella burnetii*, *Francisella tularensis*, some of the Rickettsiaceae (especially *Rickettsia prowazekii* and *Rickettsia rickettsii*), *Shigella spp.*, *Vibrio cholerae*, and *Yersinia pestis*. Many viral agents have been studied and/or weaponized, including some of the Bunyaviridae (especially Rift Valley fever virus), Ebolavirus, many of the Flaviviridae (especially Japanese encephalitis virus), Machupo virus, Marburg virus, Variola virus, and Yellow fever virus. Fungal agents that have been studied include *Coccidioides spp.*.[32][33]

Toxins that can be used as weapons include ricin, staphylococcal enterotoxin B, botulinum toxin, saxitoxin, and many mycotoxins. These toxins and the organisms that produce them are sometimes referred to as select agents. In the United States, their possession, use, and transfer are regulated by the Centers for Disease Control and Prevention's Select Agent Program.

The former US biological warfare program categorized its weaponized anti-personnel bio-agents as either **Lethal Agents** (*Bacillus anthracis*, *Francisella tularensis*, Botulinum toxin) or **Incapacitating Agents** (*Brucella suis*, *Coxiella burnetii*, Venezuelan equine encephalitis virus, Staphylococcal enterotoxin B).

Anti-agriculture

Anti-crop/anti-vegetation/anti-fisheries

The United States developed an anti-crop capability during the Cold War that used plant diseases (bioherbicides, or mycoherbicides) for destroying enemy agriculture. Biological weapons also target fisheries as well as water-based vegetation. It was believed that destruction of enemy agriculture on a strategic scale could thwart Sino-Soviet aggression in a general war. Diseases such as wheat blast and rice blast were weaponized in aerial spray tanks and cluster bombs for delivery to enemy watersheds in agricultural regions to initiate epiphytotics (epidemics among plants). When the United States renounced its offensive biological warfare program in 1969 and 1970, the vast majority of its biological arsenal was composed of these plant diseases. Enterotoxins and Mycotoxins were not affected by Nixon's order.

Though herbicides are chemicals, they are often grouped with biological warfare and chemical warfare because they

may work in a similar manner as biotoxins or bioregulators. The Army Biological Laboratory tested each agent and the Army's Technical Escort Unit was responsible for transport of all chemical, biological, radiological (nuclear) materials. Scorched earth tactics or destroying livestock and farmland were carried out in the Vietnam war (cf. Agent Orange)[34] and Eelam War in Sri Lanka.

Biological warfare can also specifically target plants to destroy crops or defoliate vegetation. The United States and Britain discovered plant growth regulators (i.e., herbicides) during the Second World War, and initiated a herbicidal warfare program that was eventually used in Malaya and Vietnam in counterinsurgency operations.

Anti-livestock

In 1980s Soviet Ministry of Agriculture had successfully developed variants of foot-and-mouth disease, and rinderpest against cows, African swine fever for pigs, and psittacosis to kill chicken. These agents were prepared to spray them down from tanks attached to airplanes over hundreds of miles. The secret program was code-named "Ecology".[32]

Attacking animals is another area of biological warfare intended to eliminate animal resources for transportation and food. In the First World War, German agents were arrested attempting to inoculate draft animals with anthrax, and they were believed to be responsible for outbreaks of glanders in horses and mules. The British tainted small feed cakes with anthrax in the Second World War as a potential means of attacking German cattle for food denial, but never employed the weapon. In the 1950s, the United States had a field trial with hog cholera. During the Mau Mau Uprising in 1952, the poisonous latex of the African milk bush was used to kill cattle.[35]

Outside the context of war, humans have deliberately introduced the rabbit disease Myxomatosis, originating in South America, to Australia and Europe, with the intention of reducing the rabbit population – which had devastating but temporary results, with wild rabbit populations reduced to a fraction of their former size but survivors developing immunity and increasing again.

Entomological warfare

Main article: Entomological warfare

Entomological warfare (EW) is a type of biological warfare that uses insects to attack the enemy. The concept has existed for centuries and research and development have continued into the modern era. EW has been used in battle by Japan and several other nations have developed and been accused of using an entomological warfare program. EW may employ insects in a direct attack or as vectors to deliver a biological agent, such as plague. Essentially, EW exists in three varieties. One type of EW involves infecting insects with a pathogen and then dispersing the insects over target areas.*[36] The insects then act as a vector, infecting any person or animal they might bite. Another type of EW is a direct insect attack against crops; the insect may not be infected with any pathogen but instead represents a threat to agriculture. The final method uses uninfected insects, such as bees, wasps, etc., to directly attack the enemy.*[37]

30.3.2 Defensive

Main article: Biodefense

Research and development into medical countermeasures

In 2010 at The Meeting of the States Parties to the Convention on the Prohibition of the Development, Production and Stockpiling of Bacteriological (Biological) and Toxin Weapons and Their Destruction in Geneva*[38] the sanitary epidemiological reconnaissance was suggested as well-tested means for enhancing the monitoring of infections and parasitic agents, for practical implementation of the International Health Regulations (2005). The aim was to prevent and minimize the consequences of natural outbreaks of dangerous infectious diseases as well as the threat of alleged use of biological weapons against BTWC States Parties.

Role of public health and disease surveillance

It is important to note that most classical and modern biological weapons' pathogens can be obtained from a plant or an animal which is naturally infected.*[39]

Indeed, in the largest biological weapons accident known—the anthrax outbreak in Sverdlovsk (now Yekaterinburg) in the Soviet Union in 1979, sheep became ill with anthrax as far as 200 kilometers from the release point of the organism from a military facility in the southeastern portion of the city and still off limits to visitors today, see Sverdlovsk Anthrax leak).*[40]

Thus, a robust surveillance system involving human clinicians and veterinarians may identify a bioweapons attack early in the course of an epidemic, permitting the prophylaxis of disease in the vast majority of people (and/or animals) exposed but not yet ill.

For example, in the case of anthrax, it is likely that by 24–36 hours after an attack, some small percentage of individuals (those with compromised immune system or who had received a large dose of the organism due to proximity to the release point) will become ill with classical symptoms and signs (including a virtually unique chest X-ray finding, often recognized by public health officials if they receive timely reports).*[41] The incubation period for humans is estimated to be about 11.8 days to 12.1 days. This suggested period is the first model that is independently consistent with data from the largest known human outbreak. These projections refines previous estimates of the distribution of early onset cases after a release and supports a recommended 60-day course of prophylactic antibiotic treatment for individuals exposed to low doses of anthrax.*[42] By making these data available to local public health officials in real time, most models of anthrax epidemics indicate that more than 80% of an exposed population can receive antibiotic treatment before becoming symptomatic, and thus avoid the moderately high mortality of the disease.*[41]

Common epidemiological clues that may signal biological attack

From most specific to least specific:*[43]

1. Single cause of a certain disease caused by an uncommon agent, with lack of an epidemiological explanation.

2. Unusual, rare, genetically engineered strain of an agent.

3. High morbidity and mortality rates in regards to patients with the same or similar symptoms.

4. Unusual presentation of the disease.

5. Unusual geographic or seasonal distribution.

6. Stable endemic disease, but with an unexplained increase in relevance.

7. Rare transmission (aerosols, food, water).

8. No illness presented in people who were/are not exposed to "common ventilation systems (have separate closed ventilation systems) when illness is seen in persons in close proximity who have a common ventilation system."

9. Different and unexplained diseases coexisting in the same patient without any other explanation.

10. Rare illness that affects a large, disparate population (respiratory disease might suggest the pathogen or agent was inhaled).

11. Illness is unusual for a certain population or age-group in which it takes presence.

12. Unusual trends of death and/or illness in animal popu-

lations, previous to or accompanying illness in humans.

13. Many affected reaching out for treatment at the same time.

14. Similar genetic makeup of agents in effected individuals.

15. Simultaneous collections of similar illness in non-contiguous areas, domestic, or foreign.

16. An abundance of cases of unexplained diseases and deaths.

Identification of bioweapons

The goal of biodefense is to integrate the sustained efforts of the national and homeland security, medical, public health, intelligence, diplomatic, and law enforcement communities. Health care providers and public health officers are among the first lines of defense. In some countries private, local, and provincial (state) capabilities are being augmented by and coordinated with federal assets, to provide layered defenses against biological weapon attacks. During the first Gulf War the United Nations activated a biological and chemical response team, Task Force Scorpio, to respond to any potential use of weapons of mass destruction on civilians.

The traditional approach toward protecting agriculture, food, and water: focusing on the natural or unintentional introduction of a disease is being strengthened by focused efforts to address current and anticipated future biological weapons threats that may be deliberate, multiple, and repetitive.

The growing threat of biowarfare agents and bioterrorism has led to the development of specific field tools that perform on-the-spot analysis and identification of encountered suspect materials. One such technology, being developed by researchers from the Lawrence Livermore National Laboratory (LLNL), employs a "sandwich immunoassay", in which fluorescent dye-labeled antibodies aimed at specific pathogens are attached to silver and gold nanowires.[44]

In the Netherlands, the company TNO has designed Bioaerosol Single Particle Recognition eQuipment (BiosparQ). This system would be implemented into the national response plan for bioweapon attacks in the Netherlands.[45]

Researchers at Ben Gurion University in Israel are developing a different device called the BioPen, essentially a "Lab-in-a-Pen", which can detect known biological agents in under 20 minutes using an adaptation of the ELISA, a similar widely employed immunological technique, that in this case incorporates fiber optics.[46]

30.3.3 Genetic Warfare

Theoretically, novel approaches in biotechnology, such as synthetic biology could be used in the future to design novel types of biological warfare agents.[47][48][49][50] Special attention has to be laid on future experiments (of concern) that:[51]

1. Would demonstrate how to render a vaccine ineffective;

2. Would confer resistance to therapeutically useful antibiotics or antiviral agents;

3. Would enhance the virulence of a pathogen or render a nonpathogen virulent;

4. Would increase transmissibility of a pathogen;

5. Would alter the host range of a pathogen;

6. Would enable the evasion of diagnostic/detection tools;

7. Would enable the weaponization of a biological agent or toxin

Most of the biosecurity concerns in synthetic biology, however, are focused on the role of DNA synthesis and the risk of producing genetic material of lethal viruses (e.g. 1918 Spanish flu, polio) in the lab.[52][53][54] Recently, the CRISPR/Cas system has emerged as a promising technique for gene editing. It was hailed by The Washington Post as "the most important innovation in the synthetic biology space in nearly 30 years."[55] While other methods take months or years to edit gene sequences, CRISPR speeds that time up to weeks.[55] However, due to its ease of use and accessibility, it has raised a number of ethical concerns, especially surrounding its use in the biohacking space.[55][56][57]

30.4 List of BW institutions, programs, projects and sites by country

30.4.1 United States

Main article: United States biological weapons program

- Fort Detrick, Maryland

  - U.S. Army Biological Warfare Laboratories (1943–69)

Researchers working in Class III cabinets at the U.S. Biological Warfare Laboratories, Camp Detrick, Maryland (1940s).

- Building 470
- One-Million-Liter Test Sphere
- Operation Whitecoat (1954–73)
- U.S. entomological warfare program
  - Operation Big Itch
  - Operation Big Buzz
  - Operation Drop Kick
  - Operation May Day
- Project Bacchus
- Project Clear Vision
- Project SHAD
- Project 112
- Horn Island Testing Station
- Fort Terry
- Granite Peak Installation
- Vigo Ordnance Plant

30.4.2 United Kingdom

Main article: United Kingdom and weapons of mass destruction § Biological weapons

- Porton Down
- Gruinard Island
- Nancekuke

- Operation Vegetarian (1942-1944)
- **Open-air field tests:**
  - Operation Harness off Antigua, 1948–1950.
  - Operation Cauldron off Stornoway, 1952.
  - Operation Hesperus off Stornoway, 1953.
  - Operation Ozone off Nassau, 1954.
  - Operation Negation off Nassau, 1954-5.

30.4.3 Soviet Union and Russia

Main article: Soviet biological weapons program

- Biopreparat (18 labs and production centers)
  - Stepnagorsk Scientific and Technical Institute for Microbiology, Stepnogorsk, northern Kazakhstan
  - Institute of Ultra Pure Biochemical Preparations, Leningrad, a weaponized plague center
  - Vector State Research Center of Virology and Biotechnology (VECTOR), a weaponized smallpox center
  - Institute of Applied Biochemistry, Omutninsk
  - Kirov bioweapons production facility, Kirov, Kirov Oblast
  - Zagorsk smallpox production facility, Zagorsk
  - Berdsk bioweapons production facility, Berdsk
  - Bioweapons research facility, Obolensk
  - Sverdlovsk bioweapons production facility (Military Compound 19), Sverdlovsk, a weaponized anthrax center
- Institute of Virus Preparations
- Poison laboratory of the Soviet secret services
- Vozrozhdeniya
- Project Bonfire
- Project Factor

30.4.4 Japan

Main article: Special Research Units

- Unit 731

- Zhongma Fortress

- Kaimingjie germ weapon attack

- Khabarovsk War Crime Trials

- Epidemic Prevention and Water Purification Department

30.4.5 Iraq

Main articles: Iraqi biological weapons program and Iraq and weapons of mass destruction

(passim)

- Al Hakum

- Salman Pak facility

- Al Manal facility

30.4.6 South Africa

Main article: South Africa and weapons of mass destruction § Biological and chemical weapons

- Project Coast

- Delta G Scientific Company

- Roodeplaat Research Laboratories

- Protechnik

30.4.7 Canada

- Grosse Isle. Quebec, site (1939–45) of research into anthrax and other BW agents

- Experimental Station Suffield, Suffield, Alberta

30.5 List of people associated with BW

Bioweaponeers:

Includes scientists and administrators

- Shyh-Ching Lo*[58]*[59]

- Kanatjan Alibekov, known as Ken Alibek*[60]

- Ira Baldwin*[61]

- Wouter Basson

- Kurt Blome*[62]

- Eugen von Haagen*[63]

- Anton Dilger*[64]

- Paul Fildes*[65]

- Arthur Galston (unwittingly)

- Kurt Gutzeit*[66]

- Riley D. Housewright

- Shiro Ishii

- Elvin A. Kabat

- George W. Merck

- Frank Olson

- Vladimir Pasechnik*[67]

- William C. Patrick III*[68]

- Sergei Popov*[69]

- Theodor Rosebury

- Rihab Rashid Taha*[70]

- Prince Tsuneyoshi Takeda

- Huda Salih Mahdi Ammash

- Nassir al-Hindawi

- Erich Traub*[71]

- Auguste Trillat

- Baron Otto von Rosen*[72]

- Yujiro Wakamatsu

- Yazid Sufaat*[73]

Writers and activists:

- Daniel Barenblatt

- Leonard A. Cole

- Stephen Endicott

- Arthur Galston

- Jeanne Guillemin*[74]

- Edward Hagerman

- Sheldon H. Harris*[75]

- Nicholas D. Kristof

- Joshua Lederberg*[76]

- Matthew Meselson*[77]

- Richard Preston

- Ed Regis

- Mark Wheelis

- David Willman

- Johnston Atoll Chemical Agent Disposal System

- Plum Island Animal Disease Center

- Project 112

- Project AGILE

- Project SHAD

- McNeill's law

- Mycotoxin

- Ten Threats

- Trichothecene

- Yellow rain

30.6 In popular culture

Main article: Biological warfare in popular culture

30.7 See also

- Animal-borne bomb attacks

- Antibiotic resistance

- Asymmetric warfare

- Baker Island

- Bioaerosol

- Biological contamination

- Biosecurity

- Chemical weapon

- Counterinsurgency

- Discredited AIDS origins theories

- Enterotoxin

- Entomological warfare

- Ethnic bioweapon

- Exotic pollution

- Herbicidal warfare

- Human experimentation in the United States

- John W. Powell

30.8 References

[1] Wheelis, Mark; Rózsa, Lajos; Dando, Malcolm (2006). Deadly Cultures: Biological Weapons Since 1945. Harvard University Press. pp. 284–293, 301–303. ISBN 0-674-01699-8

[2] Gray, Colin. (2007). Another Bloody Century: Future Warfare. Page 265 to 266. Phoenix. ISBN 0-304-36734-6.

[3] Biological Weapons Convention

[4] Archived 30 April 2011 at the Wayback Machine.

[5] "Informaworld link". Retrieved 24 October 2014.

[6] Mayor, Adrienne (2003). Greek Fire, Poison Arrows & Scorpion Bombs: Biological and Chemical Warfare in the Ancient World. Woodstock, N.Y.: Overlook Duckworth. ISBN 978-1-58567-348-3.

[7] Wheelis, Mark (2002). "Biological warfare at the 1346 siege of Caffa". Emerg Infect Dis. Center for Disease Control. 8 (9): 971–5. doi:10.3201/eid0809.010536. PMC 2732530. PMID 12194776.

[8] Barras, Vincent; Greub, Gilbert (2014). "History of biological warfare and bioterrorism". Clinical Microbiology and Infection. 20 (6): 497–502. doi:10.1111/1469-0691.12706.

[9] Andrew G. Robertson, and Laura J. Robertson. "From asps to allegations: biological warfare in history," Military medicine (1995) 160#8 pp: 369-373.

[10] Rakibul Hasan. "Biological Weapons: covert threats to Global Health Security." Asian Journal of Multidisciplinary Studies (2014) 2#9 p 38. online

[11] Fenn, Elizabeth A. "Biological Warfare in Eighteenth-Century North America: Beyond Jeffery Amherst". Journal of American History. 86 (4): 1552–1580. doi:10.2307/2567577. JSTOR 2567577.

[12] Christopher. Warren (2013). "Smallpox at Sydney Cove - Who. When. Why". *Journal of Australian Studies.* doi:10.1080/14443058.2013.849750#preview. See also History of biological warfare#New South Wales, First Fleet#First Fleet smallpox, and History wars#Controversy over smallpox in Australia.

[13] Koenig. Robert (2006). *The Fourth Horseman: One Man's Secret Campaign to Fight the Great War in America.* PublicAffairs.

[14] Prasad, S.K. (2009). *Biological Agents. Volume 2.* Discovery Publishing House. p. 36. ISBN 9788183563819.

[15] Garrett, Laurie. *Betrayal of Trust: The Collapse of Global Public Health.* (Google Books). Oxford University Press. 2003, p. 340-41. (ISBN 0198526830).

[16] Covert. Norman M. (2000). "A History of Fort Detrick. Maryland". 4th Edition: 2000.

[17] Guillemi n. J. (2006). "Scientists and the history of biological weapons: A brief historical overview of the development of biological weapons in the twentieth century". *EMBO Reports.* **7** (Spec No): S45–S49. doi:10.1038/sj.embor.7400689. PMC 1490304⊙. PMID 16819450.

[18] Williams. Peter; Wallace, David (1989). *Unit 731: Japan's Secret Biological Warfare in World War II.* Free Press. ISBN 0-02-935301-7.

[19] Hal Gold. *Unit 731 testimony.* 1996. p.64-66

[20] Barenblatt. Daniel (2004). "A Plague upon Humanity". HarperCollins: 220–221.

[21] Hudson. Christopher (2 March 2007). "Doctors of Depravity". Daily Mail.

[22] Chevrier & Chomiczewski & Garrigue 2004. p. 19.

[23] Croddy & Wirtz 2005. p. 171.

[24] Naomi Baumslag, *Murderous Medicine: Nazi Doctors, Human Experimentation, and Typhus,* 2005, p.207

[25] "Weapons of Mass Destruction: Plague as Biological Weapons Agent". GlobalSecurity.org. Retrieved 21 December 2014.

[26] Amy Stewart (25 April 2011). "Where To Find The World's Most 'Wicked Bugs': Fleas". National Public Radio.

[27] Russell Working (5 June 2001). "The trial of Unit 731". The Japan Times.

[28] Ken Alibek and K Handelman (1999). *Biohazard: The Chilling True Story of the Largest Covert Biological Weapons Program in the World Trade From the Inside by the Man Who Ran It,* New York. NY: Random House.

[29] "26 Countries' WMD Programs: A Global History of WMD Use - US - Iraq War - ProCon.org". Usiraq.procon.org. 2009-05-29. Retrieved 2013-09-05.

[30] "Loner Likely Sent Anthrax. FBI Says". *Los Angeles Times.* Archived from the original on 7 April 2008. Retrieved 30 March 2008.

[31] "Anthrax Facts | UPMC Center for Health Security". Upmc-biosecurity.org. Archived from the original on 2 March 2013. Retrieved 2013-09-05.

[32] Kenneth Alibek and S. Handelman. *Biohazard: The Chilling True Story of the Largest Covert Biological Weapons Program in the World – Told from Inside by the Man Who Ran it.* 1999. Delta (2000) ISBN 0-385-33496-6 .

[33] "Potential bioweapons" . *Clinical Immunology.* **111**: 1–15. doi:10.1016/j.clim.2003.09.010. Retrieved 24 October 2014.

[34] "Vietnam's war against Agent Orange". *BBC News.* 14 June 2004. Retrieved 17 April 2010.

[35] Verdourt, Bernard; Trump, E.C.; Church, M.E. (1969). "Common poisonous plants of East Africa" . London: Collins: 254

[36] "An Introduction to Biological Weapons. Their Prohibition, and the Relationship to Biosafety". *The Sunshine Project,* April 2002. Retrieved 25 December 2008. Archived 12 May 2013 at the Wayback Machine.

[37] Lockwood, Jeffrey A. *Six-legged Soldiers: Using Insects as Weapons of War.* Oxford University Press, USA, 2008, pp. 9–26. (ISBN 0195333055).

[38] http://www.opbw.org/new_process/msp2010/BWC_MSP_2010_WP8_E.pdf

[39] Ouagrham-Gormley S. Dissuading Biological Weapons Proliferation. Contemporary Security Policy [serial online]. December 2013;34(3):473-500. Available from: Humanities International Complete. Ipswich. MA. Accessed 28 January 2015.

[40] Guillemin. J. (2013). The Soviet Biological Weapons Program: A History. Politics & The Life Sciences. 32(1), 102-105. doi:10.2990/32_1_102

[41] Wilkening D. Modeling the incubation period of inhalational anthrax. Medical Decision Making [serial online]. 1 July 2008;28(4):593-605. Available from: Scopus®. Ipswich. MA. Accessed 28 January 2015.

[42] Toth D, Gundlapalli A. Adler F, et al. Quantitative Models of the Dose-Response and Time Course of Inhalational Anthrax in Humans. Plos Pathogens [serial online]. August 2013;9(8):1-18. Available from: Academic Search Complete. Ipswich. MA. Accessed 28 January 2015.

[43] Treadwell. Tracee (March–April 2003). "Epidemiological Clues to Bioterrorism". *Public Health Reports.* **118**: 93–94. doi:10.1093/phr/118.2.92.

[44] 'Physorg.com. "Encoded Metallic Nanowires Reveal Bioweapons". 12:50 EST. 10 August 2006." . Retrieved 24 October 2014.

[45] "BiosparQ features" . Retrieved 24 October 2014.

[46] Genuth, Iddo ; Fresco-Cohen, Lucille (13 November 2006). 'BioPen Senses BioThreats", *The Future of Things* Archived 30 April 2007 at the Wayback Machine.

[47] Kelle A (2009) Security issues related to synthetic biology. Chapter 7. In: Schmidt M, Kelle A, Ganguli-Mitra A, de Vriend H (eds) Synthetic biology. The technoscience and its societal conse- quences. Springer, Berlin

[48] Garfinkel, M., Endy, D., Epstein, G., and Friedman, R. (2007). In Synthetic Genomics: Options for Governance. Available at: http://www.jcvi.org/cms/research/projects/syngen-options/overview/.

[49] National Security Advisory Board on Biotechnology (NSABB) (2010). Addressing Biosecurity Concerns Related to Synthetic Biology. Available at: http://oba.od.nih.gov/biosecurity/pdf/NSABB%20SynBio%20-DRAFT%20Report-FINAL%20(2)_6-7-10.pdf. Retrieved 4 September 2010.

[50] M.Buller, The potential use of genetic engineering to enhance orthopox viruses as bioweapons. Presentation at the International Conference 'Smallpox Biosecurity. Preventing the Unthinkable' (21–22 October 2003) Geneva, Switzerland

[51] Kelle A. 2007. Synthetic Biology & Biosecurity Awareness In Europe . Bradford Science and Technology Report No.9

[52] Tumpej TM et al. 2005. Characterization of the Reconstructed 1918 Spanish Influenza Pandemic Virus. Science Vol. 310(5745):77–80

[53] Cello, J.; Paul, A. V.; Wimmer, E. (2002). "Chemical synthesis of poliovirus cDNA: generation of infectious virus in the absence of natural template". *Science*. **297**: 1016–1018. doi:10.1126/science.1072266. PMID 12114528.

[54] Wimmer, E.; Mueller, S.; Tumpey, T. M.; Taubenberger, J. K. (2009). "Synthetic viruses: a new opportunity to understand and prevent viral disease" . *Nat. Biotechnol.* **27**: 1163–1172. doi:10.1038/nbt.1593. PMC 2819212⊖. PMID 20010599.

[55] Basulto, Dominic (2015-11-04). "Everything you need to know about why CRISPR is such a hot technology" . *The Washington Post*. ISSN 0190-8286. Retrieved 2016-01-24.

[56] Kahn, Jennifer (2015-11-09). "The Crispr Quandary". *The New York Times*. ISSN 0362-4331. Retrieved 2016-01-24.

[57] "CRISPR, the disruptor" . *Nature News & Comment*. Retrieved 2016-01-24.

[58] Shyh-Ching Lo

[59] http://www.google.com/patents/US5242820

[60] "Interview: Dr Kanatjan Alibekov" . *Frontline*. PBS. Retrieved 8 March 2010.

[61] "Dr. Ira Baldwin: Biological Weapons Pioneer". American History. Retrieved 8 March 2009.

[62] Ute Deichmann, Biologists under Hitler, trans Thomas Dunlap (Harvard 1996). http://books.google.com.bz/books?id=gPrtE4K0WC8C&pg=PA173&dq=kurt+blome&hl=en&ei=P3o3TOLMBMKCnQe39rTVAw&sa=X&oi=book_result&ct=result&resnum=1&ved=0CCYQ6AEwAA#v=onepage&q=kurt%20blome&f=false

[63] Leyendecker, B.; Klapp, F. (1989). "Human hepatitis experiments in the 2d World War" . *Zeitschrift für die gesamte Hygiene und ihre Grenzgebiete*. **35** (12): 756–760. PMID 2698560.

[64] Maksel, Rebecca (14 January 2007). "An American waged germ warfare against U.S. in WWI" . SF Gate. Retrieved 7 March 2010.

[65] Chauhan, Sharad S. (2004). *Biological Weapons*. APH Publishing. p. 194. ISBN 81-7648-732-5

[66] Office of U.S. Chief of Counsel for the American Military Tribunals at Nurember, 1946. http://www.mazal.org/NO-series/NO-0124-000.htm

[67] "Obituary: Vladimir Pasechnik" . *The Daily Telegraph*. London. 29 November 2001. Retrieved 8 March 2010.

[68] "Anthrax attacks" . *Newsnight*. BBC. 14 March 2002. Retrieved 16 March 2010.

[69] "Interviews With Biowarriors: Sergei Popov". (2001) *NOVA Online*.

[70] "US welcomes 'Dr Germ' capture" . BBC. 13 May 2003. Retrieved 8 March 2010.

[71] Paul Maddrell, "Operation Matchbox and the Scientific Containment of the USSR" . in Peter J. Jackson and Jennifer L. Siegel (eds) Intelligence and Statecraft: The Use and Limits of Intelligence in International Society. Praeger, 2005.https://books.google.com/books?id=I3Q3_Ww-5SMC&pg=PA194&dq=erich+traub&hl=en&ei=DyJ_TPDPI4vEsAOvq_nwCg&sa=X&oi=book_result&ct=result&resnum=10&ved=0CE4Q6AEwCQ#v=onepage&q=erich%20traub&f=false

[72] "Jamie Bisher, "Baron von Rosen's 1916 Anthrax Mission," 2014". *Baron von Rosen's 1916 Anthrax Mission*. Retrieved 24 October 2014.

[73] Yazid Sufaat works on anthrax for al-Qaeda, GlobalSecurity.org

[74] "MIT Security Studies Program (SSP): Jeanne Guillemin" . MIT. Retrieved 8 March 2010.

[75] Lewis, Paul (4 September 2002). "Sheldon Harris, 74, Historian Of Japan's Biological Warfare". *The New York Times*. Retrieved 8 March 2010.

[76] Miller, Judith (2001). *Biological Weapons and America's Secret War*. New York: Simon & Schuster. p. 67. ISBN 0-684-87158-0.

[77] "Matthew Meselson – Harvard – Belfer Center for Science and International Affairs". Harvard. Retrieved 8 March 2010.

- Chevrier, Marie Isabelle; Chomiczewski, Krzysztof; Garrigue, Henri, eds. (2004). *The Implementation of Legally Binding Measures to Strengthen the Biological and Toxin Weapons Convention: Proceedings of the NATO Advanced Study Institute, Held in Budapest, Hungary, 2001*. Volume 150 of NATO science series: Mathematics, physics, and chemistry (illustrated ed.). Springer. ISBN 140202097X. Retrieved 10 March 2014.

- Croddy, Eric A.; Wirtz, James J., eds. (2005). *Weapons of Mass Destruction*. Jeffrey A. Larsen, Managing Editor. ABC-CLIO. ISBN 1851094903. Retrieved 10 March 2014.

30.9   Further reading

- Alibek, K. and S. Handelman. *Biohazard: The Chilling True Story of the Largest Covert Biological Weapons Program in the World– Told from Inside by the Man Who Ran it*. Delta (2000) ISBN 0-385-33496-6

- Appel, J. M. Is all fair in biological warfare? The controversy over genetically engineered biological weapons, *Journal of Medical Ethics*, Volume 35, pp. 429–432 (2009).

- Crosby, Alfred W., *Ecological Imperialism: The Biological Expansion of Europe, 900–1900* (New York, 1986).

- Dembek, Zygmunt (editor), *Medical Aspects of Biological Warfare*; Washington, DC: Borden Institute (2007).

- Endicott, Stephen and Edward Hagerman. *The United States and Biological Warfare: Secrets from the Early Cold War and Korea*, Indiana University Press (1998). ISBN 0-253-33472-1

- Fenn, Elizabeth A. (2000). "Biological Warfare in Eighteenth-Century North America: Beyond Jeffery Amherst". *Journal of American History*. **86** (4): 1552–1580. doi:10.2307/2567577. JSTOR 2567577.

- Keith, Jim (1999). *Biowarfare In America*. Illuminet Press. ISBN 1-881532-21-6

- Knollenberg, Bernhard. "General Amherst and Germ Warfare." *Mississippi Valley Historical Review* (1954), 41#3 489–494. British war against Indians in 1763 in JSTOR

- Leitenberg, Milton, and Raymond A. Zilinskas. *The Soviet Biological Weapons Program: A History* (Harvard University Press, 2012) 921 pp

- Mangold, Tom & Goldberg, Jeff (1999). *Plague Wars: a true story of biological warfare*. Macmillan, London. ISBN 0-333-71614-0

- Maskiell, Michelle, and Adrienne Mayor. "Killer Khilats: Legends of Poisoned Robes of Honour in India. Parts 1 & 2." Folklore [London] 112 (Spring and Fall 2001): 23–45, 163–82.

- Mayor, Adrienne. Greek Fire, Poison Arrows & Scorpion Bombs: Biological and Chemical Warfare in the Ancient World. Overlook. 2003, rev. ed. 2009. ISBN 1-58567-348-X.

- Orent, Wendy (2004). *Plague, The Mysterious Past and Terrifying Future of the World's Most Dangerous Disease*. Simon & Schuster, Inc., New York, NY. ISBN 0-7432-3685-8

- Pala, Christopher (19??), *Anthrax Island*

- Preston, Richard (2002), *The Demon in the Freezer*, New York: Random House.

- Rózsa, Lajos (2009). "The motivation for biological aggression is an inherent and common aspect of the human behavioural repertoire" (PDF). *Medical Hypotheses*. **72**: 217–219. doi:10.1016/j.mehy.2008.06.047.

- Warner, Jerry; Ramsbotham, James; Tunia, Ewelina; Vadez, James J. (May 2011). *Analysis of the Threat of Genetically Modified Organisms for Biological Warfare*. Washington, DC: National Defense University. Retrieved 8 March 2015.

- Woods, Lt Col Jon B. (ed.), *USAMRIID's Medical Management of Biological Casualties Handbook*, 6th edition, U.S. Army Medical Institute of Infectious Diseases, Fort Detrick, Maryland (April 2005).

- Zelicoff, Alan & Bellomo, Michael (2005). *Microbe: Are we Ready for the Next Plague?*. AMACOM Books, New York, NY. ISBN 0-8144-0865-6

30.10 External links

- Biological weapons and international humanitarian law, ICRC

- WHO: Health Aspects of Biological and Chemical Weapons

- "Biological Warfare". National Library of Medicine. Retrieved 2013-05-28.

- U.S Army site

Chapter 31

Chemical warfare

For other uses, see Chemical warfare (disambiguation).

Chemical warfare (**CW**) involves using the toxic properties of chemical substances as weapons. This type of warfare is distinct from nuclear warfare and biological warfare, which together make up NBC, the military acronym for nuclear, biological, and chemical (warfare or weapons), all of which are considered "weapons of mass destruction" (WMDs). None of these fall under the term conventional weapons which are primarily effective due to their destructive potential. With proper protective equipment, training, and decontamination measures, the primary effects of chemical weapons can be overcome. Many nations possess vast stockpiles of weaponized agents in preparation for wartime use. The threat and the perceived threat have become strategic tools in planning both measures and countermeasures.

31.1 Definition

Chemical warfare is different from the use of conventional weapons or nuclear weapons because the destructive effects of chemical weapons are not primarily due to any explosive force. The offensive use of living organisms (such as anthrax) is considered biological warfare rather than chemical warfare; however, the use of nonliving toxic products produced by living organisms (e.g. toxins such as botulinum toxin, ricin, and saxitoxin) *is* considered chemical warfare under the provisions of the Chemical Weapons Convention (CWC). Under this Convention, any toxic chemical, regardless of its origin, is considered a chemical weapon unless it is used for purposes that are not prohibited (an important legal definition known as the General Purpose Criterion).[1]

About 70 different chemicals have been used or stockpiled as chemical warfare agents during the 20th century. The entire class known as Lethal Unitary Chemical Agents and Munitions have been scheduled for elimination by the CWC.[2]

Under the Convention, chemicals that are toxic enough to be used as chemical weapons, or that may be used to manufacture such chemicals, are divided into three groups according to their purpose and treatment:

- **Schedule 1** – Have few, if any, legitimate uses. These may only be produced or used for research, medical, pharmaceutical or protective purposes (i.e. testing of chemical weapons sensors and protective clothing). Examples include nerve agents, ricin, lewisite and mustard gas. Any production over 100 g must be reported to the OPCW and a country can have a stockpile of no more than one tonne of these chemicals.

- **Schedule 2** – Have no large-scale industrial uses, but may have legitimate small-scale uses. Examples include dimethyl methylphosphonate, a precursor to sarin also used as a flame retardant, and thiodiglycol, a precursor chemical used in the manufacture of mustard gas but also widely used as a solvent in inks.

- **Schedule 3** – Have legitimate large-scale industrial uses. Examples include phosgene and chloropicrin. Both have been used as chemical weapons but phosgene is an important precursor in the manufacture of plastics and chloropicrin is used as a fumigant. The OPCW must be notified of, and may inspect, any plant producing more than 30 tonnes per year.

31.2 History

31.2.1 Ancient times

Chemical weapons have been used for millennia in the form of poisoned spears and arrows, but evidence can be found for the existence of more advanced forms of chemical weapons in ancient and classical times.

Ancient Greek myths about Hercules poisoning his arrows with the venom of the Hydra monster are the earliest references to toxic weapons in western literature. Homer's

214

epics. the *Iliad* and the *Odyssey*, allude to poisoned arrows used by both sides in the legendary Trojan War (Bronze Age Greece).[3]

Some of the earliest surviving references to toxic warfare appear in the Indian epics *Ramayana* and *Mahabharata*.[4] The "Laws of Manu," a Hindu treatise on statecraft (c. 400 BC) forbids the use of poison and fire arrows, but advises poisoning food and water. Kautilya's "Arthashastra", a statecraft manual of the same era, contains hundreds of recipes for creating poison weapons, toxic smokes, and other chemical weapons. Ancient Greek historians recount that Alexander the Great encountered poison arrows and fire incendiaries in India at the Indus basin in the 4th century BC.[3]

The Art of War *described the use of fire weapons against the enemy.*

Arsenical smokes were known to the Chinese as far back as c. 1000 BC[5] and Sun Tzu's "Art of War" (c. 200 BC) advises the use of fire weapons. In the second century BC, writings of the Mohist sect in China describe the use of bellows to pump smoke from burning balls of mustard and other toxic vegetables into tunnels being dug by a besieging army. Other Chinese writings dating around the same period contain hundreds of recipes for the production of poisonous or irritating smokes for use in war along with numerous accounts of their use. These accounts describe an arsenic-containing "soul-hunting fog", and the use of finely divided lime dispersed into the air to suppress a peasant revolt in 178 AD.

The earliest recorded use of gas warfare in the West dates back to the fifth century BC, during the Peloponnesian War between Athens and Sparta. Spartan forces besieging an Athenian city placed a lighted mixture of wood, pitch, and sulfur under the walls hoping that the noxious smoke would incapacitate the Athenians, so that they would not be able to resist the assault that followed. Sparta was not alone in its use of unconventional tactics in ancient Greece; Solon of Athens is said to have used hellebore roots to poison the water in an aqueduct leading from the River Pleistos around 590 BC during the siege of Kirrha.[3]

There is archaeological evidence that the Sassanians deployed chemical weapons against the Roman army in the Siege of Dura Europos in the third century AD. Research carried out on the collapsed tunnels at Dura-Europos in Syria suggests that the Persians used bitumen and sulfur crystals to get it burning. When ignited, the materials gave off dense clouds of choking sulfur dioxide gases which killed 20 Roman soldiers in a matter of two minutes. This is the earliest evidence of gas warfare.[6][7][8][9]

In the late 15th century. Spanish conquistadors encountered a rudimentary type of chemical warfare on the island of Hispaniola. The Taíno threw gourds filled with ashes and ground hot peppers at the Spaniards to create a blinding smoke screen before launching their attack.[10]

31.2.2 Early modern era

Christoph Bernhard von Galen tried to use toxic fumes during the siege of the city of Groningen in 1672.

Historian and philosopher David Hume, in his history of England, recounts how in the reign of Henry III (r.1216 - 1272) the English Navy destroyed an invading French fleet, by blinding the enemy fleet with "quicklime," the old name for calcium oxide. D'Albiney employed a stratagem against them, which is said to have contributed to the victory: Having gained the wind of the French, he came down upon them with violence; and throwing in their faces a great quantity of quicklime, which he purposely carried on board, he so blinded them, that they were disabled from defending themselves.*[11]

Leonardo da Vinci proposed the use of a powder of sulfide, arsenic and verdigris in the 15th century:

> *throw poison in the form of powder upon galleys. Chalk, fine sulfide of arsenic, and powdered verdegris may be thrown among enemy ships by means of small mangonels, and all those who, as they breathe, inhale the powder into their lungs will become asphyxiated.*

It is unknown whether this powder was ever actually used.

In the 17th century during sieges, armies attempted to start fires by launching incendiary shells filled with sulfur, tallow, rosin, turpentine, saltpeter, and/or antimony. Even when fires were not started, the resulting smoke and fumes provided a considerable distraction. Although their primary function was never abandoned, a variety of fills for shells were developed to maximize the effects of the smoke.

In 1672, during his siege of the city of Groningen, Christoph Bernhard von Galen, the Bishop of Münster, employed several different explosive and incendiary devices, some of which had a fill that included Deadly Nightshade, intended to produce toxic fumes. Just three years later, August 27, 1675, the French and the Holy Roman Empire concluded the Strasbourg Agreement, which included an article banning the use of "perfidious and odious" toxic devices.

31.2.3   Industrial era

The modern notion of chemical warfare emerged from the mid-19th century, with the development of modern chemistry and associated industries. The first proposal for the use of chemical warfare was made by Lyon Playfair, Secretary of the Science and Art Department, in 1854 during the Crimean War. He proposed a cacodyl cyanide artillery shell for use against enemy ships as way to solve the stalemate during the siege of Sevastopol. The proposal was backed by Admiral Thomas Cochrane of the Royal Navy. It was considered by the Prime Minister, Lord Palmerston, but the British Ordnance Department rejected the proposal as "as bad a mode of warfare as poisoning the wells of the enemy." Playfair's response was used to justify chemical warfare into the next century:*[12]

> *There was no sense in this objection. It is considered a legitimate mode of warfare to fill shells with molten metal which scatters among the enemy, and produced the most frightful modes of*

Lyon Playfair proposed the industrial manufacture of cyanide artillery shells for use during the Crimean War.

> *death.   Why a poisonous vapor which would kill men without suffering is to be considered illegitimate warfare is incomprehensible.   War is destruction, and the more destructive it can be made with the least suffering the sooner will be ended that barbarous method of protecting national rights.   No doubt in time chemistry will be used to lessen the suffering of combatants, and even of criminals condemned to death.*

Later, during the American Civil War, New York school teacher John Doughty proposed the offensive use of chlorine gas, delivered by filling a 10-inch (254 millimeter) artillery shell with two to three quarts (two to three liters) of liquid chlorine, which could produce many cubic feet (a few cubic meters) of chlorine gas. Doughty's plan was apparently never acted on, as it was probably*[13] presented to Brigadier General James Wolfe Ripley, Chief of Ordnance, who was described as being congenitally immune to new ideas.

A general concern over the use of poison gas manifested itself in 1899 at the Hague Conference with a proposal prohibiting shells filled with asphyxiating gas. The proposal was passed, despite a single dissenting vote from the United States. The American representative, Navy Captain Alfred Thayer Mahan, justified voting against the measure on the grounds that "the inventiveness of Americans should not be restricted in the development of new weapons."

World War I

Main article: Chemical weapons in World War I
The Hague Declaration of 1899 and the Hague Convention

Tear gas casualties from the Battle of Estaires, April 10, 1918.

of 1907 forbade the use of "poison or poisoned weapons" in warfare, yet more than 124,000 tons of gas were produced by the end of World War I. The French were the first to use chemical weapons during the First World War, using the tear gases ethyl bromoacetate and chloroacetone.

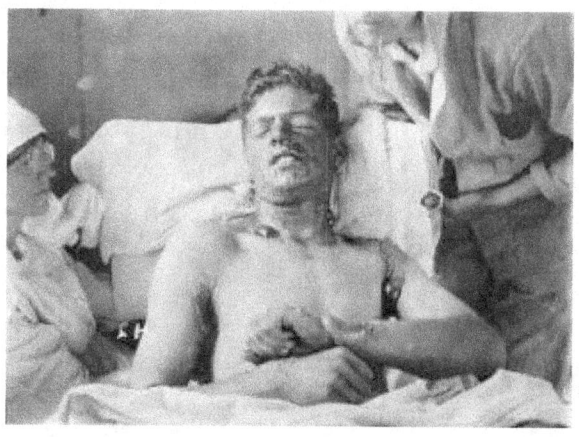

A Canadian soldier with mustard gas burns, ca. 1917–1918.

One of Germany's earliest uses of chemical weapons occurred on October 27, 1914 when shells containing the irritant dianisidine chlorosulfonate were fired at British troops near Neuve-Chapelle, France.[5] Germany used another irritant, xylyl bromide, in artillery shells that were fired in January 1915 at the Russians near Bolimów, in present-day Poland.[14] The first full-scale deployment of deadly chemical warfare agents during World War I was at the Second Battle of Ypres, on April 22, 1915, when the Germans attacked French, Canadian and Algerian troops with

chlorine gas. Deaths were light, though casualties were relatively heavy.

A total 50,965 tons of pulmonary, lachrymatory, and vesicant agents were deployed by both sides of the conflict, including chlorine, phosgene, and mustard gas. Official figures declare about 1.3 million casualties directly caused by chemical warfare agents during the course of the war. Of these, an estimated 100,000-260,000 casualties were civilians. Nearby civilian towns were at risk from winds blowing the poison gases through. Civilians rarely had a warning system put into place to alert their neighbors of the danger. In addition to poor warning systems, civilians often did not have access to effective gas masks.[15][16]

Football team of British soldiers with gas masks, Western Front, 1916.

To this day, unexploded World War I-era chemical ammunition is still uncovered when the ground is dug in former battle or depot areas and continues to pose a threat to the civilian population in Belgium and France and less commonly in other countries.

After the war, most of the unused German chemical warfare agents were dumped into the Baltic Sea, a common disposal method among all the participants in several bodies of water. Over time, the salt water causes the shell casings to corrode, and mustard gas occasionally leaks from these containers and washes onto shore as a wax-like solid resembling ambergris.

Interwar years

In 1919, the Royal Air Force dropped arsenic gas on Bolshevik troops during the British intervention in the Russian Civil War.[17]

After World War I chemical agents were occasionally used to subdue populations and suppress rebellion.

In 1920, the Arab and Kurdish people of Mesopotamia revolted against the British occupation, which cost the British dearly. As the Mesopotamian resistance gained strength, the British resorted to increasingly repressive measures. Much speculation was made about aerial bombardment of major cities with gas in Mesopotamia, with Winston Churchill, then-Secretary of State at the British War Office, arguing in favor of gas.[18][19]

The Bolsheviks also employed poison gas in 1921 during the Tambov Rebellion. An order signed by military commanders Tukhachevsky and Vladimir Antonov-Ovseyenko stipulated: *"The forests where the bandits are hiding are to be cleared by the use of poison gas. This must be carefully calculated, so that the layer of gas penetrates the forests and kills everyone hiding there."* [20]

During the Rif War in Spanish Morocco in 1921–1927, combined Spanish and French forces dropped mustard gas bombs in an attempt to put down the Berber rebellion. (*See also: Chemical weapons in the Rif War*)

In 1925, 16 of the world's major nations signed the Geneva Protocol, thereby pledging never to use gas in warfare again. Notably, while the United States delegation under Presidential authority signed the Protocol, it languished in the U.S. Senate until 1975, when it was finally ratified.

Beginning in the final week of 1935 and continuing into 1936, Fascist Italy used mustard gas during its invasion of Ethiopia in the Second Italo-Abyssinian War. Ignoring the Geneva Protocol, which it signed seven years earlier, the Italian military dropped mustard gas in bombs, sprayed it from airplanes, and spread it in powdered form on the ground. The Italians inflicted a reported 150,000 chemical casualties on the Ethiopians, mostly from mustard gas.

by aircraft in support of mobile warfare. Also in 1923, at the behest of the German army, poison gas expert Dr. Hugo Stoltzenberg negotiated with the USSR to build a huge chemical weapons plant at Trotsk, on the Volga river.

Collaboration between Germany and the USSR in poison gas continued on and off through the 1920s. In 1924, German officers debated the use of poison gas versus non-lethal chemical weapons against civilians.

Chemical warfare was revolutionized by Nazi Germany's discovery of the nerve agents tabun (in 1937) and sarin (in 1939) by Gerhard Schrader, a chemist of IG Farben.

IG Farben was Germany's premier poison gas manufacturer during World War II, so the weaponization of these agents cannot be considered accidental.[21] Both were turned over to the German Army Weapons Office prior to the outbreak of the war.

The nerve agent soman was later discovered by Nobel Prize laureate Richard Kuhn and his collaborator Konrad Henkel at the Kaiser Wilhelm Institute for Medical Research in Heidelberg in spring 1944.[22][23] The Germans developed and manufactured large quantities of several agents, but chemical warfare was not extensively used by either side. Chemical troops were set up (in Germany since 1934) and delivery technology was actively developed.

World War II

Japanese Special Naval Landing Force wearing gas masks and rubber gloves during a chemical attack near Chapei in the Battle of Shanghai. [24]

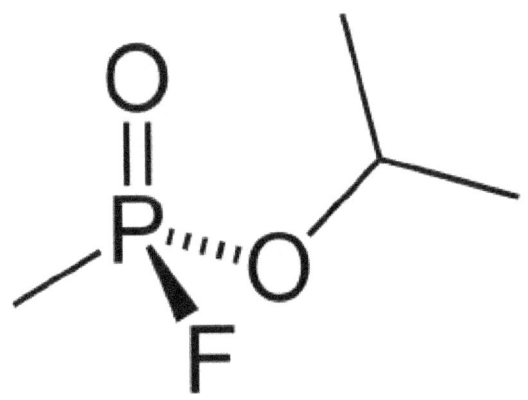

The chemical structure of sarin nerve gas, developed in Germany in 1939.

Nerve agents   Shortly after the end of World War I, Germany's General Staff enthusiastically pursued a recapture of their preeminent position in chemical warfare. In 1923, Hans von Seeckt pointed the way, by suggesting that German poison gas research move in the direction of delivery

Imperial Japanese Army   Despite the 1899 Hague Declaration *IV, 2 - Declaration on the Use of Projectiles the Object of Which is the Diffusion of Asphyxiating or Deleterious Gases*,[25] Article 23 (a) of the 1907 Hague Convention *IV - The Laws and Customs of War on Land*,[26] and a resolution adopted against Japan by the League of Nations on May 14, 1938, the Imperial Japanese Army frequently used chemical weapons. Because of fear of retaliation, however, those weapons were never used against Westerners, but against other Asians judged "inferior" by imperial propaganda. According to historians Yoshiaki Yoshimi and Kentaro Awaya, gas weapons, such as tear gas, were used only sporadically in 1937 but in early 1938, the Imperial Japanese Army began full-scale use of sneeze and nausea gas (red), and from mid-1939, used mustard gas (yellow) against both Kuomintang and Communist Chinese troops.[27]

According to historians Yoshiaki Yoshimi and Seiya Matsuno, the chemical weapons were authorized by specific orders given by Emperor Hirohito himself, transmitted by the chief of staff of the army. For example, the Emperor authorized the use of toxic gas on 375 separate occasions during the Battle of Wuhan from August to October 1938.[28] They were also profusely used during the invasion of Changde. Those orders were transmitted either by prince Kotohito Kan'in or general Hajime Sugiyama.[29] The Imperial Japanese Army had used mustard gas and the US-developed (CWS-1918) blister agent lewisite against Chinese troops and guerrillas. Experiments involving chemical weapons were conducted on live prisoners (Unit 731 and Unit 516).

The Japanese also carried chemical weapons as they swept through Southeast Asia towards Australia. Some of these items were captured and analyzed by the Allies. Historian Geoff Plunkett has recorded how Australia covertly imported 1,000,000 chemical weapons from the United Kingdom from 1942 onwards and stored them in many storage depots around the country, including three tunnels in the Blue Mountains to the west of Sydney. They were to be used as a retaliatory measure if the Japanese first used chemical weapons.[30] Buried chemical weapons have been recovered at Marrangaroo and Columboola.[31][32]

Nazi Germany Recovered documents suggest that German intelligence incorrectly thought that the Allies also knew of the nerve agent compounds, interpreting their lack of mention in the Allies' scientific journals as evidence that information about them was being suppressed. Germany ultimately decided not to use the new nerve agents, fearing a potentially devastating Allied retaliatory nerve agent deployment.[33]

Stanley P. Lovell, Deputy Director for Research and Development of the Office of Strategic Services, reports in his book *Of Spies and Stratagems* that the Allies knew the Germans had quantities of Gas Blau available for use in the defense of the Atlantic Wall. The use of nerve gas on the

The Germans may have used poison gas on survivors from the Battle of Kerch, May 1942

Normandy beachhead would have seriously impeded the Allies and possibly caused the invasion to fail altogether. He submitted the question "Why was nerve gas not used in Normandy?" to be asked of Hermann Göring during his interrogation. Göring answered that the reason gas was not used had to do with horses. The Wehrmacht was dependent upon horse-drawn transport to move supplies to their combat units, and had never been able to devise a gas mask horses could tolerate; the versions they developed would not pass enough pure air to allow the horses to pull a cart. Thus, gas was of no use to the German Army under most conditions.[34]

One reported incident indicates the German army eventually used poison gas on survivors of the Battle of Kerch on the Eastern Crimean peninsula. After the battle in mid-May 1942, roughly 3000 soldiers and civilians not evacuated by sea were besieged in a series of caves and tunnels in the nearby Adzhimuskai quarry. After holding out for approximately three months, "poison gas was released into the tunnels, killing all but a few score of the Soviet defenders."[35]

In February 1943, German troops stationed in Kuban received a telegram.[36]

> ...Russians should be eventually cleared out of the mountain range with gas.

The troops also received two wagons of toxin antidotes.[36]

Western Allies The Western Allies did not use chemical weapons during the Second World War. The British planned to use mustard gas and phosgene to help repel a German invasion in 1940-1941,[37][38] and had there been an invasion may have also deployed it against German cities.[39] General Alan Brooke, Commander-in-Chief,

The British stockpiled chemical weapons to use in case of a German invasion. Pictured is a chemical warfare bulk contamination vehicle.

Home Forces, in command of British anti-invasion preparations of the Second World War said that he "...had every intention of using sprayed mustard gas on the beaches" in an annotation in his diary.[40] The British manufactured mustard, chlorine, lewisite, phosgene and Paris Green and stored them at airfields and depots for use on the beaches.[39]

The mustard gas stockpile was enlarged in 1942-1943 for possible use by RAF Bomber Command against German cities, and in 1944 for possible retaliatory use if German forces used chemical weapons against the D-Day landings.[37]

Winston Churchill, the British Prime Minister, issued a memorandum advocating a chemical strike on German cities using poison gas and possibly anthrax. Although the idea was rejected, it has provoked debate.[41] In July 1944, fearing that rocket attacks on London would get even worse, and saying he would only use chemical weapons if it were "life or death for us" or would "shorten the war by a year",[42] Churchill wrote a secret memorandum asking his military chiefs to "think very seriously over this question of using poison gas." He stated "it is absurd to consider morality on this topic when everybody used it in the last war without a word of complaint..."

The Joint Planning Staff, however, advised against the use of gas because it would inevitably provoke Germany to retaliate with gas. They argued that this would be to the Allies' disadvantage in France both for military reasons and

because it might "seriously impair our relations with the civilian population when it became generally known that chemical warfare was first employed by us."[43]

In 1945, the U.S. Army's Chemical Warfare Service standardized improved chemical warfare rockets intended for the new M9 and M9A1 'Bazooka' launchers, adopting the M26 Gas Rocket, a cyanogen chloride (CK)-filled warhead for the 2.36-in rocket launcher.[44] CK, a deadly blood agent, was capable of penetrating the protective filter barriers in some gas masks,[45] and was seen as an effective agent against Japanese forces (particularly those hiding in caves or bunkers), whose gas masks lacked the impregnants that would provide protection against the chemical reaction of CK.[44][46][47] While stockpiled in US inventory, the CK rocket was never deployed or issued to combat personnel.[44]

Accidental release   On the night of December 2, 1943, German Ju 88 bombers attacked the port of Bari in Southern Italy, sinking several American ships—among them the SS *John Harvey*, which was carrying mustard gas intended for use in retaliation by the Allies if German forces initiated gas warfare. The presence of the gas was highly classified, and authorities ashore had no knowledge of it, which increased the number of fatalities since physicians, who had no idea that they were dealing with the effects of mustard gas, prescribed treatment improper for those suffering from exposure and immersion.

The whole affair was kept secret at the time and for many years after the war. According to the U.S. military account, "Sixty-nine deaths were attributed in whole or in part to the mustard gas, most of them American merchant seamen"[48] out of 628 mustard gas military casualties.[49]

The large number of civilian casualties among the Italian population was not recorded. Part of the confusion and controversy derives from the fact that the German attack was highly destructive and lethal in itself, also apart from the accidental additional effects of the gas (the attack was nicknamed "The Little Pearl Harbor"), and attribution of the causes of death between the gas and other causes is far from easy.[50][51] Rick Atkinson, in his book *The Day of Battle*, describes the intelligence that prompted Allied leaders to deploy mustard gas to Italy. This included Italian intelligence that Adolf Hitler had threatened to use gas against Italy if the state changed sides, and prisoner of war interrogations suggesting that preparations were being made to use a "new, egregiously potent gas" if the war turned decisively against Germany. Atkinson concludes, "No commander in 1943 could be cavalier about a manifest threat by Germany to use gas."

31.2.4 Post-war

After World War II, the Allies recovered German artillery shells containing the three German nerve agents of the day (tabun, sarin, and soman), prompting further research into nerve agents by all of the former Allies.

Although the threat of global thermonuclear war was foremost in the minds of most during the Cold War, both the Soviet and Western governments put enormous resources into developing chemical and biological weapons.

Britain

Porton Down was first established in 1916 and became the centre for the development of chemical weapons.

In the late 1940s and early 1950s, British postwar chemical weapons research was based at the Porton Down facility. Research was aimed at providing Britain with the means to arm itself with a modern nerve agent based capability and to develop specific means of defence against these agents.

Ranajit Ghosh, a chemist at the Plant Protection Laboratories of Imperial Chemical Industries was investigating a class of organophosphate compounds (organophosphate esters of substituted aminoethanethiols),[52] for use as a pesticide. In 1954, ICI put one of them on the market under the trade name Amiton. It was subsequently withdrawn, as it was too toxic for safe use.

The toxicity did not go unnoticed, and samples of it were sent to the research facility at Porton Down for evaluation. After the evaluation was complete, several members of this class of compounds were developed into a new group of much more lethal nerve agents, the V agents. The best-known of these is probably VX, assigned the UK Rainbow Code *Purple Possum*, with the Russian V-Agent coming a close second (Amiton is largely forgotten as VG).[53]

On the defensive side, there were years of difficult work to develop the means of prophylaxis, therapy, rapid detection and identification, decontamination and more effective protection of the body against nerve agents, capable of exerting effects through the skin, the eyes and respiratory tract.

Tests were carried out on servicemen to determine the effects of nerve agents on human subjects, with one recorded death due to a nerve gas experiment. There have been persistent allegations of unethical human experimentation at Porton Down, such as those relating to the death of Leading Aircraftman Ronald Maddison, aged 20, in 1953. Maddison was taking part in sarin nerve agent toxicity tests. Sarin was dripped onto his arm and he died shortly afterwards.[54]

In the 1950s the Chemical Defence Experimental Establishment became involved with the development of CS, a riot control agent, and took an increasing role in trauma and wound ballistics work. Both these facets of Porton Down's work had become more important because of the situation in Northern Ireland.[55]

In the early 1950s, nerve agents such as sarin were produced in small quantities—about 20 tons were made from 1954 until 1956. CDE Nancekuke was an important factory for stockpiling chemical weapons. Small amounts of VX were produced there, mainly for laboratory test purposes, but also to validate plant designs and optimise chemical processes for potential mass production. However, full-scale mass production of VX agent never took place, with the 1956 decision to end the UK's offensive chemical weapons programme.[56] In the late 1950s, the chemical weapons production plant at Nancekuke was mothballed, but was maintained through the 1960s and 1970s in a state whereby production of chemical weapons could easily re-commence if required.[56]

United States

In 1952, the U.S. Army patented a process for the "Preparation of Toxic Ricin", publishing a method of producing this powerful toxin. In 1958 the British government traded their VX technology with the United States in exchange for information on thermonuclear weapons. By 1961 the U.S. was producing large amounts of VX and performing its own nerve agent research. This research produced at least three more agents; the four agents (VE, VG, VM, VX) are collectively known as the "V-Series" class of nerve agents.

Between 1951 and 1969, Dugway Proving Ground was the site of testing for various chemical and biological agents, including an open-air aerodynamic dissemination test in 1968 that accidentally killed, on neighboring farms, approximately 6,400 sheep by an unspecified nerve agent.[57]

From 1962 to 1973, the Department of Defense planned

134 tests under Project 112, a chemical and biological weapons "vulnerability-testing program." In 2002, the Pentagon admitted for the first time that some of tests used real chemical and biological weapons, not just harmless simulants.[58]

Specifically under Project SHAD, 37 secret tests were conducted in California, Alaska, Florida, Hawaii, Maryland and Utah. Land tests in Alaska and Hawaii used artillery shells filled with sarin and VX, while Navy trials off the coasts of Florida, California and Hawaii tested the ability of ships and crew to perform under biological and chemical warfare, without the crew's knowledge. The code name for the sea tests was Project Shipboard Hazard and Defense —"SHAD" for short.[58]

In October 2002, the Senate Armed Forces Subcommittee on Personnel held hearings as the controversial news broke that chemical agents had been tested on thousands of American military personnel. The hearings were chaired by Senator Max Cleland, former VA administrator and Vietnam War veteran.

United States chemical respiratory protection standardization

In December 2001, the United States Department of Health and Human Services, Centers for Disease Control and Prevention (CDC), National Institute for Occupational Safety and Health (NIOSH), and National Personal Protective Technology Laboratory (NPPTL), along with the U.S. Army Research, Development and Engineering Command (RDECOM), Edgewood Chemical and Biological Center (ECBC), and the U.S. Department of Commerce National Institute for Standards and Technology (NIST) published the first of six technical performance standards and test procedures designed to evaluate and certify respirators intended for use by civilian emergency responders to a chemical, biological, radiological, or nuclear weapon release, detonation, or terrorism incident.

To date NIOSH/NPPTL has published six new respirator performance standards based on a tiered approach that relies on traditional industrial respirator certification policy, next-generation emergency response respirator performance requirements, and special live chemical warfare agent testing requirements of the classes of respirators identified to offer respiratory protection against chemical, biological, radiological, and nuclear (CBRN) agent inhalation hazards. These CBRN respirators are commonly known as open-circuit self-contained breathing apparatus (CBRN SCBA), air-purifying respirator (CBRN APR), air-purifying escape respirator (CBRN APER), self-contained escape respirator (CBRN SCER) and loose- or tight-fitting powered air-purifying respirators (CBRN PAPR).

Soviet Union

There were numerous reports of chemical weapons being used during the Soviet war in Afghanistan, sometimes against civilians.[59][60]

Due to the secrecy of the Soviet Union's government, very little information was available about the direction and progress of the Soviet chemical weapons until relatively recently. After the fall of the Soviet Union, Russian chemist Vil Mirzayanov published articles revealing illegal chemical weapons experimentation in Russia.

In 1993, Mirzayanov was imprisoned and fired from his job at the State Research Institute of Organic Chemistry and Technology, where he had worked for 26 years. In March 1994, after a major campaign by U.S. scientists on his behalf, Mirzayanov was released.[61]

Among the information related by Vil Mirzayanov was the direction of Soviet research into the development of even more toxic nerve agents, which saw most of its success during the mid-1980s. Several highly toxic agents were developed during this period; the only unclassified information regarding these agents is that they are known in the open literature only as "Foliant" agents (named after the program under which they were developed) and by various code designations, such as A-230 and A-232.[62]

According to Mirzayanov, the Soviets also developed weapons that were safer to handle, leading to the development of the binary weapons, in which precursors for the nerve agents are mixed in a munition to produce the agent just prior to its use. Because the precursors are generally significantly less hazardous than the agents themselves, this technique makes handling and transporting the munitions a great deal simpler.

Additionally, precursors to the agents are usually much easier to stabilize than the agents themselves, so this technique also made it possible to increase the shelf life of the agents a great deal. During the 1980s and 1990s, binary versions of several Soviet agents were developed and are designated as "Novichok" agents (after the Russian word for "newcomer").[63] Together with Lev Fedorov, he told the secret Novichok story exposed in the newspaper *The Moscow News*.[64]

31.2.5 Use in post-WWII conflicts

Stalag 13 prison camp

The earliest successful use of chemical agents in a noncombat setting was in 1946. Motivated by a desire to obtain revenge on Germans for the Holocaust, three members of a Jewish group calling themselves Dahm Y'Israel Nokeam

("Avenging Israel's Blood") hid in a bakery in the Stalag 13 prison camp near Nuremberg, Germany, where several thousand SS troops were being detained. The three applied an arsenic-containing mixture to loaves of bread, sickening more than 2,000 Nazi troops, of whom more than 200 required hospitalization.

North Yemen

The International Red Cross hospital at Uqd, North Yemen, where the use of chemical weapons was alleged to have occurred.

The first attack of the North Yemen Civil War took place on June 8, 1963 against Kawma, a village of about 100 inhabitants in northern Yemen, killing about seven people and damaging the eyes and lungs of 25 others. This incident is considered to have been experimental, and the bombs were described as "home-made, amateurish and relatively ineffective" . The Egyptian authorities suggested that the reported incidents were probably caused by napalm, not gas.

There were no reports of gas during 1964, and only a few were reported in 1965. The reports grew more frequent in late 1966. On December 11, 1966, fifteen gas bombs killed two people and injured thirty-five. On January 5, 1967, the biggest gas attack came against the village of Kitaf, causing 270 casualties, including 140 fatalities. The target may have been Prince Hassan bin Yahya, who had installed his headquarters nearby. The Egyptian government denied using poison gas, and alleged that Britain and the US were using the reports as psychological warfare against Egypt. On February 12, 1967, it said it would welcome a UN investigation. On March 1, U Thant, the then Secretary-General of the United Nations, said he was "powerless" to deal with the matter.

On May 10, 1967 the twin villages of Gahar and Gadafa in Wadi Hirran, where Prince Mohamed bin Mohsin was in command, were gas bombed, killing at least seventy-five. The Red Cross was alerted and on June 2, 1967, it issued a statement in Geneva expressing concern. The Institute of Forensic Medicine at the University of Berne made a statement, based on a Red Cross report, that the gas was likely to have been halogenous derivatives—phosgene, mustard gas, lewisite, chloride or cyanogen bromide.

The gas attacks stopped for three weeks after the Six-Day War of June, but resumed in July, against all parts of royalist Yemen. Casualty estimates vary, and an assumption, considered conservative, is that the mustard-and-phosgene-filled aerial bombs caused approximately 1,500 fatalities and 1,500 injuries.

Vietnamese border raids in Thailand

There is some evidence suggesting that Vietnamese troops used phosgene gas against Cambodian resistance forces in Thailand during the 1984-1985 dry-season offensive on the Thai-Cambodian border.[65][66][67]

Iran–Iraq War

See also: Iraqi chemical warfare

Chemical weapons employed by Saddam Hussein killed and injured numerous Iranians and Iraqi Kurds. According to Iraqi documents, assistance in developing chemical weapons was obtained from firms in many countries, including the United States, West Germany, the Netherlands, the United Kingdom, and France.[68]

About 100,000 Iranian soldiers were victims of Iraq's chemical attacks. Many were hit by mustard gas. The official estimate does not include the civilian population contaminated in bordering towns or the children and relatives of veterans, many of whom have developed blood, lung and skin complications, according to the Organization for Veterans. Nerve gas agents killed about 20,000 Iranian soldiers immediately, according to official reports. Of the 80,000 survivors, some 5,000 seek medical treatment regularly and about 1,000 are still hospitalized with severe, chronic conditions.[69][70][71]

Halabja

Main article: Halabja poison gas attack

Shortly before the war ended in 1988, the Iraqi Kurdish village of Halabja was exposed to multiple chemical agents, killing about 5,000 of the town's 50,000 residents.*[72]

During the Gulf War in 1991, Coalition forces began a ground war in Iraq. Despite the fact that they did possess chemical weapons, Iraq did not use any chemical agents against coalition forces. The commander of the Allied Forces, General Norman Schwarzkopf, suggested this may have been due to Iraqi fear of retaliation with nuclear weapons.

Angola

During the Cuban intervention in Angola, United Nations toxicologists certified that residue from both VX and sarin nerve agents had been discovered in plants, water, and soil where Cuban units were conducting operations against National Union for the Total Independence of Angola (UNITA) insurgents.*[73] In 1985, UNITA made the first of several claims that their forces were the target of chemical weapons, specifically organophosphates. The following year guerrillas reported being bombarded with an unidentified greenish-yellow agent on three separate occasions. Depending on the length and intensity of exposure, victims suffered blindness or death. The toxin was also observed to have killed plant life.*[74] Shortly afterwards, UNITA also sighted strikes carried out with a brown agent which it claimed resembled mustard gas.*[75] As early as 1984 a research team dispatched by the University of Ghent had examined patients in UNITA field hospitals showing signs of exposure to nerve agents, although it found no evidence of mustard gas.*[76]

The UN first accused Cuba of deploying chemical weapons against Angolan civilians and partisans in 1988.*[73] Wouter Basson later disclosed that South African military intelligence had long verified the use of unidentified chemical weapons on Angolan soil; this was to provide the impetus for their own biological warfare programme, Project Coast.*[73] During the Battle of Cuito Cuanavale, South African troops then fighting in Angola were issued with gas masks and ordered to rehearse chemical weapons drills. Although the status of its own chemical weapons program remained uncertain, South Africa also deceptively bombarded Cuban and Angolan units with coloured smoke in an attempt to induce hysteria or mass panic.*[75] According to Defence Minister Magnus Malan, this would force the Cubans to share the inconvenience of having to take preventative measures such as donning NBC suits, which would cut combat effectiveness in half. The tactic was effective: beginning in early 1988 Cuban units posted to Angola were issued with full protective gear in anticipation of a South African chemical strike.*[75]

On 29 October 1988, personnel attached to Angola's 59 Brigade, accompanied by six Soviet military advisors, reported being struck with chemical weapons on the banks of the Mianei River.*[77] The attack occurred shortly after one in the afternoon. Four Angolan soldiers lost consciousness while the others complained of violent headaches and nausea. That November the Angolan representative to the UN accused South Africa of employing poison gas near Cuito Cuanavale for the first time.*[77]

Falklands War

Technically, the reported employment of tear gas by Argentine forces during the 1982 invasion of the Falkland Islands constitutes chemical warfare.*[78] However, the tear gas grenades were employed as nonlethal weapons to avoid British casualties. The barrack buildings the weapons were used on proved to be deserted in any case. The British claim that more lethal, but legally justifiable as they are not considered chemical weapons under the Chemical Weapons Convention, white phosphorus grenades were used.*[79]

Syrian Civil War

Some of the victims of the Ghouta, Syria attack, 21 August 2013

Main article: Syrian Civil War § Chemical weapons

31.2.6 Terrorism

See also: Chemical terrorism

For many terrorist organizations, chemical weapons might be considered an ideal choice for a mode of attack, if they are available: they are cheap, relatively accessible, and easy to transport. A skilled chemist can readily synthesize most chemical agents if the precursors are available.

In July 1974, a group calling themselves the Aliens of America successfully firebombed the houses of a judge, two police commissioners, and one of the commissioner's cars, burned down two apartment buildings, and bombed the Pan Am Terminal at Los Angeles International Airport, killing three people and injuring eight. The organization, which turned out to be a single resident alien named Muharem Kurbegovic, claimed to have developed and possessed a supply of sarin, as well as four unique nerve agents named AA1, AA2, AA3, and AA4S. Although no agents were found at the time Kurbegovic was arrested in August 1974, he had reportedly acquired "all but one" of the ingredients required to produce a nerve agent. A search of his apartment turned up a variety of materials, including precursors for phosgene and a drum containing 25 pounds of sodium cyanide.[80]

U.S. Navy Seabees don their M40 Field Protective Mask

The first successful use of chemical agents by terrorists against a general civilian population was on June 27, 1994, when Aum Shinrikyo, an apocalyptic group based in Japan that believed it necessary to destroy the planet, released sarin gas in Matsumoto, Japan, killing eight and harming 200. The following year, Aum Shinrikyo released sarin into the Tokyo subway system killing 12 and injuring over 5,000.

On 29 December 1999, four days after Russian forces began an assault of Grozny, Chechen terrorists exploded two chlorine tanks in the town. Because of the wind conditions, no Russian soldiers were injured.[81]

Following the September 11, 2001 attacks on the U.S. cities of New York City and Washington, D.C., the organization Al-Qaeda responsible for the attacks announced that they were attempting to acquire radiological, biological, and chemical weapons. This threat was lent a great deal of credibility when a large archive of videotapes was obtained by the cable television network CNN in August 2002 showing, among other things, the killing of three dogs by an apparent nerve agent.[82]

On October 26, 2002, Russian special forces used a chemical agent (presumably KOLOKOL-1, an aerosolized fentanyl derivative), as a precursor to an assault on Chechen terrorists, ending the Moscow theater hostage crisis. All 42 of the terrorists and 120 out of 850 hostages were killed during the raid. Of the hostages who died, all but one or two died from the effects of the agent.

In early 2007, multiple terrorist bombings had been reported in Iraq using chlorine gas. These attacks wounded or sickened more than 350 people. Reportedly the bombers were affiliated with Al-Qaeda in Iraq,[83] and they have used bombs of various sizes up to chlorine tanker trucks.[84] United Nations Secretary-General Ban Ki-moon condemned the attacks as "clearly intended to cause panic and instability in the country."[85]

31.2.7 Chemical weapons treaties

See also: Destruction of chemical weapons and Chemical Weapons Convention

The *Protocol for the Prohibition of the Use in War of Asphyxiating, Poisonous or other Gases, and the Bacteriological Methods of Warfare,* or the Geneva Protocol, is an international treaty which prohibits the use of chemical and biological weapons in warfare. Signed into international Law at Geneva on June 17, 1925 and entered into force on February 8, 1928, this treaty states that chemical and biological weapons are "justly condemned by the general opinion of the civilised world."[86]

Chemical Weapons Convention

Main article: Chemical Weapons Convention
The most recent arms control agreement in International

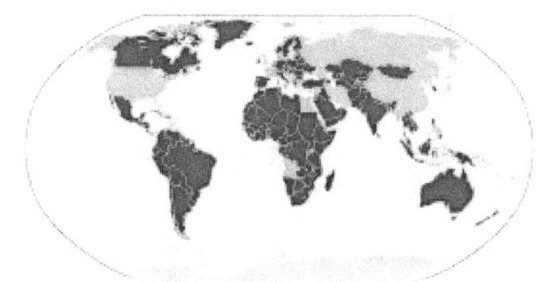

States parties to the Chemical Weapons Convention in 2015. Light colored territories are those states parties that have declared stockpiles of chemical weapons and/or have known production facilities for chemical weapons.

Law, the *Convention of the Prohibition of the Development, Production, Stockpiling and Use of Chemical Weapons and on their Destruction,* or the Chemical Weapons Convention, outlaws the production, stockpiling, and use of chemical weapons. It is administered by the Organisation for the Prohibition of Chemical Weapons (OPCW), an intergovernmental organisation based in The Hague.[87]

31.3 Technology

A Swedish Army soldier wearing a chemical agent protective suit (C-vätskeskydd) and protection mask (skyddsmask 90).

Although crude chemical warfare has been employed in many parts of the world for thousands of years.[88] "modern" chemical warfare began during World War I - see Chemical weapons in World War I.

Initially, only well-known commercially available chemicals and their variants were used. These included chlorine and phosgene gas. The methods used to disperse these agents during battle were relatively unrefined and inefficient. Even so, casualties could be heavy, due to the mainly static troop positions which were characteristic features of trench warfare.

Germany, the first side to employ chemical warfare on the battlefield.[89] simply opened canisters of chlorine upwind of the opposing side and let the prevailing winds do the dissemination. Soon after, the French modified artillery munitions to contain phosgene – a much more effective method that became the principal means of delivery.[90]

Since the development of modern chemical warfare in World War I, nations have pursued research and development on chemical weapons that falls into four major categories: new and more deadly agents; more efficient methods of delivering agents to the target (dissemination); more reli-

able means of defense against chemical weapons; and more sensitive and accurate means of detecting chemical agents.

31.3.1 Chemical warfare agents

See also: List of chemical warfare agents

A chemical used in warfare is called a *chemical warfare agent (CWA)*. About 70 different chemicals have been used or stockpiled as chemical warfare agents during the 20th and 21st centuries. These agents may be in liquid, gas or solid form. Liquid agents that evaporate quickly are said to be *volatile* or have a *high vapor pressure*. Many chemical agents are made volatile so they can be dispersed over a large region quickly.

The earliest target of chemical warfare agent research was not toxicity, but development of agents that can affect a target through the skin and clothing, rendering protective gas masks useless. In July 1917, the Germans employed mustard gas. Mustard gas easily penetrates leather and fabric to inflict painful burns on the skin.

Chemical warfare agents are divided into *lethal* and *incapacitating* categories. A substance is classified as incapacitating if less than 1/100 of the lethal dose causes incapacitation, e.g., through nausea or visual problems. The distinction between lethal and incapacitating substances is not fixed, but relies on a statistical average called the LD_{50}.

Persistency

Chemical warfare agents can be classified according to their *persistency*, a measure of the length of time that a chemical agent remains effective after dissemination. Chemical agents are classified as *persistent* or *nonpersistent*.

Agents classified as *nonpersistent* lose effectiveness after only a few minutes or hours or even only a few seconds. Purely gaseous agents such as chlorine are nonpersistent, as are highly volatile agents such as sarin. Tactically, nonpersistent agents are very useful against targets that are to be taken over and controlled very quickly.

Apart from the agent used, the delivery mode is very important. To achieve a nonpersistent deployment, the agent is dispersed into very small droplets comparable with the mist produced by an aerosol can. In this form not only the gaseous part of the agent (around 50%) but also the fine aerosol can be inhaled or absorbed through pores in the skin.

Modern doctrine requires very high concentrations almost instantly in order to be effective (one breath should contain a lethal dose of the agent). To achieve this, the primary weapons used would be rocket artillery or bombs and large ballistic missiles with cluster warheads. The contamination in the target area is only low or not existent and after four hours sarin or similar agents are not detectable anymore.

By contrast, *persistent* agents tend to remain in the environment for as long as several weeks, complicating decontamination. Defense against persistent agents requires shielding for extended periods of time. Non-volatile liquid agents, such as blister agents and the oily VX nerve agent, do not easily evaporate into a gas, and therefore present primarily a contact hazard.

The droplet size used for persistent delivery goes up to 1 mm increasing the falling speed and therefore about 80% of the deployed agent reaches the ground, resulting in heavy contamination. Deployment of persistent agents is intended to constrain enemy operations by denying access to contaminated areas.

Possible targets include enemy flank positions (averting possible counterattacks), artillery regiments, commando posts or supply lines. Because it is not necessary to deliver large quantities of the agent in a short period of time, a wide variety of weapons systems can be used.

A special form of persistent agents are thickened agents. These comprise a common agent mixed with thickeners to provide gelatinous, sticky agents. Primary targets for this kind of use include airfields, due to the increased persistency and difficulty of decontaminating affected areas.

Classes

Chemical weapons are inert agents that come in four categories: choking, blister, blood and nerve.[91] The agents are organized into several categories according to the manner in which they affect the human body. The names and number of categories varies slightly from source to source, but in general, types of chemical warfare agents are as follows:

There are other chemicals used militarily that are not scheduled by the Chemical Weapons Convention, and thus are not controlled under the CWC treaties. These include:

- Defoliants and herbicides that destroy vegetation, but are not immediately toxic or poisonous to human beings. Their use is classified as herbicidal warfare. Some batches of Agent Orange, for instance, used by the British during the Malayan Emergency and the United States during the Vietnam War, contained dioxins as manufacturing impurities. Dioxins, rather than Agent Orange itself, have long-term cancer effects and for causing genetic damage leading to serious birth deformities.

- Incendiary or explosive chemicals (such as napalm, extensively used by the United States during the Korean War and the Vietnam War, or dynamite) because their destructive effects are primarily due to fire or explosive force, and not direct chemical action. Their use is classified as conventional warfare.

- Viruses, bacteria, or other organisms. Their use is classified as biological warfare. Toxins produced by living organisms are considered chemical weapons, although the boundary is blurry. Toxins are covered by the Biological Weapons Convention.

Designations

For more details on this topic, see chemical weapon designation.

Most chemical weapons are assigned a one- to three-letter "NATO weapon designation" in addition to, or in place of, a common name. Binary munitions, in which precursors for chemical warfare agents are automatically mixed in shell to produce the agent just prior to its use, are indicated by a "−2" following the agent's designation (for example, GB-2 and VX-2).

Some examples are given below:

31.3.2 Delivery

The most important factor in the effectiveness of chemical weapons is the efficiency of its delivery, or dissemination, to a target. The most common techniques include munitions (such as bombs, projectiles, warheads) that allow dissemination at a distance and spray tanks which disseminate from low-flying aircraft. Developments in the techniques of filling and storage of munitions have also been important.

Although there have been many advances in chemical weapon delivery since World War I, it is still difficult to achieve effective dispersion. The dissemination is highly dependent on atmospheric conditions because many chemical agents act in gaseous form. Thus, weather observations and forecasting are essential to optimize weapon delivery and reduce the risk of injuring friendly forces.

Dispersion

Dispersion is placing the chemical agent upon or adjacent to a target immediately before dissemination, so that the material is most efficiently used. Dispersion is the simplest technique of delivering an agent to its target. The most common

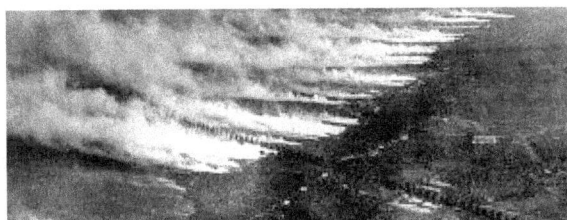

Dispersion of chlorine in World War I

Aerial photograph of a German gas attack on Russian forces circa 1916

techniques are munitions, bombs, projectiles, spray tanks and warheads.

World War I saw the earliest implementation of this technique. The actual first chemical ammunition was the French 26 mm cartouche suffocante rifle grenade, fired from a flare carbine. It contained 35g of the tear-producer ethyl bromoacetate, and was used in autumn 1914 – with little effect on the Germans.

The Germans on the other hand tried to increase the effect of 10.5 cm shrapnel shells by adding an irritant – dianisidine chlorosulfonate. Its use went unnoticed by the British when it was used against them at Neuve Chapelle in October 1914. Hans Tappen, a chemist in the Heavy Artillery Department of the War Ministry, suggested to his brother, the Chief of the Operations Branch at German General Headquarters, the use of the tear-gases benzyl bromide or xylyl bromide.

Shells were tested successfully at the Wahn artillery range near Cologne on 9 January 1915, and an order was placed for 15 cm howitzer shells, designated 'T-shells' after Tappen. A shortage of shells limited the first use against the Russians at Bolimów on 31 January 1915; the liquid failed to vaporize in the cold weather, and again the experiment went unnoticed by the Allies.

The first effective use were when the German forces at the Second Battle of Ypres simply opened cylinders of chlorine and allowed the wind to carry the gas across enemy lines. While simple, this technique had numerous disadvantages. Moving large numbers of heavy gas cylinders to the front-line positions from where the gas would be released was a lengthy and difficult logistical task.

Stockpiles of cylinders had to be stored at the front line, posing a great risk if hit by artillery shells. Gas delivery depended greatly on wind speed and direction. If the wind was fickle, as at Loos, the gas could blow back, causing friendly casualties.

Gas clouds gave plenty of warning, allowing the enemy time to protect themselves, though many soldiers found the sight of a creeping gas cloud unnerving. This made the gas doubly effective, as, in addition to damaging the enemy physically, it also had a psychological effect on the intended victims.

Another disadvantage was that gas clouds had limited penetration, capable only of affecting the front-line trenches before dissipating. Although it produced limited results in World War I, this technique shows how simple chemical weapon dissemination *can* be.

Shortly after this "open canister" dissemination, French forces developed a technique for delivery of phosgene in a non-explosive artillery shell. This technique overcame many of the risks of dealing with gas in cylinders. First, gas shells were independent of the wind and increased the effective range of gas, making any target within reach of guns vulnerable. Second, gas shells could be delivered without warning, especially the clear, nearly odorless phosgene— there are numerous accounts of gas shells, landing with a "plop" rather than exploding, being initially dismissed as dud high explosive or shrapnel shells, giving the gas time to work before the soldiers were alerted and took precautions.

The major drawback of artillery delivery was the difficulty of achieving a killing concentration. Each shell had a small gas payload and an area would have to be subjected to saturation bombardment to produce a cloud to match cylinder delivery. A British solution to the problem was the Livens Projector. This was effectively a large-bore mortar, dug into the ground that used the gas cylinders themselves as projectiles - firing a 14 kg cylinder up to 1500 m. This combined the gas volume of cylinders with the range of artillery.

Over the years, there were some refinements in this technique. In the 1950s and early 1960s, chemical artillery rockets and cluster bombs contained a multitude of submunitions, so that a large number of small clouds of the chemical agent would form directly on the target.

Thermal dissemination

An American-made MC-1 gas bomb

Thermal dissemination is the use of explosives or pyrotechnics to deliver chemical agents. This technique, developed in the 1920s, was a major improvement over earlier dispersal techniques, in that it allowed significant quantities of an agent to be disseminated over a considerable distance. Thermal dissemination remains the principal method of disseminating chemical agents today.

Most thermal dissemination devices consist of a bomb or projectile shell that contains a chemical agent and a central "burster" charge; when the burster detonates, the agent is expelled laterally.

Thermal dissemination devices, though common, are not particularly efficient. First, a percentage of the agent is lost by incineration in the initial blast and by being forced onto the ground. Second, the sizes of the particles vary greatly because explosive dissemination produces a mixture of liquid droplets of variable and difficult to control sizes.

The efficacy of thermal detonation is greatly limited by the flammability of some agents. For flammable aerosols, the cloud is sometimes totally or partially ignited by the disseminating explosion in a phenomenon called *flashing*. Explosively disseminated VX will ignite roughly one third of the time. Despite a great deal of study, flashing is still not fully understood, and a solution to the problem would be a major technological advance.

Despite the limitations of central bursters, most nations use this method in the early stages of chemical weapon development, in part because standard munitions can be adapted to carry the agents.

Soviet chemical weapons canisters from a stockpile in Albania

Aerodynamic dissemination

Aerodynamic dissemination is the non-explosive delivery of a chemical agent from an aircraft, allowing aerodynamic stress to disseminate the agent. This technique is the most recent major development in chemical agent dissemination, originating in the mid-1960s.

This technique eliminates many of the limitations of thermal dissemination by eliminating the flashing effect and theoretically allowing precise control of particle size. In actuality, the altitude of dissemination, wind direction and velocity, and the direction and velocity of the aircraft greatly influence particle size. There are other drawbacks as well: ideal deployment requires precise knowledge of aerodynamics and fluid dynamics, and because the agent must usually be dispersed within the boundary layer (less than 200–300 ft above the ground), it puts pilots at risk.

Significant research is still being applied toward this technique. For example, by modifying the properties of the liquid, its breakup when subjected to aerodynamic stress can be controlled and an idealized particle distribution achieved, even at supersonic speed. Additionally, advances in fluid dynamics, computer modeling, and weather forecasting allow an ideal direction, speed, and altitude to be calculated, such that warfare agent of a predetermined particle size can predictably and reliably hit a target.

31.3.3　Protection against chemical warfare

Israel Defense Forces "Yanshuf" battalion soldiers at chemical warfare defense exercise

Ideal protection begins with nonproliferation treaties such as the Chemical Weapons Convention, and detecting, very early, the *signatures* of someone building a chemical weapons capability. These include a wide range of intelligence disciplines, such as economic analysis of exports of dual-use chemicals and equipment, human intelligence (HUMINT) such as diplomatic, refugee, and agent reports; photography from satellites, aircraft and drones (IMINT); examination of captured equipment (TECHINT); communications intercepts (COMINT); and detection of chemical manufacturing and chemical agents themselves (MASINT).

If all the preventive measures fail and there is a clear and present danger, then there is a need for detection of chemical attacks,[92] collective protection,[93][94][95] and decontamination. Since industrial accidents can cause dangerous chemical releases (e.g., the Bhopal disaster), these activities are things that civilian, as well as military, organizations must be prepared to carry out. In civilian situations in developed countries, these are duties of HAZMAT organizations, which most commonly are part of fire departments.

Detection has been referred to above, as a technical MASINT discipline; specific military procedures, which are usually the model for civilian procedures, depend on the equipment, expertise, and personnel available. When chemical agents are detected, an alarm needs to sound, with specific warnings over emergency broadcasts and the like. There may be a warning to expect an attack.

If, for example, the captain of a US Navy ship believes there is a serious threat of chemical, biological, or radiological attack, the crew may be ordered to set Circle William, which means closing all openings to outside air, running breathing

air through filters, and possibly starting a system that continually washes down the exterior surfaces. Civilian authorities dealing with an attack or a toxic chemical accident will invoke the Incident Command System, or local equivalent, to coordinate defensive measures.[95]

Individual protection starts with a gas mask and, depending on the nature of the threat, through various levels of protective clothing up to a complete chemical-resistant suit with a self-contained air supply. The US military defines various levels of MOPP (mission-oriented protective posture) from mask to full chemical resistant suits; Hazmat suits are the civilian equivalent, but go farther to include a fully independent air supply, rather than the filters of a gas mask.

Collective protection allows continued functioning of groups of people in buildings or shelters, the latter which may be fixed, mobile, or improvised. With ordinary buildings, this may be as basic as plastic sheeting and tape, although if the protection needs to be continued for any appreciable length of time, there will need to be an air supply, typically an enhanced gas mask.[94][95]

Members of the Ukrainian Army's 19th Nuclear, Biological and Chemical Battalion practice decontamination drill, at Camp Arifjan, Kuwait

Decontamination

Decontamination varies with the particular chemical agent used. Some *nonpersistent* agents, including most pulmonary agents (chlorine, phosgene, and so on), blood gases, and nonpersistent nerve gases (e.g., GB), will dissipate from open areas, although powerful exhaust fans may be needed to clear out buildings where they have accumulated.

In some cases, it might be necessary to neutralize them chemically, as with ammonia as a neutralizer for hydrogen cyanide or chlorine. Riot control agents such as CS will dissipate in an open area, but things contaminated with CS powder need to be aired out, washed by people wearing protective gear, or safely discarded.

Mass decontamination is a less common requirement for

people than equipment, since people may be immediately affected and treatment is the action required. It is a requirement when people have been contaminated with persistent agents. Treatment and decontamination may need to be simultaneous, with the medical personnel protecting themselves so they can function."[96]

There may need to be immediate intervention to prevent death, such as injection of atropine for nerve agents. Decontamination is especially important for people contaminated with persistent agents; many of the fatalities after the explosion of a WWII US ammunition ship carrying mustard gas, in the harbor of Bari, Italy, after a German bombing on 2 December 1943, came when rescue workers, not knowing of the contamination, bundled cold, wet seamen in tight-fitting blankets.

For decontaminating equipment and buildings exposed to persistent agents, such as blister agents, VX or other agents made persistent by mixing with a thickener, special equipment and materials might be needed. Some type of neutralizing agent will be needed; e.g. in the form of a spraying device with neutralizing agents such as Chlorine, Fichlor, strong alkaline solutions or enzymes. In other cases, a specific chemical decontaminant will be required."[95]

31.4 Sociopolitical climate

The study of chemicals and their military uses was widespread in China and India. The use of toxic materials has historically been viewed with mixed emotions and moral qualms in the West. The practical and ethical problems surrounding poison warfare appeared in ancient Greek myths about Hercules' invention of poison arrows and Odysseus's use of toxic projectiles. There are many instances of the use of chemical weapons in battles documented in Greek and Roman historical texts; the earliest example was the deliberate poisoning of Kirrha's water supply with hellebore in the First Sacred War, Greece, about 590 BC."[97]

One of the earliest reactions to the use of chemical agents was from Rome. Struggling to defend themselves from the Roman legions, Germanic tribes poisoned the wells of their enemies, with Roman jurists having been recorded as declaring "armis bella non venenis geri", meaning "war is fought with weapons, not with poisons." Yet the Romans themselves resorted to poisoning wells of besieged cities in Anatolia in the 2nd century BCE."[3]

Before 1915 the use of poisonous chemicals in battle was typically the result of local initiative, and not the result of an active government chemical weapons program. There are many reports of the isolated use of chemical agents in individual battles or sieges, but there was no true tradition

of their use outside of incendiaries and smoke. Despite this tendency, there have been several attempts to initiate large-scale implementation of poison gas in several wars, but with the notable exception of World War I, the responsible authorities generally rejected the proposals for ethical reasons.

For example, in 1854 Lyon Playfair (later 1st Baron Playfair, GCB, PC, FRS (1 May 1818 – 29 May 1898), a British chemist, proposed using a cacodyl cyanide-filled artillery shell against enemy ships during the Crimean War. The British Ordnance Department rejected the proposal as "as bad a mode of warfare as poisoning the wells of the enemy."

31.4.1 Efforts to eradicate chemical weapons

See also: List of chemical arms control agreements

- August 27, 1874: The Brussels Declaration Concerning the Laws and Customs of War is signed, specifically forbidding the "employment of poison or poisoned weapons", although the treaty was not adopted by any nation whatsoever and it never went into effect.

- September 4, 1900: The First Hague Convention, which includes a declaration banning the "use of projectiles the object of which is the diffusion of asphyxiating or deleterious gases," enters into force.

- January 26, 1910: The Second Hague Convention enters into force, prohibiting the use of "poison or poisoned weapons" in warfare.

- February 6, 1922: After World War I, the Washington Arms Conference Treaty prohibited the use of asphyxiating, poisonous or other gases. It was signed by the United States, Britain, Japan, France, and Italy, but France objected to other provisions in the treaty and it never went into effect.

- February 8, 1928: The Geneva Protocol enters into force, prohibiting the use of "asphyxiating, poisonous or other gases, and of all analogous liquids, materials or devices" and "bacteriological methods of warfare"

.

31.4.2 Chemical weapon proliferation

Main article: Chemical weapon proliferation

Despite numerous efforts to reduce or eliminate them, some nations continue to research and/or stockpile chemical warfare agents. To the right is a summary of the nations that have either declared weapon stockpiles or are suspected of secretly stockpiling or possessing CW research programs. Notable examples include United States and Russia.

In 1997, future US Vice President Dick Cheney opposed the signing ratification of a treaty banning the use chemical weapons, a recently unearthed letter shows. In a letter dated April 8, 1997, then Halliburton-CEO Cheney told Sen. Jesse Helms, the chairman of the Senate Foreign Relations Committee, that it would be a mistake for America to join the Convention. "Those nations most likely to comply with the Chemical Weapons Convention are not likely to ever constitute a military threat to the United States. The governments we should be concerned about are likely to cheat on the CWC, even if they do participate," reads the letter.[101] published by the Federation of American Scientists.

The CWC was ratified by the Senate that same month. Since then, Albania, Libya, Russia, the United States, and India have declared over 71,000 metric tons of chemical weapon stockpiles, and destroyed about a third of them. Under the terms of the agreement, the United States and Russia agreed to eliminate the rest of their supplies of chemical weapons by 2012. Not having met its goal, the U.S. government estimates remaining stocks will be destroyed by 2017.

31.5 Chemical weapons destruction

31.5.1 India

In June 1997, India declared that it had a stockpile of 1044 tonnes of sulphur mustard in its possession. India's declaration of its stockpile came after its entry into the Chemical Weapons Convention, that created the Organisation for the Prohibition of Chemical Weapons, and on January 14, 1993 India became one of the original signatories to the Chemical Weapons Convention. By 2005, from among six nations that had declared their possession of chemical weapons, India was the only country to meet its deadline for chemical weapons destruction and for inspection of its facilities by the Organisation for the Prohibition of Chemical Weapons.[102][103] By 2006, India had destroyed more than 75 percent of its chemical weapons and material stockpile and was granted an extension to complete a 100 percent destruction of its stocks by April 2009. On May 14, 2009 India informed the United Nations that it has completely destroyed its stockpile of chemical weapons.[104]

31.5.2 Iraq

See also: Iraqi chemical warfare

The Director-General of the Organisation for the Prohibition of Chemical Weapons, Ambassador Rogelio Pfirter, welcomed Iraq's decision to join the OPCW as a significant step to strengthening global and regional efforts to prevent the spread and use of chemical weapons. The OPCW announced "The government of Iraq has deposited its instrument of accession to the Chemical Weapons Convention with the Secretary General of the United Nations and within 30 days, on 12 February 2009, will become the 186th State Party to the Convention". Iraq has also declared stockpiles of chemical weapons, and because of their recent accession is the only State Party exempted from the destruction timeline.[105]

31.5.3 Japan

During the Second Sino-Japanese War (1937–1945) Japan stored chemical weapons on the territory of mainland China. The weapon stock mostly containing mustard gas-lewisite mixture.[106] The weapons are classified as abandoned chemical weapons under the Chemical Weapons Convention and from September 2010 Japan has started their destruction in Nanjing using mobile destruction facilities in order to do so.[107]

31.5.4 Russia

Russia signed into the Chemical Weapons Convention on January 13, 1993 and ratified it on November 5, 1995. Declaring an arsenal of 39,967 tons of chemical weapons in 1997, by far the largest arsenal, consisting of blister agents: Lewisite, Sulfur mustard, Lewisite-mustard mix, and nerve agents: Sarin, Soman, and VX. Russia met its treaty obligations by destroying 1 percent of its chemical agents by the 2002 deadline set out by the Chemical Weapons Convention, but requested an extension on the deadlines of 2004 and 2007 due to technical, financial, and environmental challenges of chemical disposal. Since, Russia has received help from other countries such as Canada which donated C$100,000, plus a further C$100,000 already donated, to the Russian Chemical Weapons Destruction Program. This money will be used to complete work at Shchuch'ye and support the construction of a chemical weapons destruction facility at Kizner (Russia), where the destruction of nearly 5,700 tonnes of nerve agent, stored in approximately 2 million artillery shells and munitions, will be undertaken. Canadian funds are also being used for the operation of a Green Cross Public Outreach Office, to keep the civilian population informed on the progress made in chemical weapons destruction activities.[108]

As of July 2011, Russia has destroyed 48 percent (18,241 tonnes) of its stockpile at destruction facilities located in

Gorny (Saratov Oblast) and Kambarka (Udmurt Republic) - where operations have finished - and Schuch'ye (Kurgan Oblast), Maradykovsky (Kirov Oblast), Leonidovka (Penza Oblast) whilst installations are under construction in Pochep (Bryansk Oblast) and Kizner (Udmurt Republic).*[109] As August 2013, 76 percent (30,500 tonnes) were destroyed,*[110] and Russia leaves the Cooperative Threat Reduction (CTR) Program, which partially funded chemical weapons destruction.*[111]

31.5.5 United States

See also: United States and weapons of mass destruction §
Chemical weapons

On November 25, 1969, President Richard Nixon unilaterally renounced the use of chemical weapons and renounced all methods of biological warfare. He issued a decree halting the production and transport of all chemical weapons which remains in effect. From May 1964 to the early 1970s the USA participated in Operation CHASE, a United States Department of Defense program that aimed to dispose of chemical weapons by sinking ships laden with the weapons in the deep Atlantic. After the Marine Protection, Research, and Sanctuaries Act of 1972, Operation Chase was scrapped and safer disposal methods for chemical weapons were researched, with the U.S. destroying several thousand tons of mustard gas by incineration at the Rocky Mountain Arsenal, and nearly 4,200 tons of nerve agent by chemical neutralisation at Tooele Army Depot.*[112]

The U.S. ratified the Geneva Protocol which banned the use of chemical and biological weapons on January 22, 1975. In 1989 and 1990, the U.S. and the Soviet Union entered an agreement to both end their chemical weapons programs, including binary weapons. In April 1997, the United States ratified the Chemical Weapons Convention, this banned the possession of most types of chemical weapons. It also banned the development of chemical weapons, and required the destruction of existing stockpiles, precursor chemicals, production facilities, and their weapon delivery systems.

The U.S. began stockpile reductions in the 1980s with the removal of outdated munitions and destroying its entire stock of 3-Quinuclidinyl benzilate (BZ or Agent 15) at the beginning of 1988. In June 1990 the Johnston Atoll Chemical Agent Disposal System began destruction of chemical agents stored on the Johnston Atoll in the Pacific, seven years before the Chemical Weapons Treaty came into effect. In 1986 President Ronald Reagan made an agreement with the Chancellor, Helmut Kohl to remove the U.S. stockpile of chemical weapons from Germany. In 1990, as part of Operation Steel Box, two ships were loaded with over 100,000 shells containing Sarin and VX were taken from the U.S. Army weapons storage depots such as Miesau and then-classified FSTS (Forward Storage / Transportation Sites) and transported from Bremerhaven, Germany to Johnston Atoll in the Pacific, a 46-day nonstop journey.*[113]

In May 1991, President George H. W. Bush committed the United States to destroying all of its chemical weapons and renounced the right to chemical weapon retaliation. In 1993, the United States signed the Chemical Weapons Treaty, which required the destruction of all chemical weapon agents, dispersal systems, and production facilities by April 2012. The U.S. prohibition on the transport of chemical weapons has meant that destruction facilities had to be constructed at each of the U.S.'s nine storage facilities. The U.S. met the first three of the four deadlines set out in the treaty, destroying 45% of its stockpile of chemical weapons by 2007. Due to the destruction of chemical weapons, under the United States policy of Proportional Response, an attack upon the United States or its Allies would trigger a force-equivalent counter-attack. Since the United States only maintains nuclear Weapons of Mass Destruction, it is the stated policy that the United States will regard all WMD attacks (Biological, chemical, or nuclear) as a nuclear attack and will respond to such an attack with a nuclear strike.*[114]

As of 2012, stockpiles have been eliminated at 7 of the 9 chemical weapons depots and 89.75% of the 1997 stockpile has been destroyed by the treaty deadline of April 2012.*[115] Destruction will not begin at the two remaining depots until after the treaty deadline and will use neutralization, instead of incineration.

31.6   See also

- 1990 Chemical Weapons Accord

- Ali Hassan al-Majid

- Area denial weapon

- Biological warfare

- Chemical Weapons Convention

- Chemical weapon designation

- Chemical weapons and the United Kingdom

- Lethal Unitary Chemical Agents and Munitions

- List of chemical warfare agents

- List of highly toxic gases

- Ronald Maddison

- Psychochemical weapon

- Saint Julien Memorial

- Sardasht (A town attacked with chemical weapons during the Iran–Iraq War.)

- Stink bomb

- United States Army Medical Research Institute of Chemical Defense

- Weapon of mass destruction

- Zyklon B

31.7   Notes

[1]   "Convention on the Prohibition of the Development, Production, Stockpiling and Use of Chemical Weapons and on Their Destruction (CWC): Annexes and Original Signatories". Bureau of Arms Control, Verification and Compliance. Retrieved 19 January 2012.

[2]   Disarmament lessons from the Chemical Weapons Convention Archived June 6, 2013, at the Wayback Machine.

[3]   Mayor 2003

[4]   Michael Bothe; Natalino Ronzitti; Allan Rosas, eds. (1998). *The New Chemical Weapons Convention - Implementation and Prospects*, Martinus Nijhoff Publishers, p. 17, ISBN 90-411-1099-2

[5]   Richardt, Andre (2013). *CBRN Protection: Managing the Threat of Chemical, Biological, Radioactive and Nuclear Weapons*. Germany: Wiley-VCH Verlag & Co., p. 4, ISBN 978-3-527-32413-2

[6]   Science Daily, dated January 19, 2009 Sciencedaily.com

[7]   "Death Underground: Gas Warfare at Dura-Europos," Current Archaeology, November 26, 2009 (online feature) http://www.archaeology.co.uk/cwa/world-news/death-underground-gas-warfare-at-dura-europos.htm (accessed October 3, 2014)

[8]   Samir S. Patel. "Early Chemical Warfare – Dura-Europos, Syria," Archaeology, Vol. 63, No. 1, January/February 2010. http://www.archaeology.org/1001/topten/syria.html (accessed October 3, 2014)

[9]   Stephanie Pappas, "Buried Soldiers May Be Victims of Ancient Chemical Weapon," LiveScience, March 8, 2011. http://www.livescience.com/13113-ancient-chemical-warfare-romans-persians.html (accessed October 3, 2014).

[10]   Charles C. Mann (2011), *1493: Uncovering the New World Columbus Created*, Random House Digital. p. 9, ISBN 978-0-307-59672-7

[11]   David Hume, History of England, Volume II.

[12]   Eric Croddy (2002). *Chemical and Biological Warfare: A Comprehensive Survey for the Concerned Citizen*. Springer. p. 131.

[13]   Smart, Jeffery K. "CHEMICAL AND BIOLOGICAL WARFARE RESEARCH AND DEVELOPMENT DURING THE CIVIL WAR" (PDF). *United States Army*. US Army Soldier and Biological Chemical Command. Retrieved 7 November 2014.

[14]   "The First World War" (a Channel 4 documentary based on the book by Hew Strachan)

[15]   D. Hank Ellison (August 24, 2007). *Handbook of Chemical and Biological Warfare Agents, Second Edition*. CRC Press. pp. 567–570. ISBN 0-8493-1434-8.

[16]   Max Boot (August 16, 2007). *War Made New: Weapons, Warriors, and the Making of the Modern World*. Gotham. pp. 245–250. ISBN 1-5924-0315-8.

[17]   Walter E. Grunden: *Secret Weapons And World War II: Japan In The Shadow Of Big Science*. Lawrence (Kansas) 2005 , S. 172.

[18]   Was Winston Churchill really "strongly in favor of using poisoned gas against uncivilized tribes" from the Churchill Papers 16/16, 12 May 1919 at www.winstonchurchill.org , accessed 10 September 2013

[19]   Libcom.org, Libcom 1804-2003: History of Iraq

[20]   Nicolas Werth, Karel Bartošek, Jean-Louis Panné, Jean-Louis Margolin, Andrzej Paczkowski, Stéphane Courtois, *The Black Book of Communism: Crimes, Terror, Repression*, Harvard University Press, 1999, hardcover, 858 pages, ISBN 0-674-07608-7

[21]   Corum, James S., *The Roots of Blitzkrieg*, University Press of Kansas, USA, 1992, pp.106-107.

[22]   Schmaltz, Florian (2005), *Kampfstoff-Forschung im Nationalsozialismus Zur Kooperation von Kaiser-Wilhelm-Instituten, Militär und Industrie*, Wallstein Verlag

[23]   Schmaltz, Florian (2006), "Neurosciences and Research on Chemical Weapons of Mass Destruction in Nazi Germany", *Journal of the History of the Neurosciences*, **15** (3): 186–209, doi:10.1080/09647040600658229, PMID 16887760

[24]   "Chemical warfare". *World War II*. DesertWar.net.

[25]   "Laws of War: Declaration on the Use of Projectiles the Object of Which is the Diffusion of Asphyxiating or Deleterious Gases; July 29, 1899". Avalon.law.yale.edu. Retrieved 2014-01-18.

[26]   "Convention (IV) respecting the Laws and Customs of War on Land and its annex: Regulations concerning the Laws and Customs of War on Land. The Hague, 18 October 1907.". International Committee of the Red Cross. Retrieved 2014-01-18.

[27] Yuki Tanaka, *Poison Gas, the Story Japan Would Like to Forget*, Bulletin of the Atomic Scientists, October 1988, p. 16-17

[28] Y. Yoshimi and S. Matsuno, *Dokugasusen Kankei Shiryô II, Kaisetsu, Jugonen Sensô Gokuhi Shiryoshu*, 1997, p.27-29

[29] Yoshimi and Matsuno, idem, Herbert Bix, *Hirohito and the Making of Modern Japan*, 2001, p.360-364

[30] Army History Unit, *Chemical Warfare in Australia (2nd Edn) 2013 (Army Military History Series)*

[31] Geoff Plunkett, *Chemical Warfare in Australia*

[32] Geoff Plunkett, *Death By Mustard Gas*

[33] Fisk, Robert (December 30, 2000), "Poison gas from Germany", *Independent*

[34] Stanley P. Lovell, *Of Spies & Stratagems* (Englewood Cliffs, New Jersey: Prentice-Hall, 1963), p. 78.

[35] Merridale, Catherine, Ivan's War, Faber & Faber: pp. 148-150.

[36] Гречко, p. 468.

[37] Bellamy, Christopher (4 June 1996). "Sixty secret mustard gas sites uncovered". *The Independent*.

[38] "Chemical Warfare -Suffolk". *Anti-Invasion defences Suffolk World War II*. Retrieved 18 June 2008.

[39] Pears, Brian. "Chapter 5 Invasion". *Rowlands Gill and the North-East 1939 - 1945*.

[40] Alanbrooke, 2001. Entry 22 July 1940.

[41] BBC2 *Newsnight*, 1/5/81; *The Guardian*, 7,9,13,20,30/5, 2/6/81; *The Times*, 11/5/81, 20/5/81, 15/6/81; *The Listener*, 25/6, 2/7, 17/8/81; *Daily Telegraph*, 18,21,25,29/5, 2,11/6/81; *Encounter* magazine, Vol.58-9 no.2; *New Society*, Vol.60; *Bulletin of the Atomic Scientists*, Vol.4 no.4 and 'Churchill's Anthrax Bombs - a debate', Vol.4 no.12, November 1987.

[42] Paxman, Jeremy; Harris, Robert (2002-08-06) [1982]. "The War That Never Was". *A higher form of killing: the secret history of chemical and biological warfare*. p. 128. ISBN 978-0-8129-6653-4. OCLC 268949025. I want the matter studied in cold blood by sensible people

[43] "Paxman and Harris", p132-35.

[44] Smart, Jeffrey (1997). "2". *History of Chemical and Biological Warfare: An American Perspective*, Aberdeen, MD. USA: Army Chemical and Biological Defense Command, p. 32.

[45] http://www.cdc.gov/niosh/ershdb/ EmergencyResponseCard_29750039.html

[46] "Characteristics and Employment of Ground Chemical Munitions". *Field Manual 3-5*, Washington, DC: War Department, 1946, pp. 108–19.

[47] Skates, John R (2000), *The Invasion of Japan: Alternative to the Bomb*, University of South Carolina Press, pp. 93–96, ISBN 978-1-57003-354-4

[48] US Naval Historical Center, *Naval Armed Guard Service: Tragedy at Bari, Italy on 2 December 1943*, archived from the original on 2008-01-12

[49] Niderost, Eric, *World War II: German Raid on Bari* (Full text), HistoryNet.com

[50] Infield, Glenn B. Infield, *Disaster at Bari*

[51] Reminick, Gerald. *Nightmare in Bari: The World War II Liberty Ship Poison Gas Disaster and Coverup*

[52] Ghosh, R.; Newman, J.E. (Jan 29, 1955). "A new group of organophosphorus pesticides". *Chemistry and Industry*: 118.

[53] G B Carter. *Porton Down: a brief history*.

[54] G B Carter (2000). *Chemical and Biological Defence at Porton Down 1916–2000*. The Stationery Office.

[55] Robert Bud; Philip Gummett (2002). *Cold War, Hot Science: Applied Research in Britain's Defence Laboratories. 1945-1990*. Science museum.

[56] "Nancekuke Remediation Project". Ministry of Defence (Archived by The National Archives). Retrieved 25 April 2012.

[57] Staff, Committee on Veterans' Affairs, US Senate (December 8, 1994). *Is Military Research Hazardous to Veterans' Health? Lessons spanning half a century*. 103d Congress, 2d Session - Committee Print - S. Prt. 103-97

[58] Philip Huang (October 17, 2002), "Sickening strategy", *Oregon Daily Emerald*

[59] The Story of Genocide in Afghanistan Hassan Kakar

[60] Report from Afghanistan Claude Malhuret

[61] Yevgenia Albats and Catherine A. Fitzpatrick. *The State Within a State: The KGB and Its Hold on Russia - Past, Present, and Future*, 1994. ISBN 0-374-18104-7 (see pages 325–328)

[62] Fedorov, Lev (27 July 1994), *Chemical Weapons in Russia: History, Ecology, Politics*, Center of Ecological Policy of Russia

[63] Birstein, Vadim J. (2004), *The Perversion Of Knowledge: The True Story of Soviet Science*, Westview Press, ISBN 0-8133-4280-5

[64] Fedorov, Lev; Mirzayanov, Vil (1992), "A Poisoned Policy", *Moscow News* (weekly No. 39)

[65] "KPNLF says Vietnamese Using Suffocant Gas." *Bangkok World*, January 4, 1985, p. 1.

[66] "Viets Accused of Using Gas Against Rebels." *Associated Press*, Feb 19, 1985.

[67] "Thais Report a Clash with Vietnamese Troops." *Associated Press*, Feb 20, 1985.

[68] Lafayette, Lev (July 26, 2002). "Who armed Saddam?", *World History Archives*

[69] Fassihi, Farnaz (October 27, 2002). "In Iran, grim reminders of Saddam's arsenal", *New Jersey Star Ledger*

[70] Paul Hughes (January 21, 2003). "It's like a knife stabbing into me", *The Star (South Africa)*

[71] Sciolino, Elaine (February 13, 2003). "Iraq Chemical Arms Condemned, but West Once Looked the Other Way", *New York Times*, archived from the original on May 27, 2013

[72] "Death Clouds: Saddam Hussein's Chemical War Against the Kurds 5/1/1991". Retrieved 26 February 2016.

[73] Hawk, Kathleen Dupes; Villella, Ron; Varona, Adolfo Leyva de (30 July 2014). *Florida and the Mariel Boatlift of 1980: The First Twenty Days*. University of Alabama Press. ISBN 978-0817318376. Retrieved 11 October 2014.

[74] "Chemical weapons being used in Angola?". *Park City Daily News*. Bowling Green, Kentucky. 22 August 1986. Retrieved 28 July 2015.

[75] Helen E. Purkitt, Stephen F. Burgess: *The Rollback of South Africa's Chemical and Biological Warfare Program*. Air University, Counterproliferation Center, Maxwell Air-force Base, Alabama. 2001

[76] "Cubans using poison gas in Angola". *The Lewiston Journal*. Lewiston–Auburn, Maine. 26 August 1988. Retrieved 28 July 2015.

[77] Tokarev, Andrei; Shubin, Gennady, eds. (2011). *Bush War: The Road to Cuito Cuanavale: Soviet Soldiers' Accounts of the Angolan War*. Auckland Park: Jacana Media (Pty) Ltd. pp. 128–130. ISBN 978-1-4314-0185-7.

[78] *The Argentine Fight for The Falklands*, Lieutenant-Commander Sanchez-Sabarots

[79] *Falkland Islanders at war*, Bound, Graham, Pen and Sword Books Limited, ISBN 1-84415-429-7.

[80] "T Is for Terror: A mad bomber who stalked Los Angeles in the '70s could be the poster boy for the kind of terrorist the FBI fears today", *Newsweek Web Exclusive*, 9 July 2003

[81] Ксения Мяло. Россия и последние войны XX века: к истории падения сверхдержавы. Глава 5: Чеченский узел. М.: Вече, 2002

[82] Nic Robertson (August 19, 2002). "Disturbing scenes of death show capability with chemical gas", *CNN*, archived from the original on April 7, 2013

[83] Multi-National Force Iraq, Combined Press Information Center (20 April 2007), *Chlorine Tanks Destroyed, Terrorists Killed in Raids*, Press Release A070420a, archived from the original on August 7, 2009

[84] Multi-National Force Iraq, Combined Press Information Center (6 April 2007), *Suicide Vehicle Detonates outside Police Checkpoint*, Press Release 20070406-34, archived from the original on September 12, 2009

[85] Ban, Ki-Moon (19 March 2007), "Secretary-General Condemns Chlorine Attack in Iraq", *United Nations Radio*

[86] "Text of the Biological and Toxin Weapons Convention", Brad.ac.uk. Retrieved 2013-09-05.

[87] "UNTC". Treaties.un.org. Retrieved 2011-09-16.

[88] Syed, Tanya (2009-01-19), *Ancient Persians 'gassed Romans'*, BBC, retrieved 2009-02-21

[89] Irwin, Will (22 April 1915), "The Use of Poison Gas", *New York Tribune*

[90] Johnson, Jeffrey Allan (1990). *The Kaiser's Chemists: Science and Modernization in Imperial Germany*. University of North Carolina Press

[91] Gray, Colin. (2007). *Another Bloody Century: Future Warfare*. **Page 269**. Phoenix. ISBN 0-304-36734-6.

[92] Griffin Davis (May 24, 2006), "CBRNE - Chemical Detection Equipment", *EMedicine*, retrieved 2007-10-22

[93] US Department of Defense (2 June 2003), *Multiservice Tactics, Techniques, and Procedure for NBC Nuclear, Biological, and Chemical (NBC) Protection (FM 3-11.4 / MCWP 3-37.2 / NTTP 3-11.27 / AFTTP(I) 3-2.46)* (PDF), GlobalSecurity.org, FM 3-11.4, retrieved 2007-10-22

[94] Centers for Disease Control and Prevention (2002-09-12), *Protecting Building Environments from Airborne Chemical, Biologic, or Radiologic Attacks*, retrieved 2007-10-22

[95] US Department of Defense (29 September 2000), *Multiservice Tactics, Techniques, and Procedure for NBC Defense of Theater Fixed Sites, Ports, and Airfields (FM 3-11.34/MCRP 3-37.5/NWP 3-11.23/AFTTP(I) 3-2.33)* (PDF), GlobalSecurity.org, retrieved 2007-10-22

[96] Ciottone, Gregory R; Arnold, Jeffrey L (January 4, 2007). "CBRNE - Chemical Warfare Agents", *EMedicine*, retrieved 2007-10-22

[97] Adrienne Mayor. "Greek Fire, Poison Arrows & Scorpion Bombs: Biological and Chemical Warfare in the Ancient World" Overlook-Duckworth, 2003, rev ed with new Introduction 2008

[98] "STATUS OF PARTICIPATION IN THE CHEMICAL WEAPONS CONVENTION AS AT 14 OCTOBER 2013". *Organisation for the Prohibition of Chemical Weapons*. OPCW. 14 October 2013.

[99] "SIGNATORY STATES". *Organisation for the Prohibition of Chemical Weapons*. OPCW. 2 September 2013.

[100] "Myanmar Joins Chemical Weapons Convention". OPCW. 9 July 2015.

[101] *In Surprise Testimony Cheney Renews Opposition to CWC* (PDF). United States Senate. 1997-04-08, archived from the original (PDF) on December 17, 2008, retrieved 2009-01-04.

[102] "India declares its stock of chemical weapons". Retrieved 26 February 2016.

[103] "India to destroy chemical weapons stockpile by 2009". DominicanToday.com. Archived from the original on 2013-09-07. Retrieved 2011-09-16.

[104] "India destroys its chemical weapons stockpile". Zeenews.india.com. 2009-05-14. Retrieved 2011-09-16.

[105] "Iraq Joins the Chemical Weapons Convention". Opcw.org. Retrieved 2011-09-16.

[106] "Abandoned Chemical Weapons (ACW) in China". Nti.org. Archived from the original on August 29, 2011. Retrieved 2011-09-16.

[107] "Ceremony Marks Start of Destruction of Chemical Weapons Abandoned by Japan in China". Opcw.org. Retrieved 2011-09-16.

[108] "Canada Contributes to Russia's Chemical Weapons Destruction Programme". Opcw.org. Retrieved 2011-09-16.

[109] "Research Library: Country Profiles: China Chemical". NTI. Archived from the original on June 5, 2011. Retrieved 2011-09-16.

[110] "Russia destroys over 76 percent of its chemical weapons stockpile".

[111] Guarino, Douglas P. "New U.S.-Russian Security Deal Greatly Scales Back Scope, Experts Say | Global Security Newswire". NTI. Retrieved 2013-09-05.

[112] "Rocky Mountain Arsenal | Region 8 | US EPA". Epa.gov. Retrieved 2011-09-16.

[113] The Oceans and Environmental Security: Shared U.S. and Russian Perspectives.

[114] "Not with Impunity: Assessing US Policy for Retaliating to a Chemical or Biological Attack". Airpower.maxwell.af.mil. Retrieved 2011-09-16.

[115] Army Agency Completes Mission to Destroy Chemical Weapons, USCMA, January 21, 2012

31.8 References

• CBWInfo.com (2001). A Brief History of Chemical and Biological Weapons: Ancient Times to the 19th Century. Retrieved Nov. 24, 2004.

• Chomsky, Noam (Mar. 4, 2001). *Prospects for Peace in the Middle East*, page 2. Lecture.

• Cordette, Jessica, MPH(c) (2003). Chemical Weapons of Mass Destruction. Retrieved Nov. 29, 2004.

• Croddy, Eric (2001), *Chemical and Biological Warfare*, Copernicus, ISBN 0-387-95076-1

• Smart, Jeffery K., M.A. (1997). History of Biological and Chemical Warfare. Retrieved Nov. 24, 2004.

• United States Senate, 103d Congress, 2d Session. (May 25, 1994). The Riegle Report. Retrieved Nov. 6, 2004.

• Gerard J Fitzgerald. American Journal of Public Health. Washington: Apr 2008. Vol. 98, Iss. 4; p. 611

• Гречко, А.А. (1976). *Годы Войны*. Военное Издательство Министерства Обороны СССР.Москва.

31.9 Further reading

• Leo P. Brophy and George J. B. Fisher; *The Chemical Warfare Service: Organizing for War* Office of the Chief of Military History, 1959; L. P. Brophy, W. D. Miles and C. C. Cochrane, *The Chemical Warfare Service: From Laboratory to Field* (1959); and B. E. Kleber and D. Birdsell, *The Chemical Warfare Service in Combat* (1966), official US history;

• Gordon M. Burck and Charles C. Flowerree; *International Handbook on Chemical Weapons Proliferation* 1991

• L. F. Haber. *The Poisonous Cloud: Chemical Warfare in the First World War* Oxford University Press: 1986

• James W. Hammond Jr; *Poison Gas: The Myths Versus Reality* Greenwood Press, 1999

• Jiri Janata, Role of Analytical Chemistry in Defense Strategies Against Chemical and Biological Attack, *Annual Review of Analytical Chemistry*, 2009

- Ishmael Jones, *The Human Factor: Inside the CIA's Dysfunctional Intelligence Culture*, Encounter Books, New York 2008, revised 2010. ISBN 978-1-59403-382-7. WMD espionage.

- Benoit Morel and Kyle Olson; *Shadows and Substance: The Chemical Weapons Convention* Westview Press, 1993

- Adrienne Mayor, "Greek Fire, Poison Arrows & Scorpion Bombs: Biological and Chemical Warfare in the Ancient World" Overlook-Duckworth, 2003, rev ed with new Introduction 2008

- Geoff Plunkett, *Chemical Warfare in Australia: Australia's Involvement In Chemical Warfare 1914 - Today, (2nd Edition), 2013.*. Leech Cup Books. A volume in the Army Military History Series published in association with the Army History Unit.

- Jonathan B. Tucker. *Chemical Warfare from World War I to Al-Qaeda* (2006)

31.10   External links

- Chemical weapons and international humanitarian law

- ATSDR Case Studies in Environmental Medicine: Cholinesterase Inhibitors, Including Insecticides and Chemical Warfare Nerve Agents U.S. Department of Health and Human Services

- Russian Biological and Chemical Weapons, about the danger posed by non-state weapons transfers

- Gaddum Papers at the Royal Society

- Chemical Weapons stored in the United States

- The Organisation for the Prohibition of Chemical Weapons OPCW

- Chemical Warfare in Australia

- Classes of Chemical Agents U.S. National Library of Medicine

- Chemical warfare agent potency, logistics, human damage, dispersal, protection and types of agents (bomb-shelter.net)

- "'War of Nerves': A History of Chemical Weapons" (interview with Jonathan Tucker from National Public Radio *Talk of the Nation* program, May 8, 2006

- Chemical weapons in World War II

- "Our Army's Defense Against Poison Gas" . *Popular Science*, February 1945, pp. 106–111.

Chapter 32

Cobalt bomb

For cancer radiation treatments delivered from a device with a Cobalt-60 isotope source, see cobalt therapy.

A **cobalt bomb** is a type of "salted bomb": a nuclear weapon designed to produce enhanced amounts of radioactive fallout, intended to contaminate a large area with radioactive material. The concept of a cobalt bomb was originally described in a radio program by physicist Leó Szilárd on February 26, 1950.[1] His intent was not to propose that such a weapon be built, but to show that nuclear weapon technology would soon reach the point where it could end human life on Earth, a doomsday device.[2][3] Such "salted" weapons were requested by the U.S. Air Force and seriously investigated, but not deployed. In the 1964 edition of the U.S. Department of Defense book *The Effects of Nuclear Weapons*, a new section titled radiological warfare clarified the "Doomsday device" issue.[4]

The Russian Federation has allegedly developed cobalt warheads for use with their System-6 nuclear torpedoes.[5][6][7] However many commentators doubt that this is a real project, and see it as more likely to be a staged leak to intimidate the US. Amongst other comments on it, Edward Moore Geist wrote a paper in which he says that "Russian decision makers would have little confidence that these areas would be in the intended locations"[8] and Russian military experts are cited as saying that "Robotic torpedo shown could have other purposes, such as delivering deep-sea equipment or installing surveillance devices." [9]

The Operation Antler/Round 1 test by the British at the Tadje site in the Maralinga range in Australia on September 14, 1957, tested a bomb using cobalt pellets as a radiochemical tracer for estimating yield. This was considered a failure and the experiment was not repeated.[10] The triple "taiga" nuclear salvo test, as part of the preliminary March 1971 Pechora–Kama Canal project, produced relatively high amounts of Co-60 from the steel that surrounded the Taiga devices, with this fusion generated neutron activation product being responsible for about half of the gamma dose now (2011) at the test site. This high percentage contribution is largely because the devices did not rely much at all on fission reactions and thus the quantity of gamma emitting cesium-137 fallout, is therefore comparatively low. Photosynthesizing vegetation exists all around the lake that was formed.[11][12]

32.1 Mechanism

A cobalt bomb could be made by placing a quantity of ordinary cobalt metal (^{59}Co) around a thermonuclear bomb. When the bomb explodes, the neutrons produced by the fusion reaction in the secondary stage of the thermonuclear bomb's explosion would transmute the cobalt to the radioactive cobalt-60 (^{60}Co), which would be vaporized by the explosion. The cobalt would then condense and fall back to Earth with the dust and debris from the explosion, contaminating the ground.

The deposited cobalt-60 would have a half-life of 5.27 years, decaying into ^{60}Ni and emitting two gamma rays with energies of 1.17 and 1.33 MeV, hence the overall nuclear equation of the reaction is:

$$^{59}_{27}Co + n \rightarrow ^{60}_{27}Co \rightarrow ^{60}_{28}Ni + e^- + \text{gamma rays.}$$

Nickel-60 is a stable isotope and undergoes no further decays after emitting the gamma rays.

The 5.27 year half life of the ^{60}Co is long enough to allow it to settle out before significant decay has occurred, and to render it impractical to wait in shelters for it to decay, yet short enough that intense radiation is produced.[10] Many isotopes are more radioactive (gold-198, tantalum-182, zinc-65, sodium-24, and many more), but they would decay faster, possibly allowing some population to survive in shelters.

32.2 Fallout from cobalt bombs vs. other nuclear weapons

Fission products are more deadly than neutron-activated cobalt in the first few weeks following detonation. After one to six months, the fission products from even a large-yield thermonuclear weapon decay to levels tolerable by humans. The large-yield three-stage (fission–fusion–fission) thermonuclear weapon is thus automatically a weapon of radiological warfare, but its fallout decays much more rapidly than that of a cobalt bomb. Areas irradiated by fallout from even a large-yield thermonuclear weapon begin to increasingly become habitable again after one to six months; a cobalt bomb's fallout on the other hand would render affected areas effectively stuck in this interim state for decades of habitable, but not safely so under constant habitation, conditions.

Initially, gamma radiation from the fission products of an equivalent size fission-fusion-fission bomb are much more intense than Co-60: 15,000 times more intense at 1 hour; 35 times more intense at 1 week; 5 times more intense at 1 month; and about equal at 6 months. Thereafter fission product fallout radiation levels drop off rapidly, so that Co-60 fallout is 8 times more intense than fission at 1 year and 150 times more intense at 5 years. The very long-lived isotopes produced by fission would overtake the ^{60}Co again after about 75 years.[13]

Theoretically, a device containing 510 tons of Co-60 can spread 1 g of the material to each square km of the Earth's surface (510,000,000 km^2). Radiation output from 1 g of Co-60 over one half life is equivalent to 44,000 GBq, which is sufficient to kill any inhabitants. If one assumes that all of the material is converted to Co-60 at 100 percent efficiency and if it is spread evenly across the Earth's surface, it is possible for a single bomb to kill every person on Earth. However, in fact, complete 100% conversion into Co-60 is unlikely, as 1957 British experiment at Maralinga showed that Co-59's neutron absorption ability was much lower than predicted, resulting in a very limited formation of Co-60 isotope in practice.

In addition, another important point in considering the effects of cobalt bombs is that deposition of fallout is not even throughout the path downwind from a detonation, so that there are going to be areas relatively unaffected by fallout and places where there is unusually intense fallout, so that the Earth would not be universally rendered lifeless by a cobalt bomb.[14] The fallout and devastation following a nuclear detonation does not scale upwards linearly with the explosive yield (equivalent to tons of TNT). As a re-

sult, the concept of "overkill"—the idea that one can simply estimate the destruction and fallout created by a thermonuclear weapon of the size postulated by Leo Szilard's "cobalt bomb" thought experiment by extrapolating from the effects of thermonuclear weapons of smaller yields—is fallacious.[15]

32.3 Example of radiation levels vs. time

Assume a cobalt bomb deposits intense fallout causing a dose rate of 10 sieverts (Sv) per hour. At this dose rate, any unsheltered person exposed to the fallout would receive a lethal dose in about 30 minutes (assuming a median lethal dose of 5 Sv). People in well-built shelters would be safe due to radiation shielding.

After one half-life of 5.27 years, only half of the cobalt-60 will have decayed, and the dose rate in the affected area would be 5 Sv/hour. At this dose rate, a person exposed to the radiation would receive a lethal dose in 1 hour.

After 10 half-lives (about 53 years), the dose rate would have decayed to around 10 mSv/hour. At this point, a healthy person could spend 1 to 4 days exposed to the fallout with no immediate effects.

After 20 half-lives (about 105 years), the dose rate would have decayed to around 10 μSv/hour. At this stage, humans could remain unsheltered full-time since their yearly radiation dose would be about 80 mSv. However, this yearly dose rate is on the order of 30 times greater than the peacetime exposure rate of 2.5 mSv/year. As a result, the rate of cancer incidence in the survivor population would likely increase.

After 25 half-lives (about 130 years), the dose rate from cobalt-60 would have decayed to less than 0.4 μSv/hour (natural background radiation) and could be considered negligible.

32.4 Decontamination

See also: Cactus Dome and Radioactive contamination § Decontamination

In practice it is unlikely that people would simply sit and wait for nuclear decay to go to completion, as in all historical fallout cases, decontamination of valuable land has occurred. This is most commonly done with the use of simple equipment such as lead glass covered excavators and bulldozers, similar to those employed in the Lake Chagan project.*[16] By skimming off the thin layer of fallout on the topsoil surface and burying it in the likes of a deep trench along with isolating it from ground water sources, the gamma air dose is cut by orders of magnitude.*[17]*[18] The decontamination after the Goiânia accident in Brazil 1987 and the possibility of a "dirty bomb" with Co-60, which has similarities with the environment that one would be faced with after a nuclear yielding cobalt bomb's fallout had settled, has prompted the invention of "Sequestration Coatings" and cheap liquid phase sorbents for Co-60 that would further aid in decontamination, including that of water.*[19]*[20]*[21]

32.5 Cultural references

The concept of cobalt bombs has been used in a number of works of apocalyptic fiction.

- The 1952 short story by Fritz Lieber "The Moon Is Green",*[22] describes the catastrophic consequences of a war fought with cobalt bombs.

- Similarly, the 1954 science fiction short story "Exhibit Piece" by Philip K. Dick ends with the newspaper headline "RUSSIA REVEALS COBALT BOMB: TOTAL WORLD DESTRUCTION AHEAD".

- In the 1957 novel *On the Beach* by Nevil Shute (and the films based on it), the source of a global contamination of radioactive material is the detonation of cobalt bombs in the Northern Hemisphere.

- The 1959 novel *Level 7*, by Mordecai Roshwald, involves the use of weapons intended to exterminate the populace by permanently contaminating the surface of the earth.

- In the 1960 book *On Thermonuclear War*, nuclear theorist Herman Kahn mentions cobalt weapons with the implication that they're militarily irrelevant or irresponsible (there's a chapter largely devoted to the drastically destabilizing nature of attempts to use nuclear blackmail with "doomsday machines" such as cobalt bombs or other radiological weapons). Herman Kahn was one of the main influences for Stanley Kubrick's 1964 film "Dr. Strangelove or: How I Learned to Stop Worrying and Love the Bomb", in which the Soviet Union establishes a secret nuclear deterrent comprising 50 buried cobalt bombs, more specifically the "Cobalt Thorium G doomsday machine."

- The mutant human New Yorkers in the 1970 postapocalyptic film *Beneath the Planet of the Apes* pray to an "Alpha-Omega" bomb, which is a doomsday weapon with a cobalt casing; one character detonates the bomb at the end of the film, after which a narrator states that the planet "is now dead". In the next film of the series *Escape from the Planet of the Apes* (1971), chimpanzee scientist Zira states she saw "the rim of the earth melt" from space when the bomb went off.

- In the 1968 film *Countdown*, the astronaut (Chiz) played by Robert Duvall makes reference to a "cobalt bomb" or salted bomb as it then pertained to the Cold War tensions between the United States and the former Soviet Union.

- In the 1973 film *Battle for the Planet of the Apes*, set two millennia earlier than *Beneath the Planet of the Apes*, Governor Kolp had ordered Méndez to detonate the bomb if he and his troops failed to return from their mission to destroy the ape village; instead, Méndez created a religion around the bomb.

The concept was also used in some other works of fiction as well.

- A cobalt and iodine "atomic device" is supplied by the Chinese Communist government to Auric Goldfinger in the 1964 James Bond film *Goldfinger*, where he intends to detonate the bomb inside Fort Knox, rendering the USA's gold bullion reserves radioactive for 58 years.

- In the *Star Trek* episode "Obsession", Ensign Garrovick compares the destructive power of an ounce of antimatter to "ten thousand cobalt bombs".

- In Stan Lee's comic book story featuring The Mighty Thor in *Journey Into Mystery* no. 86, a mad scientist from the year 2262 named Zarrko goes back in time to the Nevada desert in the year 1962 to steal a cobalt bomb from a military weapons test site.

- In the 1970s TV series *The Bionic Woman* ("Doomsday Is Tomorrow" (episodes 13 and 14)), professor Elijah Cooper incorporates a cobalt bomb in a doomsday device in an attempt to blackmail the world into peace.

- In the 1975 serial 5, series 13 of *Doctor Who*, "Revenge of the Cybermen", the Doctor, Harry Sullivan and Sarah Jane Smith are coerced, by the Cybermen, to carry cobalt bombs to the centre of the planet Voga, The Planet of Gold, so that they could destroy it. In a passing reference, the Doctor claims that the Cybermen and the Daleks both continue to use the devices in spite of their having been outlawed on most planets.

- Paul Erdman's 1976 novel *The Crash of '79* includes Iran using cobalt bombs to attack Middle East oilfields, rendering them "totally inaccessible for at least twenty-five years".

- In the cartoon series *Big Guy and Rusty the Boy Robot*, the Big Guy is described as having a Cobalt/Thorium G power core.

- In the 2008 series of TV programme *Ultimate Force* (Series 4, Episode 5), a "slow bomb" was stolen and set to detonate in Central London.

- In the 2009 TV series *Castle* (episodes "Setup" (16) and "Countdown" (17)), a cobalt bomb is built with the intention to destroy New York City.

- In 2015, a Russian nuclear-armed torpedo design was apparently leaked. It has been speculated that the warhead would be a cobalt bomb, designed for "creating wide areas of radioactive contamination, rendering them unusable for military, economic or other activity for a long time".[9]

32.6 See also

- Neutron bomb

32.7 References

[1] Brian Clegg. *Armageddon Science: The Science of Mass Destruction*. St. Martins Griffin. p. 77. ISBN 978-1-250-01649-2.

[2] Bhushan, K.; G. Katyal (2002). *Nuclear, Biological, and Chemical Warfare*. India: APH Publishing. pp. 75–77. ISBN 81-7648-312-5.

[3] Sublette, Carey (July 2007). "Types of nuclear weapons". FAQ. The Nuclear Weapon Archive. Retrieved 2010-02-13.

[4] Samuel Glasstone, *The Effects of Nuclear Weapons*, 1962, revised 1964. U.S. Department of Defense and U.S. Department of Energy. pp. 464–465. This section was removed from later editions, but, according to Glasstone in 1978, not because it was inaccurate or because the weapons had changed.

[5] http://nationalinterest.org/feature/russias-new-super-torpedo-carries-the-threat-nuclear-14537

[6] http://www.foxnews.com/world/2015/11/12/kremlin-controlled-tv-airs-secret-plans-for-new-submarine-launched-nuclear.html

[7] https://www.rt.com/news/321640-leaked-russian-nuclear-torpedo/

[8] Would Russia's undersea "doomsday drone" carry a cobalt bomb?

[9] "Russia reveals giant nuclear torpedo in state TV 'leak'". BBC News. November 12, 2015. Retrieved February 16, 2017.

[10] "1.6 Cobalt Bombs and other Salted Bombs". Nuclearweaponarchive.org. Retrieved February 10, 2011.

[11] "Radiological investigations at the 'Taiga' nuclear explosion site: Site description and in situ measurements". *Journal of Environmental Radioactivity*. **102**: 672–680. doi:10.1016/j.jenvrad.2011.04.003.

[12] "Radiological investigations at the 'Taiga' nuclear explosion site, part II: man-made γ-ray emitting radionuclides in the ground and the resultant kerma rate in air". *Journal of Environmental Radioactivity*. **109**: 1–12. doi:10.1016/j.jenvrad.2011.12.009.

[13] "Section 1.0 Types of Nuclear Weapons". *nuclear-weaponarchive.org*.

[14] Samuel Glasstone; Philip J. Dolan, eds. (1977). "The Effects of Nuclear Weapons" (PDF) (3rd ed.). Washington, D.C.: United States Department of Defense and Department of Energy.

[15] Martin, Brian (December 1982). "The global health effects of nuclear war". *Current Affairs Bulletin*. **59** (7): 14–26.

[16] *Born of Nuclear Blast: Russia's Lakes of Mystery*. YouTube. November 28, 2010.

[17] Joint FAO/IAEA Programme. "Joint Division Questions & Answers - Nuclear Emergency Response for Food and Agriculture, NAFA". *iaea.org*.

[18] International Atomic Energy Agency International Atomic Enmergy Agency, 2000 - Technology & Engineering - restoration of environments with radioactive residues : papers and discussions, 697 pages

[19] "Scavenging cobalt from radwaste". *neimagazine.com*.

[20] "Sequestration Coating Performance Requirements for Mitigation of Contamination from a Radiological Dispersion Device- 9067" (PDF). Wmsym.org. Retrieved 2015-11-12.

[21] John Drake. "Sequestration Coating Performance Requirements for Mitigation of Contamination from a Radiological Dispersion Device" (PDF). Cfpub.epa.gov. Retrieved 2015-11-12.

[22] Lieber, Fritz (1952). "The Moon Is Green". Gutenburg Project.

32.8 Text and image sources, contributors, and licenses

32.8.1 Text

- **Nuclear terrorism** *Source:* https://en.wikipedia.org/wiki/Nuclear_terrorism?oldid=766077202 *Contributors:* Magnus Manske, Trelvis, Eloquence, Ixfd64, Kingturtle, Ed Cormany, PBS, Romanm, Securiger, Auric, Alan Liefting, Fastfission, Wwoods, Abqwildcat, Beland, Rdsmith4, Commodore Sloat, Kevin Rector, Kingal86, D6, Discospinster, Rich Farmbrough, Bender235, Mr. Billion, SpencerWilson, Idleguy, Jakew, Mrzaius, Alansohn, Damburger, Hohum, Green slash, Bobrayner, Former user 2, Stefanomione, Paxsimius, Descendall, Rjwilmsi, Tangotango, Ground Zero, Survivor, Mmx1, Limulus, Nick, Rwalker, Wangi, Psy guy, Black Falcon, Unforgiven24, PTSE, Sallymcvegas, SmackBot, Wine Guy, Enr-v, MadCow257, John, Zapptastic, Mike1901, Fluppy, MTSbot~enwiki, Stanley011, BranStark, Iridescent, MaxHarmony, Tawkerbot2, ShelfSkewed, Mewantcookie, Anthonyhcole, Phonemonkey, Zer0faults, The machine512, Thijs!bot, Epbr123, Honeplus, KimDabelsteinPetersen, Hempfel, AntiVandalBot, Magioladitis, Renyseneb, S3000, Tobor0, Keith D, Largeassociates, Hodja Nasreddin, Lifeboatpres, Halmstad, Cecil9sm, Johnfos, Vlad fedorov, Philip Trueman, ErikWarmelink, Triesault, HiDrNick, AHMartin, TurtleShroom, La poet, RW Marloe, Megansmith18, JL-Bot, ClueBot, Plastikspork, Lampak, Megiddo1013, Coccyx Bloccyx, Redthoreau, Dubmill, ESO Fan, Lkcutler, Addbot, Krawndawg, Micahmedia, Yanking13, Blackcell16, Zorrobot, Everyme, Luckas-bot, AzureFury, AnomieBOT, Jim1138, Galoubet, Materialscientist, Dynablaster, LilHelpa, Bairdjr, JanDeFietser, Wikireader41, Born Gay, Zahkar, Ironboy11, Orenburg1, Trappist the monk, BonifaciusVIII, Grammarxxx, RjwilmsiBot, KinkyLipids, EmausBot, John of Reading, Parkywiki, Boundarylayer, Dewritech, Dcirovic, Marwatrostock, ZéroBot, Kilpazz, ClueBot NG, Helpful Pixie Bot, MusikAnimal, Hamish59, Leilis.easc, BattyBot, StarryGrandma, Makecat-bot, Timothysandole, GentleDjinn, Pdiddy1994, Acetotyce, Melonkelon, NatSecWonk, Buffbills7701, Limnalid, Thewillsterr, Monkbot, Vieque, Lucasjohansson, Sriac LEX, 0xF8E8, GeneralizationsAreBad, KSFT, DisuseKid, Antrangelos, CaptainCarlosdeCorona, Baking Soda, Spolglans, A Missing WMD, Bender the Bot and Anonymous: 149

- **Radiological weapon** *Source:* https://en.wikipedia.org/wiki/Radiological_weapon?oldid=754266609 *Contributors:* TwoOneTwo, ClaudeMuncey, The Anome, Miguel~enwiki, William Avery, Roadrunner, Maury Markowitz, Alan_d, Rsabbatini, Patrick, Eric119, Bon d'une cythare, Aarchiba, Andres, Mulad, RodC, Pstudier, Sanders muc, Securiger, Jleedev, DocWatson42, Fastfission, Mboverload, Pne, Mzajac, Neutrality, Ponder, Bender235, Alansohn, Joshbaumgartner, BRW, DV8 2XL, Kazvorpal, WadeSimMiser, Urbane Legend, Gurch, YurikBot, RussBot, Crazytales, Stephenb, Palpalpalpal, Light current, NHSavage, Dr U, HereToHelp, Sardanaphalus, SmackBot, Arniebutelt, AndySayler, Ncprm2026, A5b, Pilotguy, IgWannA, Chue03, Timetracker, Iridescent, Cnstewart, Brandizzi, Tawkerbot2, Cryptic C62, Give Peace A Chance, Dancter, Fleshwater, Honeplus, Mentifisto, K7aay, Tjmayerinsf, Sluzzelin, Deadbeef, Inks.LWC, BenB4, JimCubb, Dragonnas, Magioladitis, VoABot II, Midgrid, Cgingold, Hodja Nasreddin, Cromdog, Trumpet marietta 45750, Malinaccier, UnitedStatesian, Lamro, Falcon8765, Caltas, JabbaTheBot, RW Marloe, Francvs, Twinsday, ClueBot, Paulcmnt, Ahmed91981, TonyBallioni, Rhotel1, SkyLined, Addbot, Download, Lightbot, Legobot, Guy1890, Pganas, AnomieBOT, DemocraticLuntz, Metalhead94, Sugoi47, Lithenium, FrescoBot, Adam9389, RoyGoldsmith, BenzolBot, MastiBot, Callanecc, Dinamik-bot, Vrenator, EmausBot, AvicAWB, Arbnos, ChuispastonBot, ClueBot NG, Astatine211, Helpful Pixie Bot, Curb Chain, Marcocapelle, CitationCleanerBot, Tkbx, Cyberbot II, Jionpedia, Jray310, Limnalid, Jerryzhu2004, GreenC bot, (Walker Snarling) and Anonymous: 102

- **Radiological warfare** *Source:* https://en.wikipedia.org/wiki/Radiological_warfare?oldid=760009525 *Contributors:* Rsabbatini, Eric119, Palm dogg, La goutte de pluie, Wavelength, SmackBot, AndySayler, Cadmium, Headbomb, Republitarian, Cgingold, Hodja Nasreddin, Cromdog, EmxBot, Addbot, Vedran12, AnomieBOT, Erik9bot, Jonathandeamer, Arbnos, Marcocapelle, Limnalid, Demoniccathandler, NewYorkActuary and Anonymous: 6

- **Nuclear weapon** *Source:* https://en.wikipedia.org/wiki/Nuclear_weapon?oldid=763447030 *Contributors:* AxelBoldt, Magnus Manske, TwoOneTwo, Trelvis, The Epopt, Dreamyshade, Sodium, ClaudeMuncey, Ansible, Eloquence, Mav, Wesley, Bryan Derksen, Robert Merkel, The Anome, Tarquin, AstroNomer, Taw, Manning Bartlett, Ed Poor, Alex.tan, AdamW, Andre Engels, Ted Longstaffe, Youssefsan, Arvindn, Rmhermen, Toby Bartels, SJK, Little guru, Roadrunner, Ray Van De Walker, SimonP, Maury Markowitz, Zoe, Graft, FlorianMarquardt, Hephaestos, Soulpatch, Tedernst, Olivier, Patrick, RTC, Infrogmation, JohnOwens, Michael Hardy, GABaker, Modster, Cprompt, DopefishJustin, Dante Alighieri, Norm, Dominus, Ixfd64, Bcrowell, Frank Shearar, Cameron Dewe, TakuyaMurata, GTBacchus, Dori, Eric119, Minesweeper, Alfio, Ronabop, Mkweise, Ellywa, Ahoerstemeier, Anders Feder, Snoyes, Angela, Kingturtle, Erzengel, BigFatBuddha, Aarchiba, Ugen64, Glenn, Djmutex, Vzbs34, Susurrus, Jiang, Oliezekat, Alex756, [212] Mxn, Ilyanep, Lommer, Conti, Pizza Puzzle, Rami Neudorfer, Trevor Lawson, Ehn, Vroman, Jengod, Malbi, Ec5618, Jonadab~enwiki, Timwi, David Newton, Dino, Jefelex, Daniel Quinlan, Jfeckstein, Fuzheado, Andrewman327, WhisperToMe, Zoicon5, Jessel, DJ Clayworth, Haukurth, Tpbradbury, Maximus Rex, E23~enwiki, Pacific1982, Saltine, Kaal, Nv8200pa, Tempshill, Zero0000, Omegatron, Babbler, Thue, Bevo, Xevi~enwiki, Shizhao, Topbanana, Toreau, Vaceituno, Stormie, Raul654, Pstudier, Bcorr, Jusjih, Johnleemk, Finlay McWalter, Skybunny, Owen, Stargoat, Jni, Riddley, Robbot, Ke4roh, Sander123, Astronautics~enwiki, ChrisO~enwiki, Nabeel, Fredrik, Kizor, PBS, Chris 73, Chocolateboy, Kadin2048, Romanm, Arkuat, Securiger, Lowellian, Ukuk~enwiki, Merovingian, Sverdrup, Rfc1394, Academic Challenger, SchmuckyTheCat, Texture, Meelar, Yacht, Rhombus, Bkell, Mervyn, Hadal, Victor, LX, TPK, Tsavage, Seth Ilys, Diberri, Dina, David Gerard, SimonMayer, Ancheta Wis, Alexwcovington, Benji Franklyn, DocWatson42, Christopher Parham, Oberiko, Mat-C, Sj, Kim Bruning, Inter, Tom harrison, Lupin, Ferkelparade, Fastfission, Dersen, Zigger, Karn, Peruvianllama, Everyking, No Guru, Jacob1207, Anville, Perl, Curps, Electric goat, Bensaccount, Cantus, Mike40033, Guanaco, Tom-, Jherico, Zhen Lin, Gracefool, MRubenzahl, Steven jones, Matt Crypto, Chrissmith, Bobblewik, Deus Ex, Golbez, Kandar, DontMessWithThis, Christopherlin, Hob, Stevietheman, Barneyboo, Gadfium, Knutux, LiDaobing, Sonjaaa, Quadell, Fangz, Antandrus, Tom the Goober, Beland, Estel~enwiki, Apox~enwiki, PDH, Armaced, Jossi, CaribDigita, Rdsmith4, OwenBlacker, Mitaphane, Woofles, Tothebarricades.tk, Daniel11, Mysidia, Bk0, Tyler McHenry, Anirvan, Creidicki, Neutrality, Imjustmatthew, Karl Dickman, Deglr6328, Mtnerd, Barnaby dawson, Trevor MacInnis, Zaf, Mormegil, Rfl, Freakofnurture, N328KF, Venu62, Nimbulan, DanielCD, Discospinster, Rich Farmbrough, Rhobite, FiP, Jpk, Silence, Chowells, Prateep, Dsadinoff, Xezbeth, Ponder, Ioliver, Mani1, Pavel Vozenilek, Aardark, Paul August, Stereotek, SpookyMulder, Night Gyr, Bender235, ESkog, Kaisershatner, Danny B-), Hapsiainen, Brian0918, El C, Chairboy, Aude, Shanes, Sietse Snel, RoyBoy, Triona, Bookofjude, Deanos, DarkArctic, Jburt1, Balok, Adambro, Bobo192, Smalljim, BrokenSegue, Duk, Shenme, Viriditas, Serialized, Vortexrealm, Kormoran, Jag123, Scott Ritchie, Jojit fb, Kjkolb, BM, Townmouse, Bawolff, PeterisP, Naturenet, Daf, WikiLeon, Rje, Pschemp, MPerel, Sam Korn, Haham hanuka, Ral315, Ylwsub68, Nsaa, Jakew, HasharBot~enwiki, OGoncho, Jumbuck, Stephen G. Brown, Alansohn, Gary, Tablizer, Uncle.bungle, SnowFire, Mo0, 119, Atlant, Rd232, Mr Adequate, Keenan Pepper, Trainik, Joshbaumgartner, Rwoodsco, Andrew Gray, Lord

Pistachio, Lectonar, MarkGallagher, Zippanova, SlimVirgin, Lightdarkness, Stigocki, Garfield226, InShaneee, Dark Shikari, Hgrenbor, Hu, Maio, Idont Havaname, Bart133, GregLindahl, Hohum, Snowolf, Melaen, ClockworkSoul, Super-Magician, Evil Monkey, Ramius, Cat 1234, RainbowOfLight, Randy Johnston, Sciurinæ, Mikeo, Pethr, Vuo, Ianblair23, DV8 2XL, LordAmeth, Stepheno, Gene Nygaard, Redvers, Admiral Valdemar, HenryLi, Dan100, Tr00st, GreatGatsby, Crosbiesmith, Feezo, Itinerant, MickWest, Sylvain Mielot, Boothy443, Reinoutr, OwenX, Woohookitty, Jannex, GrouchyDan, JarlaxleArtemis, Superstring, Jersyko, Guy M, James Kemp, Nvinen, TomTheHand, Nameneko, JeremyA, MONGO, Nakos2208~enwiki, Tabletop, Cabhan, Firien, Bluemoose, GregorB, M412k, Petwil, Atomicarchive, Wayward, Volkz, Smartech~enwiki, Christopher Thomas, Dysepsion, GSlicer, Johndoe85839, Graham87, JiMidnite, Deltabeignet, Magister Mathematicae, Kalmia, MC MasterChef, Ligar~enwiki, Kbdank71, FreplySpang, Josh Parris, Gorrister, Rjwilmsi, Joefu, George Burgess, Phileas, Panoptical, Mystalic, Bill37212, Hiberniantears, Linuxbeak, JHMM13, Tawker, Mred64, Oblivious, Ligulem, CQJ, Frenchman113, The wub, Double-Blue, ATLBeer, Sango123, Yamamoto Ichiro, Lcolson, Titoxd, FlaBot, Mirror Vax, RobertG, Ground Zero, A scientist, Nihiltres, Josh~enwiki, TheMelenchukSmell, Crazycomputers, Survivor, JIMBO WALES, Subterfuge~enwiki, RexNL, Gurch, Ayla, Jimbo D. Wales, RobyWayne, SweBrainz, KFP, OrbitOne, Cause of death, Jtling, Butros, King of Hearts, Scimitar, Chobot, Hatch68, Theo Pardilla, GangofOne, Bgwhite, Digitalme, Dj Capricorn, Simesa, NSR, Gwernol, Peter Grey, Loco830, UkPaolo, The Rambling Man, YurikBot, Wavelength, TexasAndroid, Dimimimon4, Extraordinary Machine, Sceptre, Blightsoot, Hairy Dude, Jimp, RussBot, Arado, Red Slash, John Quincy Adding Machine, Majin Gojira, Anonymous editor, Splash, Alavena, Stalmannen, Anders.Warga, Anomaly1, 0nizuka the Great, Cmk5b, Akamad, CambridgeBayWeather, Shaddack, Bisqwit, Sweetwilliams, GeeJo, Bullzeye, Finbarr Saunders, David R. Ingham, PaulGarner, Shanel, NawlinWiki, SEWilcoBot, Wiki alf, Ceremony1968, Harrisale, WAS, Jaxl, Milo99, Robchurch, JDoorjam, Nick, Ragesoss, Anetode, Dmoss, PhilipO, Misza13, Grafikm fr, Lomn, LarryMac, Aaron Schulz, Karl Meier, DeadEyeArrow, Psy guy, Kander, Superiority, Essexmutant, Mtu, Mgnbar, Saric, FF2010, Newagelink, Georgewilliamherbert, Vontraginoff, Enormousdude, Ali K, Lt-wiki-bot, Gtdp, Ageekgal, Theda, Jwissick, Fang Aili, Adilch, Nemu, Dspradau, Rhallanger, GraemeL, JoanneB, TBadger, CWenger, Smurrayinchester, JLaTondre, Garion96, AGToth, Gorgan almighty, David Biddulph, Jack Upland, Junglecat, RG2, Mikedogg, GrinBot~enwiki, Airconswitch, Dkasak, Vreddy92, Nick-D, Sam Weber, BiH, Jade Knight, DVD R W, K14bdy, Marquez~enwiki, NetRoller 3D, Luk, Sycthos, Bigcheesegs, Sardanaphalus, Joshbuddy, A bit iffy, SmackBot, Looper5920, YellowMonkey, Mattarata, Tarret, Prodego, KnowledgeOfSelf, Royalguard11, Hydrogen Iodide, Melchoir, Jhartshorn, Unyoyega, Pgk, C.Fred, Bomac, Neptunius, Davewild, Wikedpedia~enwiki, CMD Beaker, Alksub, Delldot, Desk003, Sam8, Ajm81, AnOddName, Vilerage, Aivazovsky, Nscheffey, Wittylama, Alex earlier account, Gaff, Xaosflux, Yamaguchi 先 生, Zvonsully, Aksi great, Gilliam, The Gnome, Ppntori, Andy M. Wang, Psiphiorg, Afa86, The monkeyhate, Saros136, Chris the speller, Master Jay, Payam81, SlimJim, Persian Poet Gal, Lordkazan, Thumperward, Emt147, Silly rabbit, Hibernian, Imaginaryoctopus, Croquant, Sbharris, Darth Panda, A. B., Mikker, Cigale, Gsp8181, Dinnyy, Royboycrashfan, PeRshGo, Zsinj, Rogermw, Dethme0w, Can't sleep, clown will eat me, Jahiegel, Милан Јелисавчић, Jorvik, Hooded-Man, Markkasan, Skidude9950, Nixeagle, OOODDD, Korinkami, Prmacn, Addshore, Interfector, Joema, Mrdempsey, Khoikhoi, WhereAmI, Digitize, Jumping cheese, Iapetus, Khukri, Makemi, Engwar, Nakon, Savidan, Loannes, Kevlar67, Shadow1, Dreadstar, Mini-Geek, Lcarscad, Polonium, TCorp, DMacks, Dlamini, Daniel4004, Kotjze, Whiplashxe, Edgeris, Daniel.Cardenas, Pilotguy, Prasi90, ぐ- 口, Ohconfucius, Kuzaar, Nmnogueira, WikiWitch, Rory096, Harryboyles, AAA765, Zahid Abdassabur, Dbtfz, Kreb Dragonrider, Kuru, John, HellecticMojo, Buchanan-Hermit, J 1982, Xu3w3nan, Kipala, Amenzix, Lazylaces, Sir Nicholas de Mimsy-Porpington, JorisvS, Minna Sora no Shita, Captain-Vindaloo, Zarniwoot, Gevalt, JohnWittle, Scetoaux, Jaywubba1887, Syra987, Ckatz, Kkken, Slakr, Werdan7, Stwalkerster, Shangrilaista, Tasc, Mr Stephen, Nitro-X, InedibleHulk, Waggers, SandyGeorgia, Mets501, Java 109, Spook`, Unnamed01, Ryulong, Sertin, Ryanjunk, Gary Jacobsen, Balderdash707, Aktalo, Kenny&becca, Iridescent, K, Hydra Rider, Aspuar, CapitalR, Esurnir, Aeons, Dublan, Aaron DT, Civil Engineer III, Thebigone45, Morgan Wick, Tawkerbot2, Brian53199, Dlohcierekim, Chetvorno, Zaphody3k, Benfranklinlover, Penguincornguy, ERAGON, Vikram.raja, JForget, Pigstinky, Ale jrb, Sir Vicious, TheHerbalGerbil, Pools200, Scohoust, Iced Kola, JohnCD, Grimgor79, Randall1in, 5-HT8, GHe, Dgw, Toropop, Adrienhocky16, Evilhairyhamster, Avillia, Hipdog11, Qwertyman4444, Borislav Dopudja, TJDay, PC supergeek, Abeg92, Ryan, Tkloumo, Jeffdb123, UncleBubba, Gogo Dodo, HPaul, Travelbird, Deathmak, Llort, A Softer Answer, Give Peace A Chance, Pascal.Tesson, Nate74, Noohgodno, Tawkerbot4, Roberta F., Optimist on the run, Kingthwomp, Kansas Sam, Omicronpersei8, Bascombe2, UberScienceNerd, EvocativeIntrigue, FrancoGG, Mathpianist93, Epbr123, Dua89563, PolaroidKiss, Forsaken88, Tairen125, Corsair18, Jedibob5, Sagaciousuk, Mansoorhabib, Drift~enwiki, Mojo Hand, Aftillidie13, Louis Waweru, West Brom 4ever, John254, Bobblehead, Pavel from Russia, CST, Cj67, Geostar1024, Pcbene, RamanVirk, Nick Number, Mm11, SebastianSalceek, Thedarkestshadow, Dawnseeker2000, Escarbot, Bilbobjoe, Dagingsta, Supran, KrakatoaKatie, Ialsoagree, AntiVandalBot, Yonatan, Gioto, Luna Santin, Settersr, Seaphoto, Sobaka, Opelio, Chairman Meow, Quintote, Doc Tropics, Paste, Pokemeharder, Robxz, Sweart1, Dr who1975, Arclem, Jj137, Postlewaight, Dylan Lake, LibLord, Farosdaughter, EP111, MrBill, VonV, Aliwalla, MishMich, Lorethal, Canadian-Bacon, Bigjimr, JAnDbot, Najeb, Husond, Fidelfair, FidelFair, MER-C, Skomorokh, Mark Grant, Nthep, Instinct, Arnegrim, Seddon, Rearete, Hello32020, Ribonucleic, Tengfred, Andonic, Roleplayer, TAnthony, Fluffy the Cotton Fish, LittleOldMe, Yahel Guhan, Brandox1, Magioladitis, Pedro, Bongwarrior, VoABot II, Edwardmking, Nyq, Weser, Trnj2000, Carom, J mcandrews, Ben515, Redaktor, Btiene, Akmoilan, Avicennasis, Gblay, Bubba hotep, Animum, Adrian J. Hunter, Allstarecho, A3nm, Cpl Syx, Tokino, Spellmaster, Vssun, Glen, DerHexer, JaGa, Fulvius~enwiki, Khalid Mahmood, Hans Moravec, Saganaki-, Philbj, Tuviya, Stevepaget, FisherQueen, Leaderofearth, Hdt83, MartinBot, Mornock, Racepacket, Ninestrokes, Arjun01, One of them, Rettetast, Mschel, Nono64, PrestonH, Kentucho, Headmaster2008, MnM2324, Napalmdeth~enwiki, Zephyr21, RockMFR, Zarathura, Limongi, J.delanoy, Pharaoh of the Wizards, Loongyh, Bogey97, Hacbarton, Leaflet, Maurice Carbonaro, Fleiger, Mike.lifeguard, Menew22, Yucki8aby, Octevious, Thaurisil, Rahzvel, Sigamfan, Tdadamemd, Johnl1479, Aym710, Daedalus CA, Icseaturtles, Bot-Schafter, Nosfartu, Ncmvocalist, McSly, Ninjadeath, L'Aquatique, Grumpyapp, Sheahae, Reichner1000, AllanDeGroot, Pyrospirit, Detah, RenniePet, NewEnglandYankee, ChineseGoldFarmer, DadaNeem, Skrelk, Malerin, Halfvamp, Nikobro, Hrishie, Xecog, Bloodvayne, MetsFan76, Billyx1337x, Asdfasdf321, Antepenultimate, Morimura, Nukeitup2, Vanished user 39948282, Gemini1980, Natl1, Waterfox1, Cs302b, Useight, MissAtomicbomb, Stopthe-DatabaseState, Awesomeman42, Hmsbeagle, Phr0gor, Pistonhonda4, Idioma-bot, Wikieditor06, ACSE, Dansen3008, Happy guy of happyness, Eleron123, Mattybobo, X!, Cdmajava, JoshBuck123456789, Spartan 2.0, Nucwikigirl, VolkovBot, CN111111111, TreasuryTag, CWii, Johnfos, ABF, Jeff G., Indubitably, Lbunker, Stopping Power, HJ32, Soliloquial, Vulgarkid, Barneca, Flintsparkler, Jomorepinch, Oshwah, XavierGreen, GimmeBot, Solracm 021, SeanNovack, Maximillion Pegasus, Baldusi, Dj stone, TommyKiwi, Dchall1, Karmos, Something915, Qxz, DavidSaff, Noob wikipedian, Billy1223billy, Bloigen, Don4of4, Destroyer 2943, LeaveSleaves, Heidit, Seb az86556, UnitedStatesian, Arigato1, Cremepuff222, Hooduphodlum, Mazarin07, Tybluesum, Buffs, Sparkyrob, Feudonym, Cantiorix, Falcon8765, Tompkins818, Turgan, -ross616-, Burntsauce, Babilingbaboon, Dustybunny, Afonsecajames, NPguy, AlleborgoBot, Funeral, 682635q, Kampking13, MattW93, Worship cindy, LOTRrules, SieBot, Dusti, ShiftFn, Sonicology, Tiddly Tom, Scarian, SheepNotGoats, Jacotto, Awesome Truck Ramp, RJaguar3, Triwbe, Mcygan123, Calabraxthis, Lexicog, Bootha, Aprudhomme, Arda Xi, Dattebayo321, Keilana, Tiptoety, Oda Mari, Arbor to SJ, PeaveyStrat5, Chridd, Lagrange613, Games14pmw, Lanzarotemaps, Oxymoron83, Antonio Lopez, Byrialbot, Faradayplank, AngelOfSadness, Steven Crossin,

Apostrostomper, WillBo, Dr. Ronald Cutburth and Anonymous: 368

- **Nuclear fallout** *Source:* https://en.wikipedia.org/wiki/Nuclear_fallout?oldid=765259763 *Contributors:* Trelvis, Mav, Dachshund, Rmhermen, Ray Van De Walker, Kurt Jansson, Ktsquare, David spector, Rsabbatini, Edward, Patrick, Michael Hardy, Evanherk, Eric119, Angela, Aarchiba, Jasper, Julesd, Whkoh, Mulad, Charles Matthews, Maximus Rex, E23~enwiki, Furrykef, Itai, Tempshill, Topbanana, Bearcat, Robbot, Securiger, Rfc1394, DocWatson42, Fastfission, Karn, Anville, Wikibob, Stern~enwiki, Nomad~enwiki, Bobblewik, LucasVB, Antandrus, Rdsmith4, FoeNyx, Neutrality, Jocelyn@soleil.org, Lacrimosus, Discospinster, Brianhe, Rich Farmbrough, Avriette, MeltBanana, Deelkar, Bender235, Kaisershatner, JoeSmack, Evice, Shenme, Diceman, Kjkolb, Nk, Pcxbjh, Helix84, Alansohn, Gary, Malo, Rwendland, Snowolf, DV8 2XL, Gene Nygaard, BerserkerBen, OleMaster, Pekinensis, Roboshed, Woohookitty, REggert, BriskWiki, Uris, JohnC, Rtcpenguin, Ashmoo, Cuchullain, Ligar~enwiki, GrundyCamellia, BorgHunter, Shimbo, Rjwilmsi, Strait, R.O.C, Alban, SocialistMyrmadonUprising, Old Moonraker, Nihiltres, Survivor, Rbonvall, SweBrainz, AnthonyA7, WriterHound, YurikBot, JWB, I need a name, Samuel Wiki, Edward Wakelin, Vlad4599, Hede2000, RadioFan, Shaddack, EWS23, Nick, Aaron Brenneman, DuffDudeX1, Georgewilliamherbert, Tevildo, Rearden9, Bweenie, Sardanaphalus, Frankie, Hardscarf, KnightRider~enwiki, SmackBot, Kimon, Edgar181, Cazort, Gilliam, Bluebot, Cadmium, Jprg1966, Colonies Chris, Bpgreen, Laslovarga, Wonderstruck, Jdlambert, DMacks, Michael Rogers, MegaHasher, Nishkid64, Harryboyles, Calvados~enwiki, John, Euchiasmus, Javit, Naaman Brown, Civil Engineer III, Courcelles, Zaphody3k, Aquinex, CalebNoble, SkyWalker, CmdrObot, Trails, Slazenger, Cydebot, Playtime, Llort, Photocopier, Underpants, Daniel Olsen, Bolesjohnb, Thijs!bot, VKemyss, Headbomb, Mnemeson, Jtmoon, AntiVandalBot, Prolog, Kaini, Djinchao, Arch dude, OhanaUnited, Lighthope, VoABot II, CattleGirl, Soulbot, ErKURITA, Catgut, Howlingfool, Cgingold, Robotman1974, K7net, The cattr, DerHexer, Wi-king, T e r o, D.h. Tuviya, Sir Intellegence, MartinBot, Jeronen1, SB42, CommonsDelinker, Tokyogirl79, NewEnglandYankee, Monkey711, Olegwiki, Jwh335, Петър Петров, TWCarson, Banjodog, ACSE, VolkovBot, Johnfos, TallNapoleon, Minwu, Philip Trueman, TXiKiBoT, Oshwah, Mark v1.0, Templationist, Rourin bushi, Drappel, Cremepuff222, BotKung, ARUNKUMAR P.R, Mdeweydiii, Andy Dingley, AlleborgoBot, Kehrbykid, Lyinginbedmon, Oda Mari, Phil Bridger, RW Marloe, Obliterat, BenPhenicie, Johnyjohn, Anyeverybody, Escape Orbit, ImageRemovalBot, Martarius, ClueBot, Hutcher, Aohara1986, 1redrun, Wikifast1991, Michal Sobkowski, Maxtitan, Neverquick, Hectic525, Excirial, Crywalt, Three-quarter-ten, V Verweij, UncleGiggo, Redthoreau, Thingg, PCHS-NJROTC, Charles 91, Wnt, Nettings, XLinkBot, Ost316, Mitch Ames, SilvonenBot, Addbot, AkhtaBot, Micjon, Download, Glane23, Joomple, Tide rolls, Luckas-bot, Yobot, AnomieBOT, Archon 2488, Rubinbot, Killiondude, Jim1138, Greenbreen, Kingpin13, Bluerasberry, Materialscientist, Citation bot, Neurolysis, Xqbot, Benjabby, Beastmanphilmw89, DSisyphBot, Ubcule, Ewhalen, Omnipaedista, RibotBOT, Bellerophon, Brutaldeluxe, Lol22246, Biem, FrescoBot, Dendereon, RoyGoldsmith, Gesalbte, Pinethicket, December21st2012Freak, Lotje, Callanecc, I am Blue 862, Mr.98, Reaper Eternal, TheGrimReaper NS, Everyone Dies In the End, Jeffrd10, Suffusion of Yellow, Omnomnom123, Smd75jr, EmausBot, Gfoley4, Mk5384, BoundaryLayer, Atavism, Angel Rosado, Wikipelli, IsayDumbStuff, Josve05a, Infinitjest, Gamma287, Arbnos, A930913, Thine Antique Pen, Donner60, ChuispastonBot, DASHBotAV, Petrb, ClueBot NG, Gareth Griffith-Jones, Satellizer, Shaddim, Chester Markel, Wunderbread11, Cntras, Levi10101010, Asukite, Brickmack, Widr, MeggyFurEver, Lukes3281998, The big pineapple, BG19bot, Indycat222, Furkhaocean, CityOfSilver, Marcocapelle, Mark Arsten, Zaath, Bcary, Ddude1969, Mickdermack, SD Energy, BKFORTHEWINXx, Ducknish, Alpha1983, Mysterious Whisper, Thaturgenoise, Rsn00b4life, Debouch, Advocatejake, Kpax7777, Ghost7800, GeekyGamer01, Manul, Mrthedreat, Monkbot, Patient Zero, Theltzardreborn, Julietdeltalima, Atvica, Laytons332, Jackboron123, Gregarditus, Nickyonge, SirJaeline, Quinlan455555, Sethagawea, Bender the Bot and Anonymous: 303

- **Weapon of mass destruction** *Source:* https://en.wikipedia.org/wiki/Weapon_of_mass_destruction?oldid=764312679 *Contributors:* Trelvis, Eloquence, Mav, Bryan Derksen, The Anome, Guppie, Ed Poor, Rmhermen, Roadrunner, Shii, Heron, Hotlorp, Formulax~enwiki, Camembert, Ericd, Patrick, Michael Hardy, GABaker, Willsmith, Dante Alighieri, Nixdorf, MartinHarper, Gabbe, Menchi, Wapcaplet, Ixfd64, (, Bon d'une cythare, Card~enwiki, Sbuckley, Fantasy, Ronz, Jebba, Kingturtle, BigFatBuddha, Julesd, Vzbs34, Jiang, Kaihsu, Ruhrjung, Hike395, Stephenw32768, Dcoetzee, Reddi, Viajero, Paul Stansifer, Denni, Daniel Quinlan, Fuzheado, DJ Clayworth, Tpbradbury, Furrykef, Tuomas, Sabbut, SH~enwiki, Mackensen, Fvw, Scott Sanchez, Ortonmc, Bcorr, Secretlondon, Kutkuri, Lumos3, Nufy8, Robbot, Kizor, Donreed, Jmabel, Pibwl, EON, Yelyos, Lowellian, Academic Challenger, Timrollpickering, Sunray, Hadal, UtherSRG, Mushroom, Drstuey, Jleedev, Stefan Kögl, GreatWhiteNortherner, Christopher Parham, Laudaka, Nadavspi, Wolfkeeper, Lupin, Ferkelparade, Fastfission, Karn, Ido50, Cantus, Bovlb, Get-back-world-respect, Bobblewik, LennartBolks~enwiki, Alexf, Tetso, Geni, SarekOfVulcan, LiDaobing, Formeruser-81, Jdevine, Antandrus, Jossi, Pembers, MacGyverMagic, Oneiros, Mzajac, Secfan, Supadawg, Sam Hocevar, Neutrality, Mschlindwein, Jh51681, Master Of Ninja, Arminius, Mike Rosoft, Ornil, D6, ClockworkTroll, Blanchette, Discospinster, Rich Farmbrough, Vsmith, Kzzl, Mani1, Bender235, Sloppy, Brian0918, Mr. Billion, MBisanz, El C, J-Star, Workster, Aude, Art LaPella, RoyBoy, Jpgordon, Afed, Bobo192, Dreish, Maurreen, Toh, Chirag, PeterisP, Larryv, Minghong, GChriss, Idleguy, Sam Korn, Danski14, Alansohn, Neitram, CyberSkull, Philip Cross, Carbon Caryatid, Dubonbacon, Fritz Saalfeld, Echuck215, Lightdarkness, Sugaar, Mysdaao, Rwendland, KJK::Hyperion, ClockworkSoul, BRW, Vanish3, Versageek, Redvers, Axeman89, Recury, Richard Weil, Mahanga, Angr, OwenX, Woohookitty, TigerShark, Nuggetboy, Stickguy, TomTheHand, BillC, Chris Mason, Scjessey, WadeSimMiser, CiTrusD, Uris, Zorpheus, Meneth, SDC, Prashanthns, Mtloweman, Reignerok, PeregrineAY, Behun, Graham87, Deltabeignet, BD2412, Drbogdan, Rjwilmsi, Agrumer, Fieari, P3Pp3r, Koavf, PinchasC, Tangotango, Stardust8212, Bruce1ee, Darguz Parsilvan, Binkymagnus, Daniel Collins, Jehochman, Bhadani, Bubamara (usurped), Titoxd, FlaBot, Spaceman85, Survivor, Nivix, Chanting Fox, RexNL, Ewlyahoocom, Gurch, Verybighsh86, Deciusmagnus, Jeremygbyrne, NotJackhorkheimer, Butros, Chobot, Copperchair, Bighalonut, Gwernol, Bartleby, YurikBot, Wavelength, TexasAndroid, Rt66lt, Jimp, Wolfmankurd, John Callender, RussBot, Dimimimon5, Red Slash, John Quincy Adding Machine, Bhny, Wordmonkey, Gaius Cornelius, Pjanini1, Rsrikanth05, Pseudomonas, Wimt, ENeville, EWS23, Speculative catholic, Wiki alf, Daemon8666, Mike18xx, Welsh, Berend de Boer, Rjensen, Midnite Blue Ice, Ches88, Robotmannick, Robert McClenon, Coderzombie, Abb3w, Moe Epsilon, Zagalejo, CrazyLegsKC, BOT-Superzerocool, DeadEyeArrow, Marvan Hogan, Slicing, Mugunth Kumar, Zzuuzz, PTSE, Jsrduck, Chrisjj2, Mabcom, Keithd, Brz7, Ordinary Person, JLaTondre, Garion96, Archer7, Ajuk, Allens, Junglecat, Erudy, The Catfish, John Broughton, Tyomitch, Pentasyllabic, CIreland, Quadpus, Johnmarkh, Itub, TravisTX, Sardanaphalus, Sarah, SmackBot, Thaagenson, Meshach, F. Impaciente, KnowledgeOfSelf, DuncanBCS, Ze miguel, Unvoyega, Pgk, C.Fred, Jrockley, Thunder Wolf, Jrtf83, Sushisushi, Frymaster, AnOddName, Onebravemonkey, Edgar181, HalfShadow, Mauls, Gilliam, MPD01605, ERcheck, Chris the speller, Master Jay, Jprg1966, Master of Puppets, Deli nk, WeniWidiWiki, Ted87, DHN-bot~enwiki, Realityhammer, Darth Panda, A. B., Charles-Thomas, Kotra, Can't sleep, clown will eat me, AntiVan, Petersk, Zleitzen, Onorem, Avb, Whatthree16, Rrburke, Xyzzyplugh, Midnightcomm, Celarnor, Baiter, PaulBaldowski, Stevenmitchell, ConMan, Emre D., Dylanrush, Nakon, RevAladdinSane, Uriel-238, Xathaec, DMacks, Er Komandante, Ck lostsword, Kukini, Ohconfucius, Will Beback, Lambiam, Serein (renamed because of SUL), Doug Bell, Hestemand, Sina Kardar, Gobonobo, Robofish, JorisvS, Joffeloff, NYCJosh, PseudoSudo, Nobunaga24, 16@r, Hvn0413, Shangrilaista, Booksworm, Beetstra, CelloerTB, Martinp23, SimonATL, Waggers, EdC~enwiki, AEMoreira042281, DI2000, Hu12, Davecornell, Levineps, Iridescent, Joseph Solis in Aus-

Man, Vina, Karol Langner, JimWae, DragonflySixtyseven, Icairns, Sam Hocevar, Arosa, Jacooks, Oskar Sigvardsson, Discospinster, Vsmith, Paul August, Night Gyr, Bender235, Robert P. O'Shea, Edwinstearns, Femto, Bobo192, Dungodung, Kjkolb, Bobbis, Obradovic Goran, Pearle, Jumbuck, Autopilots, Grutness, Alansohn, CyberSkull, Babajobu, Wtmitchell, Rebroad, TenOfAllTrades, Vuo, DV8 2XL, Zereshk, WojciechSwiderski~enwiki, Blaxthos, Umapathy, Oleg Alexandrov, Zntrip, Simetrical, Qaddosh, Pol098, Macaddct1984, Jasonkibby, Graham87, V8rik, BD2412, DePiep, Mastatsan, Salix alba, Miserlou, Fred Bradstadt, Rangek, Mathbot, Gurch, Fresheneesz, King of Hearts, Chobot, GangofOne, DVdm, YurikBot, RobotE, JWB, Wolfmankurd, Robert A West, Stephenb, Maartend8, Terra Green, NormalAsylum, Shanel, NawlinWiki, Mipadi, R.a.f., Mlouns, Wap, Lomn, Mysid, Gadget850, Bota47, CorbieVreccan, Ms2ger, Nikkimaria, Josh3580, Scootersig, Teply, GrinBot~enwiki, Serendipodous, Cmglee, Sbyrnes321, Finell, SmackBot, Jclerman, KnowledgeOfSelf, FloNight, Davewild, Lds, TOMNORTHWALES, PeterSymonds, Gilliam, Ohnoitsjamie, Dauto, Ati3414, Chris the speller, Bluebot, Ottawakismet, MalafayaBot, Metacomet, Nbarth, DHN-bot~enwiki, Shalom Yechiel, HoodedMan, Vladislav, Stevenmitchell, Waprap, Michael Ross, DMacks, Sayden, SashatoBot, Vildricianus, KLLvr283, John, Kognyto, CaptainVindaloo, JohnWittle, Jim.belk, Slakr, Olivierd, Shoeofdeath, Zero sharp, Woodshed, Robot Chicken, Tawkerbot2, CRGreathouse, Huns0004, CBM, Rohan2kool, Laxdude, Casper2k3, Kyotodesertfox, Pce3@ij.net, DumbBOT, Thijs!bot, Epbr123, Yesiammanu, LeeG, Anupam, Headbomb, Marek69, John254, A3RO, Tofof, DaveJ7, Gierszep, Uruiamme, AntiVandalBot, Luna Santin, Orionus, Prolog, Edokter, Random user 8384993, Myanw, VoABot II, CS46, Animum, TheRealOzzy, Seba5618, MartinBot, Mermaid from the Baltic Sea, R'n'B, J.delanoy, Trusilver, Bobcheese411, Scj2315, Maurice Carbonaro, Brianhama, Katalaveno, Johnbod, Mahewa, McSly, Touch Of Light, Misbach, Петър Петров, Gtg204y, WinterSpw, AdrianCooke, Idioma-bot, Cuzkatzimhut, VolkovBot, Johan1298~enwiki, Philip Trueman, Toll booth, EzraZebra, Hqb, Qxz, Clarince63, Wiae, Ojmcdoogle, Yadada mean, Blurpeace, Dragonwish, Andiconda, Joseph A. Spadaro, Manikato, Falcon8765, MCTales, Lpag01, Blood sliver, Reaper111089, EJF, SieBot, Happysailor, Flyer22 Reborn, Oda Mari, Romsey5, Theunbelievers, OKBot, WordsExpert, ClueBot, Fyyer, The Thing That Should Not Be, Starkiller88, NPIC, Drmies, WWEslfan, SNCreal, Rockfang, Taz00, Arjayay, Razorflame, Niftanilla3ee, Jonverve, Bojan1989, Lunboks, Avoided, Mcdope1010, ZooFari, Vow011, Savegerninja, Addbot, Some jerk on the Internet, AkhtaBot, Ronhjones, Fieldday-sunday, NjardarBot, MrOllie, Favonian, Ehrenkater, VASANTH S.N., ScAvenger, HalfFonts, LuK3, Luckas-bot, Yobot, Tohd8BohaithuGh1, Ptbotgourou, Voxpolaris16, Philawinna, Eric-Wester, AnomieBOT, IRP, Stikhead, Materialscientist, Bluey4696, ArthurBot, LilHelpa, Xqbot, Qwertytrader, Capricorn42, Renaissancee, Jmundo, GrouchoBot, SassoBot, AlimanRuna, Thekillinglots, Santâr, Thehelpfulbot, GT5162, Pinethicket, JakeClay, Jonesey95, Rameshngbot, Serols, PRONIZ, YourMomLikesMeMore, Orenburg1, Lotje, Vrenator, Περίεργος, MegaSloth, DARTH SIDIOUS 2, Mean as custard, Vaypertrail, Bento00, Kyaw Zwar Lynn, EmausBot, WikitanvirBot, Immunize, Pimpsta992, Darklord5555, GoingBatty, Solarra, Tommy2010, Dcirovic, Hhhippo, JSquish, ZéroBot, BAICAN XXX, Shuipzv3, Orange Suede Sofa, RockMagnetist, 28bot, Whoop whoop pull up, ClueBot NG, Widr, Helpful Pixie Bot, Novusuna, DanDan0101, Guy vandegrift, Displacedguy, Hallows AG, MusikAnimal, Rostamy, Joydeep, Félix Wolf, Capauer, CitationCleanerBot, Tiberman, Ultraultragaben, Flutershy, GabrielWilliams, GoShow, Lightmeter, BrightStarSky, Sai102938, Subhadeep1908, Aakanksha.j.saxena, Jochen Burghardt, Nicereddy, Me, Myself, and I are Here, Reatlas, RootSword, Epicgenius, Historygeekwormkinggangstermony, Hardvardstudent123, Kharkiv07, Acc12345acc, Feartherapist, Rickdutta, Justlookdown, Balon Greyjoy, Gurbur, Trackteur, Fappy keegan, Patok332, KasparBot, Hyrum McDaniel, JJMC89, Tbushra, Kizami, Germny Himler, Diplobiont, Yoloswegmester420blaze, Jackbirda, R96001, CandyHat and Anonymous: 477

- **Dirty bomb** *Source:* https://en.wikipedia.org/wiki/Dirty_bomb?oldid=760007947 *Contributors:* AxelBoldt, Tretvis, Derek Ross, Bryan Derksen, The Anome, Wayne Hardman, Maury Markowitz, Edward, Patrick, RTC, Ixfd64, Pizza Puzzle, WhisperToMe, Wik, Selket, Riddley, Paranoid, Sanders muc, Securiger, Bkell, Xanzzibar, DocWatson42, Wolfkeeper, Jabra, Fastfission, Everyking, No Guru, Cce, Eequor, Tweenk, Tothebarricades.tk, Neutrality, Lostchicken, Grunt, Rich Farmbrough, Cacycle, ArnoldReinhold, Mani1, Bender235, Mr. Billion, El C, Spencer-Wilson, Solidus~enwiki, Whosyourjudas, R. S. Shaw, ParticleMan, Scollk, Kjkolb, Holdek, Mcknut, Jonathunder, Geo Swan, Linmhall, Axl, Zippanova, Mailer diablo, RoboAction~enwiki, DV8 2XL, Kazvorpal, Hojimachong, GregorB, Yst, Ligar~enwiki, Rjwilmsi, Qqqqqq, FlaBot, Bitolfish, Kolbasz, Bihzad, YurikBot, RussBot, Limulus, Gaius Cornelius, Shaddack, Cpuwhiz11, EWS23, Speedevil, Ospalh, Dbhrs, Bota47, Loffy, Junglecat, Groyolo, Swpmre, SmackBot, Ignotum per Ignotius, Blue520, Jab843, Sam8, KelleyCook, The Ronin, HalfShadow, Brianski, Chris the speller, Jamie C. Hibernian, D T G, Gracenotes, Lenin and McCarthy, Run!, Brian577, Weregerbil, Sfboi79, Mister Five, Chaldean, Platonides, Kuru, John, Microchip08, Regan123, Rkmlai, Publicus, Kieranmrhunt, Pauric, JYi, HisSpaceResearch, Mfrosz, Cls14, Tawkerbot2, TheHorseCollector, ScumMania1979, Themightyquill, Foregather, Otto4711, Lugnuts, Wfaxon, Armscontrol, Arwen4014, Omicronpersei8, Pustelnik, Z10x, DPdH, Denverjeffrey, AntiVandalBot, TimVickers, Qwerty Binary, JAnDbot, Roleplayer, PlazzTT, Bakilas, Republitarian, Nick Cooper, KConWiki, Bernd vdB~enwiki, Animum, Henrychrist, Allstarecho, JaGa, MartinBot, Kiyokun, Richardman84, Terry976, Thaurisil, Hodja Nasreddin, Collegebookworm, AntiSpamBot, Cugerbrant, Nikki311, BernardZ, Idarin, ACSE, Hugo999, VolkovBot, Johnfos, Jeff G., Oshwah, Malinaccier, Batbert, Arcyqwerty, Triesault, Meters, Ryder55, Starkrm, PookeyMaster, StaticGull, Leroyhurdfan, Bodhi Peace, Pinkadelica, ImageRemovalBot, Twinsday, ClueBot, Fasettle, The Thing That Should Not Be, VQuakr, Shaliya waya, Samuel 947, Socrates2008, Abrech, Thingg, Graevemoore, Spitfire, Ipla10s, Pamejudd, Addbot, Ronhjones, Reemrevnivek, Damiens.rf, MrOllie, Download, Justpassin, Numbo3-bot, Tide rolls, Lightbot, ScAvenger, Crusademedia, MiltonP Ottawa, Gail, Luckas-bot, AzureFury, Librsh, REDyellowGreenBLUE, AnomieBOT, Efa, 个仇忔, Materialscientist, Citation bot, Avocats, Sunapi386, Capricorn42, Dubai Cube, Elrohvjxhsudghhgdx, GliderMaven, 58Extraten, FrescoBot, RoyGoldsmith, Citation bot 1, Ntse, Moonraker, Robo Cop, Autumnalmonk, Salvio giuliano, EmausBot, Notoriousnoah, Trofobi, Boundarylayer, Rjl207, Arbnos, H3llBot, EWikist, Predpred, Scientific29, Rememberway, ClueBot NG, WIERDGREENMAN, Catlemur, JoshSherick, BG19bot, Vagobot, Marcocapelle, YodaRULZ, ProudIrishAspie, Minsbot, Cyberbot II, ChrisGualtieri, Embrittled, Ahmad albnawi, Jray310, Raymond1922A, Ocutarnervosa, Solaristv, Kevin12xd, FiredanceThroughTheNight, Pvpoodle, Derpinmyderp, Njol, Eurodyne, Salmin, Anarchyte, Jacob51515, Baking Soda, Parsley Man, InternetArchiveBot, FtLauderGuy, GreenC bot, SkyWarrior, Deglester-Hadunkichud, Juanesmartino, Kainweir, Althebear2, Bender the Bot, WizardMeme and Anonymous: 228

- **Salted bomb** *Source:* https://en.wikipedia.org/wiki/Salted_bomb?oldid=746574761 *Contributors:* Patrick, Itai, Bender235, Jhfrontz, Blaxthos, MONGO, Xihr, Kvn8907, HereToHelp, Whaa?, SmackBot, Gilliam, Crab182, Robofish, Chetvorno, Scooteristi, JamesAM, KrakatoaKatie, Albany NY, LorenzoB, Hqb, UnitedStatesian, Arjayay, MystBot, Addbot, Pietrow, Yobot, AnomieBOT, Bobbinz, Jagons, Fingerz, A412, Skiff, RjwilmsiBot, Razor2988, Ὁ οἶστρος, Arbnos, Nitroglycol, ClueBot NG, Helpful Pixie Bot, Marcocapelle, The Illusive Man, SNAAAAKE!!, MarchOrDie, Sushibob6, Monkbot, Ceosad, Salmin, Nøkkenbuer, Outspokenthegreat, Fantesykikachu and Anonymous: 27

- **Psychological warfare** *Source:* https://en.wikipedia.org/wiki/Psychological_warfare?oldid=766189122 *Contributors:* Bryan Derksen, The Anome, Danny, SimonP, Stevertigo, Patrick, MartinHarper, Grendelkhan, Sabbut, Wblakesx, Lowellian, Flauto Dolce, Bkell, Ruakh, GreatWhiteNortherner, Tom harrison, Fastfission, DO'Neil, Cloud200, DavidBrooks, Alexf, Popefauvexxiii, Dvavasour, MacGyverMagic, Ganymead, Rich Farmbrough, Murtasa, Tsujigiri~enwiki, Bender235, RJHall, CanisRufus, Johnkarp, Keron Cyst, Maurreen, Kjkolb, Jon-

troppo, AdjustShift, Nt204, Photographerguy, Knowledgekid87, Materialscientist, Citation bot, OllieFury, ArthurBot, Lqstuart, Xqbot, Wmerc, NeverGonnaDonate, Gloomkitten, AbigailAbernathy, Monkey0101, Damus72, Mathonius, Panda 51, Shadowjams, Fotaun, Abcdefgy2, FrescoBot, Surv1v4l1st, Vinithehat, Unomi, RicHard-59, RoyGoldsmith, Jabberdocky, Citation bot 1, HRoestBot, Tom.Reding, Mercuryiscool, Jauhienij, Double sharp, TobeBot, Trappist the monk, Lotje, Extra999, Mr.98, Micktheclick, Tbhotch, Misakubo, Samozaparola, Jessimaiso, The Utahraptor, RjwilmsiBot, Ripchip Bot, Teravolt, Salvio giuliano, DASHBot, WikitanvirBot, Dadaist6174, Pete Hobbs, Cableable2, Boundarylayer, Ibbn, Tommy2010, Alexander.enchevich, Wikipelli, Ornithikos, 6zeta2tothehalf, Hhhippo, AvicBot, ZéroBot, Prayerfortheworld, Unendingfear, H3llBot, AlphaPikachu578, Wayne Slam, Justiceformjj, Cf. Hay, Gijs.deRue, FreedomWorksHard, Hazard-Bot, HandsomeFella, Whoop whoop pull up, Uraniumproductionz, ClueBot NG, Zyrath, AerobicFox, Corbenic97, Deliveryrevited, Snotbot, XXI.Venom999, Dribrook, Helpful Pixie Bot, Calabe1992, BG19bot, Neuronz, Verity1980, Neon, MusikAnimal, Ericmackay, AdventurousSquirrel, Tsanquist, MrBill3, Klitidiplomus, Shisha-Tom, BattyBot, GoShow, TylerDurden8823, JYBot, Dexbot, Mysterious Whisper, Megabeth4, Hillbillyholiday, Greatuser, HFEO, Jodosma, Evolution and evolvability, BruceBlaus, Limnalid, Coreyemotela, Monkbot, SebastianG.FA, BrayLockBoy, Nloveladyallen, KBH96, TranquilHope, NekoKatsun, Sundayclose, HereAndThereNowAndThen, +Treker, Mcremp, Ch55x, InternetArchiveBot, Hitheremynameisjeff, GreenC bot, Bender the Bot, SammyMajed and Anonymous: 622

- **Radiation hormesis** *Source:* https://en.wikipedia.org/wiki/Radiation_hormesis?oldid=754015619 *Contributors:* Ed Poor, Maury Markowitz, Tijmz, Ewen, Michael Hardy, DopefishJustin, Andrewa, RL Kierant, JonathanDP81, Pstudier, Phil Boswell, Altenmann, Wolfkeeper, Dratman, Miya, Whitis, Tweenk, Rich Farmbrough, ESkog, Diamonddavej, O18, Arcadian, Plumbago, Wouterstomp, Utfish, Btyner, Jacj, Mandarax, Keithpickering, Rjwilmsi, JWWalker, Ground Zero, Old Moonraker, Choess, Jeremygbyrne, Wavelength, Mahahahaneapneap, Chris Capoccia, Limulus, Gaius Cornelius, Morphh, Settsu, 2over0, Jules.LT, CharlesHBennett, SmackBot, Stepa, Ottawakismet, Cadmium, IIXII, Radagast83, John, Peyre, Iridescent, Rp2006, Skapur, Dashpool, Rmallins, A876, Cuhlik, Photocopier, Headbomb, Alphachimpbot, Light Dragon, Magioladitis, Yakushima, Cgingold, Ben Ram, Ewvitz, Djma12, AstroHurricane001, Zdavatz, Cruftbane, Gamemaster0, Michelet, StAnselm, VVVBot, Likebox, Sunrise, Iknowyourider, Escape Orbit, Mlaffs, Dana boomer, DumZiBoT, Addbot, DOI bot, Download, Yobot, Electronsoup, AnomieBOT, Arjun G. Menon, Archon 2488, Citation bot, Embram, Nasa-verve, Rccapps, Biem, FrescoBot, Unomi, Alarics, Citation bot 1, PRONIZ, Trappist the monk, Robertiki, Dmtrlk, RjwilmsiBot, Corkscrew99, Boundarylayer, Eekh.eu, Ὁ οἶστρος, H3llBot, AManWithNoPlan, NanooGeek, ClueBot NG, Adam Majer, Curb Chain, Bibcode Bot, BG19bot, ConradMayhew, Shisha-Tom, Siphon06, Dexbot, Kbog, Pintoch, Faizan, Mjwood 26, Monkbot, Pommapple, Moobatman, Charlesy and Anonymous: 56

- **Radionuclide** *Source:* https://en.wikipedia.org/wiki/Radionuclide?oldid=762387521 *Contributors:* AxelBoldt, Kpjas, Trelvis, Mav, Fnielsen, Ubiquity, Lir, Shellreef, Jketola, Tannin, Looxix~enwiki, Ahoerstemeier, Aarchiba, Andres, Jordi Burguet Castell, Smack, Stone, Taxman, Donarreiskotfer, Gentgeen, Robbot, Kristof vt, Ojigiri~enwiki, Drstuey, DocWatson42, Herbee, Andycjp, Antandrus, Icairns, Sam Hocevar, CALR, Discospinster, Bender235, Joanjoc~enwiki, Bobo192, O18, Arcadian, Alansohn, Keenan Pepper, AjAldous, Mlessard, Utfish, Vuo, Zereshk, Voxadam, Mindmatrix, WadeSimMiser, Eleassar777, Isnow, Graham87, Rjwilmsi, Tangotango, Ems57fcva, Ayla, Kolbasz, Butros, Essaregee, Chobot, DaGizza, DVdm, YurikBot, Wavelength, Borgx, Wimt, Zwobot, Tweeq, Kkmurray, LeonardoRob0t, SmackBot, FocalPoint, Jclerman, Unyoyega, Chris the speller, Postoak, Sbharris, TheGerm, Vladislav, OrphanBot, Addshore, Drphilharmonic, Kukini, John, H.sand01, Gobonobo, Dicklyon, Citicat, Iridescent, PetaRZ, Gungasdindin, CmdrObot, Harej bot, Nilfanion, A876, ST47, Raomap, Optimist on the run, JLD, Epbr123, Pjvpjv, Gierszep, AntiVandalBot, Deberle, JAnDbot, Arch dude, Albany NY, Sophie means wisdom, I80and, VoABot II, AdamWalker, Animum, Cgingold, Dirac66, Squidonius, MartinBot, Rogerruo, Maurice Carbonaro, Stan J Klimas, Actarux, Philip Trueman, TXiKiBoT, Seb az86556, PGWG, Breakyunit, Keilana, Yerpo, Rhenning007, Simonbayly, Verytas, KathrynLybarger, Jessiehawkes, Nergaal, Jons63, ClueBot, The Thing That Should Not Be, Wysprgr2005, Niceguyedc, Nuclearmedzors, PixelBot, Jotterbot, Dekisugi, Johndoe616, Panos84, Vegetator, Jonverve, Jeremycenus, TNTM64, XLinkBot, Avoided, WikHead, Teslaton, Dfoxvog, Addbot, Gregisfat2, Haruth, Leszek Jańczuk, CUSENZA Mario, Tide rolls, Ben Ben, Luckas-bot, Yobot, AnomieBOT, Zhieaanm, Ciphers, Sz-iwbot, Materialscientist, Citation bot, ArthurBot, Xqbot, Braindamagehurts, Biggieyankfan, Hamburgers1212, Dan6hell66, FrescoBot, LucienBOT, Slastic, Citation bot 1, Minivip, FoxBot, Double sharp, Trappist the monk, Lotje, DARTH SIDIOUS 2, Ripchip Bot, EmausBot, WikitanvirBot, Look2See1, Demomoer, Steve123963, Donner60, Research new, ChuispastonBot, Cgt, Xrayburst1, ClueBot NG, Ecubar, Smm201`0, Reify-tech, Maripaz21, Bibcode Bot, MusikAnimal, Lazord00d, Crh23, Zedshort, BattyBot, Pratyya Ghosh, Layzeeboi, Ssscienccce, Dexbot, Burzuchius, Marqaz, CensoredScribe, Tinyscooter, Monktues, Karl Hess666, NewEnglandDr, Teddyktchan, Trackteur, SStewartGallus, liKkEe, DSCrowned, Pinnate foliage, KasparBot, Oluwa2Chainz, The Quixotic Potato, Bear-rings, Iambic Pentameter and Anonymous: 184

- **Radioactive waste** *Source:* https://en.wikipedia.org/wiki/Radioactive_waste?oldid=765700168 *Contributors:* Trelvis, Mav, Bryan Derksen, Rmhermen, William Avery, Anthere, Patrick, D. Tillwe, Ixfd64, Tomos, Mkweise, Ahoerstemeier, Andrewa, Aarchiba, Julesd, CarlKenner, Harvester, Smack, Lommer, Mulad, Dcoetzee, Fuzheado, Maximus Rex, Furrykef, Omegatron, ObjectFarm, Pstudier, Pakaran, Finlay McWalter, Francs2000, Robbot, Chealer, Pigsonthewing, Securiger, Auric, Hadal, Alan Liefting, Giftlite, Andries, Jyril, Cobaltbluetony, FleaPlus, Karn, Peruvianllama, Guanaco, Tweenk, Alvestrand, JRR Trollkien, OldakQuill, Utcursch, MikeX, CryptoDerk, Yath, Pcarbonn, Antandrus, Goog, Rdsmith4, Icairns, Sam Hocevar, Neutrality, Imjustmatthew, Frau Holle, Deglr6328, Danh, Mike Rosoft, Archer3, JTN, Discospinster, Rich Farmbrough, Guanabot, NeuronExMachina, Vsmith, Florian Blaschke, Pavel Vozenilek, Paul August, Cromis, Kenb215, Bender235, ESkog, Sietse Snel, Jpgordon, Bobo192, Nigelj, Smalljim, Viriditas, Vortexrealm, Phlake, Giraffedata, Kjkolb, MPerel, Frodet, Danski14, Alansohn, Gary, Borisborf, ChrisGlew, Eric Kvaalen, Arthena, Riana, Rwendland, Snowolf, Wtmitchell, Velella, Evil Monkey, Jheald, DV8 2XL, Gene Nygaard, Zereshk, Forteblast, Ultramarine, Lloydd, Bastin, Thryduulf, Kelly Martin, Imaginatorium, Woohookitty, CyrilleDunant, Oliphaunt, Benbest, WadeSimMiser, Duncan.france, Kgrr, Pogue, Bluemoose, X444, Crucis, Gimboid13, Teemu Leisti, Driftwoodzebulin, Mandarax, RedBLACKandBURN, BD2412, Squideshi, Martinevos~enwiki, Sjakkalle, Rjwilmsi, Joffan, Dcoryh192, Vegaswikian, CannotResolveSymbol, The wub, DirkvdM, Lcolson, Old Moonraker, Nihiltres, Crazycomputers, CarolGray, RexNL, Gurch, RobyWayne, Kolbasz, Drumguy8800, AnthonyA7, King of Hearts, DVdm, Mhking, Simesa, Roboto de Ajvol, YurikBot, Wavelength, Borgx, RobotE, JWB, Huw Powell, Phmer, Pip2andahalf, CKSemmens, RobHutten, Hede2000, Captainden, Ytrottier, BillMasen, Nesbit, Akamad, Stephenb, Shaddack, Haizum, Ugur Basak, David R. Ingham, NawlinWiki, NW036, Nirvana563, Harksaw, Tokachu, Ringford, Danlaycock, Misza13, Zwobot, Dbfirs, Scottisher, DeadEyeArrow, PS2pcGAMER, Oliverdl, Essexmutant, Veatch, Deville, Pb30, Ntouran, Abune, Sean Whitton, GraemeL, Bumpoh, Naught101, Paul D. Anderson, Katieh5584, Kungfuadam, Mdwyer, Sinus, NeilN, Vvmaks, CIreland, ChemGardener, Snalwibma, Bob.appleyard, Bluewave, Frankie, KnightRider~enwiki, SmackBot, Reedy, Prodego, FloNight, CRKingston, Pgk, Yuyudevil, Matveims, Chairman S., KVDP, Eskimbot, Mdd4696, Onebravemonkey, Man with two legs, HalfShadow, Septegram, Gilliam, Hmains, Skizzik, Scaife, Chris the speller, Cadmium, MaRoWi~enwiki, Isaacsurh, Sbharris, Darth Panda, Rlevse, Scalene, Theneokid, Royboycrashfan, Gbsx32, Can't sleep, clown will eat me, Nick Levine, VMS Mosaic, Addshore, Elendil's Heir, Theanphibian, Wen D House, Decltype, Epachamo, Bobtheguy, BWDuncan, Invincible Ninja,

Ospalh, Cedar101, Petri Krohn, Kungfuadam, Derelk, SmackBot, Jclerman, Edgar181, Aram.harrow, Chris the speller, Roomba, Sbharris, Emurphy42, Can't sleep, clown will eat me, MyNameIsVlad, Shalom Yechiel, KerathFreeman, Theodorevii, Theanphibian, PiMaster3, Nr-cprm2026, Mikaduki, John, Darktemplar, Rsquid, Smith609, Dave3141, Forwardbias, Wjejskenewr, Barticus88, Mojo Hand, Headbomb, AntiVandalBot, XyBot, Magioladitis, Rich257, WolfyB, MartinBot, Trusilver, NewEnglandYankee, Potatoswatter, HazyM, VolkovBot, Basharab, Toddy1, TXiKiBoT, Openman, Leafyplant, UnitedStatesian, Brian Huffman, Beast of traal, Dbagdi92, ClueBot, Nailedtooth, Arunsingh16, Excirial, Biochem67, SpartanIname, Inorganiker, Addbot, Snaily, Luckas-bot, Ptbotgourou, Tecatelite, AnomieBOT, Angry bee, Ciphers, Rubinbot, Galoubet, Materialscientist, Citation bot, Xqbot, Sandragerou, GrouchoBot, FrescoBot, Tubas-en, Trappist the monk, Speciman00, Jsobry, RjwilmsiBot, Codehydro, Alph Bot, Salvio giuliano, Garrett.mitchener, Solomonfromfinland, ZeroBot, ClueBot NG, This lousy T-shirt, Lndeo, Rm1271, Joshtaco, GabeIglesia, Faizan, Sukumaru, ♯, Luke J Pickett, DudeWithAFeud, FourViolas, SkyWarrior, John "Hannibal" Smith and Anonymous: 115

- **Plutonium** *Source:* https://en.wikipedia.org/wiki/Plutonium?oldid=762933031 *Contributors:* Mav, Bryan Derksen, Drj, LA2, Rmhermen, William Avery, Roadrunner, Ray Van De Walker, Robert Foley, Heron, Camembert, Bth, Modemac, Tedernst, Stevertigo, Spiff~enwiki, Patrick, RTC, JohnOwens, Alan Peakall, Fred Bauder, Wapcaplet, Yann, Gbleem, Egil, Mkweise, Ahoerstemeier, Andrewa, Aarchiba, Julesd, Marco Krohn, Andres, Jiang, Cherkash, GCarty, Ghewgill, Schneelocke, Uriber, Emperorbma, Ec5618, Stone, Lfh, David Latapie, Audin, Andrewman327, Svante~enwiki, Dragons flight, FurryKef, SEWilco, Paul-L~enwiki, Omegatron, Antatnsu, Thue, Warbeck~enwiki, Nosebud, Pstudier, Bcorr, Pakaran, Jerzy, Finlay McWalter, Pollinator, Jeffq, Twang, Chuunen Baka, Donarreiskoffer, Robbot, Chealer, Fredrik, Donreed, ZimZalaBim, Romanm, Naddy, Arkuat, Rursus, Catbar, Quadalpha, Lupo, Tea2min, Marc Venot, Centrx, Giftlite, Graeme Bartlett, DocWatson42, Ryanrs, Fastfission, Karn, Wwoods, Michael Devore, Yekrats, Mboverload, Tweenk, Gzornenplatz, VampWillow, Darrien, Tagishsimon, Utcursch, Geni, Yath, Antandrus, Mako098765, Jossi, Rdsmith4, Oneiros, Kesac, DragonflySixtyseven, Thincat, Icairns, Sam Hocevar, Tdent, Positron, Deandeto, Deglr6328, Mike Rosoft, Citizensunshine, Diagonalfish, Discospinster, Rich Farmbrough, Sladen, NeuronExMachina, Vsmith, Ardonik, ArnoldReinhold, Alistair1978, SpookyMulder, Bender235, TerraFrost, RJHall, FirstPrinciples, El C, Dnwq, Joanjoc~enwiki, Kwamikagami, Shanes, Kouhoutek, Remember, Art LaPella, Orlady, Femto, Bobo192, Smalljim, Clawson, Shenme, Apyule, Bbartlog, Kjkolb, Railgun, Yalbik, Honeycake, Alansohn, Anthony Appleyard, Qwe, Polarscribe, Sl, Pen1234567, Axl, Redfarmer, Wdfarmer, Avenue, Rwendland, Snowolf, Evil Monkey, TenOfAllTrades, H2g2bob, DV8 2XL, Kazvorpal, Pediddle, April Arcus, BerserkerBen, Kenyon, Joriki, Veemonkamiya, Woohookitty, NewbieDoo, LOL, Riffsyphon1024, ToddFincannon, MrWhipple, Benbest, Pol098, WadeSimMiser, Miss Madeline, The Jacobin, Sengkang, GregorB, Eyreland, Eras-mus, El Suizo, Xiong Chiamiov, Shanedidona, Lawrence King, Graham87, Deltabeignet, Magister Mathematicae, Keeves, DePiep, Grammarbot, Drbogdan, Saperaud~enwiki, Rjwilmsi, Jake Wartenberg, Jmcc150, Zizzybaluba, Bubba73, Durin, The wub, Lcolson, Allen Moore, RobertG, Ground Zero, Old Moonraker, Nihiltres, Rune.welsh, RexNL, Scottrainey, Kolbasz, Zotel, Physchim62, Valentinian, Chobot, Jaraalbe, Igordebraga, Bgwhite, Simesa, Whosasking, Algebraist, Roboto de Ajvol, YurikBot, Wavelength, Borgx, JWB, Hairy Dude, Phmer, Phixt, Midgley, RussBot, Open4D, Armistej, Chuck Carroll, Limulus, Hellbus, Hydrargyrum, Polluxian, Yyy, Shaddack, Kimchi.sg, Wimt, Bill-on-the-Hill, NawlinWiki, Hawkeye7, Dysmorodrepanis~enwiki, Wiki alf, Howcheng, Omega 13, Texboy, Gillis, Lomn, Hinto, Syrthiss, Brat32, DeadEyeArrow, Psy guy, Tetracube, FF2010, Lt-wiki-bot, Ninly, Bayerischermann, Closedmouth, Cedar101, Fang Aili, E Wing, Reyk, Petri Krohn, CWenger, Hurricane Devon, Geoffrey.landis, Anclation~enwiki, GrinBot~enwiki, Saikiri, That Guy, From That Show!, Kimdino, Luk, ChemGardener, Itub, Criticality, Attilios, SmackBot, H2eddsf3, Narson, Stux, KnowledgeOfSelf, Melchoir, Shoy, Pgk, C.Fred, Martylunsford, FRS, Onebravemonkey, Edgar181, Man with two legs, Commander Keane bot, Andy M. Wang, GoneAwayNowAndRetired, Chris the speller, Kurykh, Keegan, Cadmium, Thumperward, Fuzzform, Stellar-TO, Domthedude001, SchfiftyThree, IIXII, EdgeOfEpsilon, Sbharris, Rama's Arrow, Farseer, Rogermw, Tsca.bot, Can't sleep, clown will eat me, Joema, Edivorce, Ruffin' writer, Cybercobra, Nakon, TedE, RandomP, A.R., Polonium, Soarhead77, Kendrick7, A5b, Pinktulip, SashatoBot, Rockvee, ArglebargleIV, Sia15998, John, HVS, Sosodank, Stuerst, Mcshadypl, Iliev, Heimstern, JorisvS, Mgiganteus1, CredoFromStart, IronGargoyle, JHunterJ, Werdan7, Beetstra, Mr Stephen, Xiaphias, SandyGeorgia, Super8Guy, Fan-1967, Iridescent, Polymerbringer, Igoldste, Walkinglikeahuricane, Courcelles, Fullerene~enwiki, Tawkerbot2, Ouishoebean, Pootkris, Ur mom sux, JForget, Markjoseph125, CmdrObot, The Cake is a Lie, Rwflammang, Bayou Banjo, Asro940, Skoch3, Borislav Dopudja, Stebbins, Supremeknowledge, A876, Steel, Not R, HPaul, Flowerpotman, Cuhlik, Give Peace A Chance, Benjiboi, Christian75, Nabokov, Jaerik, Omicronpersei8, Quartic, Kirk Hilliard, Thijs!bot, Epbr123, Pajz, Pstanton, David from Downunder, Martin Hogbin, Bladiebla~enwiki, Mojo Hand, Headbomb, Marek69, John254, Wiki fanatic, Zachary, Ludde23, Escarbot, Mentifisto, AntiVandalBot, Barneyg, Tpth, Farosdaughter, Spencer, SkoreKeep, Myanw, Defordj, Gökhan, Edward J. Picardy, BeefRendang, JAnDbot, Deflective, Husond, NapoliRoma, MER-C, Plantsurfer, Igodard, Getaway, East718, Tstrobaugh, Adzze, Savant13, LittleOldMe, .anacondabot, Acroterion, Angelofdeath275, Magioladitis, WolfmanSF, Karlhahn, Bongwarrior, VoABot II, BigDukeSix, Bulbeck, Sdaaw4g, Cgingold, Spellmaster, JaGa, Chuckwatson, Forai, Sasper, MartinBot, Pagw, Sean Goodwin, CommonsDelinker, Brothejr, Ash, Marhault, Thirdright, Watch37264, J.delanoy, Whitneyjones, DrKay, Trusilver, Euku, Uncle Dick, MistyMorn, Tbom.fynn, Ccmdav, Dispenser, Slithymatt, NavyPunk426, Darth Krayt, Rcnet, AntiSpamBot, Margareta, Warut, ARTE, Sugarbat, Tanaats, I am Super Ryan, Mirithing, Action Jackson IV, WhickityWhite, HazyM, Ron Magic, Cs302b, Squids and Chips, Spellcast, ACSE, Frog is God, Hugo999, 28bytes, VolkovBot, CWii, Thedjatclubrock, Johnfos, Jeff G., Abatacha, Philip Trueman, Big chutty, TXiKiBoT, Dajwilkinson, Zidonuke, GimmeBot, Cosmic Latte, .:CoReHaCk:., Hqb, Rei-bot, Anonymous Dissident, Monkey Bounce, Piperh, Leafyplant, Zenswashbuckler, Jackfork, LeaveSleaves, Guest9999, Maxim, Milkbreath, Lamro, Enviroboy, Brianga, AlleborgoBot, TwoWildnCrazyKids, Boooboobear7, FlyingLeopard2014, EmxBot, Starkrm, Kbrose, SieBot, Snugggy, Tresiden, PlanetStar, Swliv, Graham Beards, Soralin, WereSpielChequers, Jauerback, Dawn Bard, Nelhowt4, MeegsC, Smsarmad, JLKrause, Black lupin, Oxymoron83, Megansmith18, Techman224, Hak-kâ-ngin, Grunkhead, Doublesuited, Moletrouser, AWeishaupt, Afernand74, OKBot, Svick, LonelyMarble, StaticGull, Anchor Link Bot, Wuhwuzdat, Mygerardromance, Anyeverybody, Kenny1678, Misiu mp, Pinkadelica, Dolphin51, Nergaal, Gderbysh, Science Focus, Atif.t2, Eoghain0708, Ratemonth, Sfan00 IMG, MBK004, ClueBot, Trojancowboy, Binksternet, Artichoker, Fasettle, Bobathon71, The Thing That Should Not Be, Spidaman23, Inver471ness, Pi zero, Polyamorph, CounterVandalismBot, Fredrock800, Piledhigheranddeeper, Neverquick, DragonBot, Deaxman, Howdyfolk, Excirial, Alexbot, Bdweiler, PixelBot, Shinkolobwe, NuclearWarfare, Jotterbot, Wikimedes, Thingg, Aitias, Versus22, Skier lad, SoxBot III, YouRang?, Bridies, XLinkBot, Ultramince, Wertuose, Gerhardvalentin, FellGleaming, Rreagan007, Mxfh, ErkinBatu, WikiDao, ZooFari, MystBot, Hosain54, Shoemaker's Holiday, HexaChord, Pyfan, Mr0t1633, Roentgenium111, DOI bot, Flbribri8788, Ocdnctx, AkhtaBot, Blethering Scot, Download, LaaknorBot, CarsracBot, Infeh, Glane23, Nesstopher, Aunva6, LinkFA-Bot, 5 albert square, Hahnium, Numbo3-bot, Evildeathmath, Joe Friendly, Tide rolls, Gail, Frehley, Luckas-bot, Yobot, EchetusXe, 2D, Rccoms, THEN WHO WAS PHONE?, Christhi, AnakngAraw, Tempodivalse, AnomieBOT, KDS4444, Jim1138, IRP, Galoubet, Piano non troppo, EHRice, Theseeker4, Ulric1313, Crystal whacker, Materialscientist, The High Fin Sperm Whale, Citation bot, Kalamkaar, Carlsotr, Sweet smell of salami, Frankenpuppy, Neurolysis, ArthurBot, LovesMacs, LilHelpa, Jay L09, Xqbot, Gigemag76, JWBE, Beeline23, Daners, Neoshero5, Srich32977, Whiteskin420, GrouchoBot, Lop7685, Nedim Ardoǧa, Stratocracy, Doulos Christos, MerlLinkBot, GainLine, Gordonrox24, Shadowjams, DaneCrazy, Ankitbhatt, MeDrewNotYou, AlimanRuna, Chaheel Riens, SchnitzelMannGreek, Einsteinmc2300, A. di M., Dougofborg, BoomerAB, 1stlegionarmy, R8R Gtrs, Eatabug1234, Abcdefgy2, FrescoBot, Dufydtaf, Bigmikeaziz, Dszerxftcghjk, Quatro dose, Gcalis, StephenHart, Wervo, Bananaman2222, Ankit555551, Buijs, Cannolis, Citation bot 1, AstaBOTh15, Pinethicket, HRoestBot, Jonesey95, Tom.Reding, Calmer Waters, Geogene, Trần Nam Ha 2001, RedBot, MastiBot, Plasticspork, Jauhienij, Orenburg1, Double sharp, TobeBot, Speciman00, Dkozza, Mono, ArthurBorges, Clarkcj12, Mr.98, Reaper Eternal, கலை, Tbhotch, DARTH SIDIOUS 2, RjwilmsiBot, DexDor, Ripchip Bot, Yaush, Korderrius, Emaus-Bot, Matalovita, Boundarylayer, GoingBatty, XinaNicole, MyFaceBeAFunnyFace, HarDNox, Wikipelli, Endicott65, Josve05a, Ninjamarmot, StringTheory11, ElationAviation, Alpha Quadrant (alt), H3llBot, SporkBot, Aschwole, Dtobin123, Sebastian barnes, ChuispastonBot, Hand-someFella, DASHBotAV, Whoop whoop pull up, Joesmith989, ClueBot NG, RaptorHunter, MelbourneStar, PaleCloudedWhite, Parcly Taxel, Widr, Newyorkadam, Helpful Pixie Bot, Janiberry, JohnSRoberts99, Me1423, Strike Eagle, Bibcode Bot, Lowercase sigmabot, Muqman 52, Vagobot, ElphiBot, Lazord00d, Jhfjdhfjhsdfkd, FormerNukeSubmariner, Keeplookin, Djmackie, Dagonking123, THEJUDGE24, Alarbus, Minsbot, Nitrobutane, BattyBot, Whydidthatpagechange, Cyberbot II, Plutonium12345, JYBot, Samcz, Dexbot, NewebNL, XXzoonamiXX, Lugia2453, CaSJer, Froshirt, Little green rosetta, Lucas0425, Vanischenu from public computers, Reatlas, Salamancer42, Ruby Murray, In-glok, Misterballs1337, Madtrolls, Thedigitalabe, Basti1basti, Ugog Nizdast, DudeWithAFeud, Hugginsian, AwesomeEvilGenius, Scinarchist, Wearyalchemist21, Trollepedia troll, Whoamiswagger, Monkbot, JuanRiley, InfoDataMonger, Sangdeboeuf, Isambard Kingdom, Pinkomega, KasparBot, ChemWarfare, Stewader91, LL212W, Linrx and Anonymous: 828

- **Plutonium-238** *Source:* https://en.wikipedia.org/wiki/Plutonium-238?oldid=763969117 *Contributors:* Bryan Derksen, Ceaser, Julesd, GCarty, Stone, Donarreiskoffer, Iain.mcclatchie, DocWatson42, Fastfission, Beland, Neko-chan, Art LaPella, Ardric47, Crust, DV8 2XL, Benbest, Bkuschel, Emerson7, Pmj, Bubba73, Kolbasz, Baszoetekouw, JWB, Robert A West, Limulus, Heltbus, Shaddack, Cstaffa, Pr1268, Johndburger, Ntouran, CharlesHBennett, Petri Krohn, Geoffrey.landis, SmackBot, Kilo-Lima, Gilliam, Skizzik, Bluebot, Vladislav, Pulu, Polonium, John, Vitall, Smith609, Spiel496, Craigboy, Zaphody3k, CmdrObot, Comrade42, ShelfSkewed, Cuhlik, Give Peace A Chance, Headbomb, Hcobb, Golgofrinchian, Swpb, Bobkeyes, BatteryIncluded, Rod57, HazyM, Mark v1.0, Piperh, Inventis, Drkarger, Flyer22 Reborn, Thomasonline, Misiu mp, Dradler, Addbot, Roentgenium111, Ronkonkaman, Fmwagner, KitemanSA, Yobot, AnomieBOT, GrouchoBot, Erik9bot, Fres-coBot, Bob Saint Clar, Dcheagle, LittleWink, IJBall, Jimantis, Bento00, Sverigekillen, FourtySix&Two, Lucas hamster, Onegumas, Dcirovic, A2soup, StringTheory11, Aschwole, MichiganY, Ffreibert, ChuispastonBot, Rjchemdoc, Northamerica1000, Badon, 220 of Borg, Poikiloblas-tic, Cyberbot II, Tony Mach, Napy65, TheWanderer1357, DudeWithAFeud, Monkbot, Hullspeed, GreenC bot, Blahvscarrots, Rdmadding and Anonymous: 44

- **Americium-241** *Source:* https://en.wikipedia.org/wiki/Americium-241?oldid=765507619 *Contributors:* Karn, BRW, BD2412, Huskydog, TimSE, Bgwhite, Ytrottier, Shaddack, SmackBot, Jushi, Joeylawn, Iridescent, Christian75, Headbomb, Electron9, Avicennasis, LaMona, CommonsDelinker, Doub, Ehnebuske, Alexbot, Luckas-bot, The Earwig, SwisterTwister, Vuerqex, Erik9bot, Diannaa, John of Reading, Neilgra-hamshaw, ClueBot NG, Lanthanum-138, BG19bot, Guy Adler, Zedshort, Comfr, Pinkomega, InternetArchiveBot, GreenC bot and Anonymous: 15

- **Isotopes of californium** *Source:* https://en.wikipedia.org/wiki/Isotopes_of_californium?oldid=761357672 *Contributors:* Mav, Bryan Derksen, Merovingian, Urhixidur, Femto, Avian, Rjwilmsi, Bgwhite, JWB, Anomalocaris, CWenger, Vina-iwbot~enwiki, Headbomb, Leyo, TXiKiBoT, UnitedStatesian, The way, the truth, and the light, Addbot, Roentgenium111, LaaknorBot, Luckas-bot, Amirobot, Floquenbeam, XZeroBot, Erik9bot, Double sharp, Trappist the monk, 777sms, DexDor, XinaNicole, ZéroBot, Quondum, Helpful Pixie Bot, Bibcode Bot, Citation-CleanerBot, ChrisGualtieri, Monkbot, Lubomir Kucera, Tnetrpm, Fterpm, Crackhorace and Anonymous: 3

- **Caesium-137** *Source:* https://en.wikipedia.org/wiki/Caesium-137?oldid=763086283 *Contributors:* AxelBoldt, Bryan Derksen, Schewek, Fx-mastermind, Edward, Julesd, Stone, Wetman, Donarreiskoffer, Alan Liefting, DocWatson42, Poupoune5, Andycjp, Beland, Vsmith, Bro-kenSegue, Brim, Jonaro~enwiki, Radical Mallard, Capecodeph, HenryLi, Forteblast, Woohookitty, Miss Madeline, Rjwilmsi, Runarb, Itinerant1, Kolbasz, Sperxios, WriterHound, JWB, Huw Powell, Shaddack, Anomie, Kantokano, Grafen, CecilWard, Wknight94, MaeseLeon, MacsBug, SmackBot, Tonyr68uk, Dr Satori, Sam8, Chris the speller, Cadmium, Dwchin, Master Bob, Giancarlo Rossi, John, Minna Sora no Shita, Jmgon-zalez, Lamiot, Rwflammang, Myscrnnm, Armscontrol, Headbomb, Uruiamme, AntiVandalBot, AndreasWittenstein, Demonkey36, TV4Fun, Sophie means wisdom, Yakushima, Cathsherguy, KylieTastic, HazyM, KudzuVine, ACSE, Mark v1.0, Piperh, Falcon8765, SieBot, GlassCo-bra, Avidallred, Yamaka122, Wiki-ny-2007, Shawks2003, Binksternet, Palantir-palantir, Dpmuk, Awickert, Wprlh, FellGleaming, MystBot, Addbot, Arthur to, Internet lord, Mvkushnir, Pince Nez, Ehrenkater, Luckas-bot, Yobot, AnomieBOT, Citation bot, IAINATOR, ArthurBot, Kevin chen2003, Pontificalibus, Gumok, Xasodfuih, Wilsonchas, Some standardized rigour, Haploidavey, Erik9bot, Toddbailey, Bob Saint Clar, Tubas-en, Citation bot 1, Spume, Robertiki, Лте Лой, EmausBot, WikitanvirBot, Dadaist6174, ZéroBot, Redhanker, SporkBot, Wiki user 4~enwiki, וגיל, Bomazi, 28bot, ClueBot NG, Jack Greenmaven, RaptorHunter, MelbourneStar, Theolos, Mark Zelinka, Nao1958, Mean-Standev, Ramaksoud2000, BG19bot, Joshua hykes, BattyBot, ChrisGualtieri, Khazar2, Basroil, TwoTwoHello, Was 203.27.72.5, ChemTerm, Babitaarora, Finnusertop, Jarash, Nanapanners, DudeWithAFeud, Monkbot, Yikkayaya, WikicleanerB, Arnon81, KCGrimes, Liechtenstein96, Benjamin Redd, GreenC bot and Anonymous: 112

- **Cobalt-60** *Source:* https://en.wikipedia.org/wiki/Cobalt-60?oldid=760825370 *Contributors:* Bryan Derksen, Edward, Darkwind, Andrewa, SEWilco, Donarreiskoffer, Sho Uemura, Poupoune5, Beland, Canterbury Tail, Rich Farmbrough, NeuronExMachina, Vsmith, Arcadian, Burzum, DV8 2XL, Gene Nygaard, Bryan986, The Nameless, BD2412, Ketiltrout, Rjwilmsi, Strait, FlaBot, Kolbasz, Scalytail, Mrnatural, RussBot, Limulus, Shaddack, Eleassar, Bill-on-the-Hill, Johantheghost, CharlesHBennett, Ray Chason, Algae, Rtc, Arniebuteft, Edgar181, Grandmartin11, Chris the speller, Bluebot, Trekphiler, RedHillian, Giancarlo Rossi, Muhammad Hamza, John, Aussie Alchemist, Kevin W., Joseph Solis in Australia, Chetvorno, Rwflammang, HPaul, Guitardemon666, Aldis90, Epbr123, Headbomb, Rosarinagazo, Escarbot, Gavia immer, Jarekt, CommonsDelinker, Leyo, Meam5555, J.delaney, Jakebathman, Trumpet marietta 45750, Adam Schwing, HazyM, Metro67, RingtailedFox, Mark v1.0, Smptq, Falcon8765, Pijuvwy, Billy Huang, EoGuy, Robadavis, VQuakr, Eiland, SuperHamster, NuclearWarfare, Zdravljica, Mifter, Noctibus, Antewolf, Addbot, Mjamja, Debresser, Bonapace, Legobot, आशीष भटनागर, Luckas-bot, Yobot, Brad Razner, Churchofcheese, Dfe6543, Redakie, Materialscientist, Xqbot, Itoughnuke, Erik9bot, Bob Saint Clar, Tubas-en, DMKTirpitz, Mikespedia, Full-date unlinking bot, Ripchip Bot, Orangwiki, John of Reading, WikitanvirBot, Wikipelli, SporkBot, JulioLS, One.Ouch.Zero, Sundarnut, Gray-Fullbuster, Whoop whoop pull up, ClueBot NG, Widr, Frze, Joshua hykes, Justanonymous, WheresTristan, Dexbot, Tristan Surtel, UW Dawgs, Qwh, EvergreenFir, Babitaarora, Jfung1999, JaconaFrere, Dcmjim, Timtam463, Rovingrobert, Pinkomega, Stewader91, SireWonton, InternetArchiveBot, Sillysaturn and Anonymous: 127

- **Isotopes of iridium** *Source:* https://en.wikipedia.org/wiki/Isotopes_of_iridium?oldid=741245873 *Contributors:* Bryan Derksen, Julesd, Donarreiskoffer, Graeme Bartlett, Rchandra, Femto, Edison, Rjwilmsi, Vegaswikian, Limulus, Anomalocaris, Itub, SmackBot, John, Zanhsieh, Headbomb, Jimjamjak, Leyo, Piperh, Gerakibot, Dolphin51, Sv1xv, Addbot, LaaknorBot, Jasper Deng, TaBOT-zerem, Xqbot, GrouchoBot, Erik9bot, 777sms, GoingBatty, XinaNicole, ZéroBot, Remux, Whoop whoop pull up, ClueBot NG, Lanthanum-138, Helpful Pixie Bot, Excerpted31, Benzband, Makecat-bot, Manul, Kaspersmith1992 and Anonymous: 6

- **Isotopes of polonium** *Source:* https://en.wikipedia.org/wiki/Isotopes_of_polonium?oldid=765643903 *Contributors:* Bryan Derksen, Schewek, Julesd, Donarreiskoffer, PBS, Urhixidur, Femto, Deicas, BRW, LukeSurl, Benhocking, Magister Mathematicae, DePiep, Rjwilmsi, Bubba73, Kolbasz, Bgwhite, Chris Capoccia, Shaddack, Goodbyemoff, Anomalocaris, SmackBot, Chris the speller, DocKrin, Sbharris, Vladislav, DrGeneNelson, Zaphody3k, Eastlaw, Max sang, Zanhsieh, Wikid77, Headbomb, R'n'B, Leyo, J.delanoy, Maurice Carbonaro, Athomic69, Mosmof, SieBot, I Like Cheeseburgers, Friendly person, Senor Cuete, Trojancowboy, SoxBot, DumZiBoT, MystBot, Addbot, Luckas-bot, Yobot, Krgeqewrjsif, AnomieBOT, Jim1138, Carlsotr, Yeoman Scrap, Erik9bot, Double sharp, 777sms, XinaNicole, ZéroBot, DavidMCEddy, ClueBot NG, Latifahphysics, BG19bot, IronOak, Gknapp1, 3.14159265358pi, Cyberbot II, Darian2, Anythingcouldhappen, Thisisisaax, GreenC bot and Anonymous: 29

- **Isotopes of radium** *Source:* https://en.wikipedia.org/wiki/Isotopes_of_radium?oldid=686451087 *Contributors:* Bryan Derksen, Urhixidur, Femto, Rjwilmsi, Hellbus, Anomalocaris, Zanhsieh, Robertsteadman, Headbomb, Leyo, TXiKiBoT, SoxBot, Addbot, Luckas-bot, Erik9bot, Banak, Double sharp, TobeBot, Speciman00, 777sms, EmausBot, XinaNicole, ZéroBot, StringTheory11, Latifahphysics, Helpful Pixie Bot, 3.14159265358pi, Batgontork and Anonymous: 7

- **Strontium-90** *Source:* https://en.wikipedia.org/wiki/Strontium-90?oldid=733003132 *Contributors:* Bryan Derksen, Fxmastermind, Ewen, Emperor, DocWatson42, Rchandra, Andycjp, Eyrian, Sladen, Alansohn, Miranche, Stillnotelf, Vuo, TheGoblin, Graham87, BD2412, ScottJ, Mortice, Kolbasz, Wavelength, Quentin X, JWB, Limulus, Lar, Stephenb, Shaddack, Doctorsundar, Anomie, Trovatore, Black Falcon, Bosco13, Geoffrey.landis, SmackBot, Twerges, Chris the speller, Cadmium, Sangrolu, GeeksHaveFeelings, TenPoundHammer, Moeburn, Pat Payne, Jimvin, IronGargoyle, Smith609, Mike Doughney, Phatom87, Headbomb, James086, Nick Number, Northumbrian, Goldenband, Mack2, JAnDbot, Jespley, Ambrosia-, Alistairbell, Nono64, Rod57, Trumpet marietta 45750, Juliancolton, HazyM, ACSE, Mark v1.0, Una Smith, Chimino, Lazyrussian, NellieBly, Harlock81, MystBot, Addbot, Element16, Milepost53, Blah28948, Luckas-bot, Yobot, Amirobot, AnomieBOT, Rubinbot, Bluerasberry, Materialscientist, The High Fin Sperm Whale, ArthurBot, Eugene-elgato, Erik9, Erik9bot, GliderMaven, FrescoBot, Bob Saint Clar, RedBot, Smatrese, MrX, RjwilmsiBot, EmausBot, Boundarylayer, Wikipelli, ZéroBot, Johnny Beta, ClueBot NG, Widr, Latifahphysics, BG19bot, Pedroaznarez, Mongrel1956, BattyBot, DudeWithAFeud, Monkbot, Graemem56, CV9933, UncertaintyPrinciple1, NukePro and Anonymous: 49

- **Goiânia accident** *Source:* https://en.wikipedia.org/wiki/Goi%C3%A2nia_accident?oldid=762093078 *Contributors:* AxelBoldt, Bryan Derksen, The Anome, Rmhermen, Shii, TomCerul, Leandrod, Edward, Delirium, Eric119, Aarchiba, Ineuw, Ehn, Mulad, RodC, Warmfuzzygrrl, WhisperToMe, LMB, JonathanDP81, AnonMoos, Scott Sanchez, Wetman, Hajor, Mjmcb1, Robbot, Tlogmer, Korath, Sanders muc, Cecropia, Mattflaschen, Carnildo, Alan Liefting, Inkling, Wwoods, Mboverload, Solipsist, Pne, Dvavasour, LucasVB, DragonflySixtyseven, Clarknova, Quota, Deglr6328, Abdull, Adambondy, Eyrian, Brianhe, Pjacobi, Adam850, MeltBanana, Sam Derbyshire, Pavel Vozenilek, Quietly, Mashford, Neko-chan, MisterSheik, Alereon, Orlady, Jpgordon, Reinyday, Vortexrealm, Giraffedata, Helix84, Caeruleancentaur, Hooperbloob, Blahnia, Keenan Pepper, Axl, Hu, Dental, Dhartung, Velella, Evil Monkey, Vuo, Dave.Dunford, ShawnVW, Gene Nygaard, Dan100, Falcorian, Feezo, Firsfron, Ylem, Oliphaunt, Crackerbelly, Ikescs, Ardfern, Cbustapeck, Jacj, Stevey7788, Graham87, Mucky Duck, Tim!, Urbane Legend, XP1, Rillian, BlueMoonlet, SLi, Patrickr, Kolbasz, Wgfcrafty, JonathanFreed, Pstevens, Dylan Thurston, EamonnPKeane, YurikBot, Wavelength, NTBot~enwiki, Huw Powell, Kiscica, Brandmeister (old), Guslacerda, Epolk, Shawn81, Shaddack, Lusanaherandraton, G@iffen~enwiki, Howcheng, FreelanceWizard, Mgcsinc, Supten, Chris S, Tetracube, Leviramsey, EdX20, Nippoo, Dsyzdek, A bit iffy, SmackBot, YellowMonkey, Fireworks, Franny Wentzel, Septegram, Yellowbounder, AxelHarvey, Alias777, Cadmium, Moshe Constantine Hassan Al-Silverburg, Rolypolyman, Scwlong, Tsca.bot, Racklever, Thatnewguy, BrianTung, Fuhghettaboutit, Jumping cheese, BWDuncan, Dogosaurus, Acdx, Captainbeefart, John, Microchip08, Minna Sora no Shita, Beta34, Tonsa, The-Pope, TJ Spyke, Lapinbleu, Joseph Solis in Australia, Driegendre, Rocketman768, Nixxonvaldez, CmdrObot, Rwflammang, Gyopi, Mateus Hidalgo, Lentower, Deusnoctum, Location, Rifleman 82, Headbomb, Sijarvis, Mdotley, Ingolfson, Serpent's Choice, Dariosity, WolfmanSF, JamesBWatson, Schoowru, Swpb, The Anomebot2, Nick Cooper, Jimjamjak, ZackTheJack, Diotime, Liontish0, Welshleprechaun, Cfrydj, Akronym, RockMFR, Pharaoh of the Wizards, McSly, Trumpet marietta 45750, Woodega, Nwbeeson, Tvbrichmond, DeFaultRyan, Widders, GrahamHardy, Hugo999, VolkovBot, Johnfos, Lear's Fool, Mazarin07, Truthanado, Ian Glenn, Gorpik, Rob.bastholm, Afernand74, Johnnywiggle, Anyeverybody, ClueBot, Badger Drink, Palantir-palantir, Bbalasub, Roxport, Redrocketred, Aaroncorey, Ktr101, AssegaiAli, PixelBot, Shinkolobwe, NuclearWarfare, InternetMeme, FellGleaming, Ltmboy, Eleman, Cabayi, Addbot, Jacopo Werther, CanadianLinuxUser, Ashanda, LaaknorBot, Doniago, 84user, Bwrs, Verbal, Lightbot, The Bushranger, Luckas-bot, Yobot, Yngvadottir, Amy Baily, AnomieBOT, KDS4444, Archon 2488, Materialscientist, Aff123a, Citation bot, Xqbot, Anonymous from the 21st century, Johndcurry, Tibidibtibo, Intrepid-NY, Kgrad, RjwilmsiBot, EmausBot, Trofobi, ElTorbe, Dadaist6174, Dazman83, A2soup, SporkBot, Demiurge1000, Prandr, Whoop whoop pull up, ClueBot NG, Jnorton7558, DieSwartzPunkt, LucaNevski, CopperSquare, Reify-tech, Avecmonami, Gabriel Yuji, MeanMotherJr, Cyberbot II, Zbjornson, Zerabat, Tigraan, Agnichuck, Opencooper, Fer48, InternetArchiveBot, GreenC bot and Anonymous: 171

- **Biological warfare** *Source:* https://en.wikipedia.org/wiki/Biological_warfare?oldid=766188489 *Contributors:* AxelBoldt, Magnus Manske, Trelvis, LC~enwiki, Mav, Bryan Derksen, Timo Honkasalo, Koyaanis Qatsi, Malcolm Farmer, Andre Engels, Roadrunner, Heron, Ram-Man, Lorenzarius, Zocky, Lexor, DopefishJustin, Dante Alighieri, Gabbe, TakuyaMurata, Bon d'une cythare, Mbessey, Minesweeper, NuclearWinner, Ihcoyc, Kingturtle, Darkwind, Astudent, Ghewgill, Darkonc, Adam Bishop, PaulinSaudi, Andrevan, Viajero, Ike9898, Sue D. Nymme, GulDan, Morwen, Thue, Raul654, Francs2000, Kuikuri, Robbot, Astronautics~enwiki, Moriori, Korath, RedWolf, Psychonaut, Lowellian, SEKIUCHI, Mirv, Nilmerg, Hadal, JesseW, Wikibot, Wereon, Vikingstad, Reytan, Wayland, Dave6, Giftlite, DocWatson42, FleaPlus, Everyking, Varlaam, Dmmaus, Get-back-world-respect, Matthead, Deus Ex, JRR Trollkien, Wmahan, Chowbok, Geni, Antandrus, Jossi, Aequo, Ellsworth, Imjustmatthew, Discospinster, ElTyrant, Rich Farmbrough, Pjacobi, Vsmith, Zen-master, ArnoldReinhold, Paul August, Stereotek, Bender235, Terrapin, ReallyNiceGuy, Dpotter, CanisRufus, El C, Walden, Kross, Chairboy, Aude, RoyBoy, Palm dogg, Richard Cane, Thu, Bobo192, Stesmo, Sentience, Harald Hansen, Infocidal, Smalljim, Nectarflowed, Clawson, MITalum, Sriram sh, Roy da Vinci, Obradovic Goran, Nsaa, OGoncho, Jumbuck, Preuninger, Danski14, Alansohn, Eleland, LtNOWIS, Great Scott, SlimVirgin, Rwendland, Kelson Vibber, Wtmitchell, Velella, TaintedMustard, Evil Monkey, Netkinetic, Kelba, Ceyockey, Dtobias, Bacteria, Guy M, Madchester, Pol098, Ruud Koot, Firien, JRHorse,

Stephenb, Gaius Cornelius, Ksyrie, NawlinWiki, Complainer, Borbrav, Grafen, Badagnani, Sekhui, Rjensen, Dmoss, Rmky87, Lockesdonkey, BOT-Superzerocool, Karl Meier, Bota47, Wknight94, SamuelRiv, Yummy123, Theda, NHSavage, De Administrando Imperio, Tevildo, Daschtrois, CWenger, Fram, Nick-D, DVD R W, Sardanaphalus, Vanka5, SmackBot, Roger Davies, Falustra77, Pgk, Ikip, KocjoBot~enwiki, Thunderboltz, Bwiihh, Delldot, Cla68, Edgar181, Cool3, Peter Isotalo, Gilliam, Folajimi, Hmains, Betacommand, YMB29, Chris the speller, Master Jay, Keegan, Persian Poet Gal, MalafayaBot, Hibernian, Moshe Constantine Hassan Al-Silverburg, Deli nk, Sadads, DHN-bot~enwiki, MercZ, Rcbutcher, Can't sleep, clown will eat me, OrphanBot, Buttered Bread, TKD, Ryanluck, El guero, Mayooresan, Jwy, MartinCollin, Sammy1339, A5b, Ligulembot, Zeamays, Pilotguy, SashatoBot, Nishkid64, Mcgowan30, General Ization, Sina Kardar, Soumya92, Evan Robidoux, CaptainVindaloo, Peterlewis, Nobunaga24, RandomCritic, Beetstra, Darry2385, JYi, TW2, Butcherbird52, IvanLanin, Provocateur, RGrimmig, Audiosmurt, Tawkerbot2, Dlohcierekim, JForget, Unreal128, Adam Keller, Liam Skoda, CmdrObot, Edward Vielmetti, Jesse Viviano, Sarveshkarkhanis, Bnwwf91, WeggeBot, Pyro13368, Danrok, Bellerophon5685, Flowerpotman, Sa.vakilian, DavidMcCabe, Dynaflow, Nabokov, Narayanese, Kozuch, Weirdo59, Omicronpersei8, Gimmetrow, TH3 W1R3D, BetacommandBot, Rjm656s, Hypnosadist, Wandalstouring, Epbr123, Andyjsmith, Hcberkowitz, PanAndScan, Marek69, Bobblehead, Yettie0711, Davidhorman, Escarbot, Hajji Piruz, AntiVandalBot, Majorly, Marokwitz, TimVickers, Noroton, Dylan Lake, Tashtastic, Lost Boy, Gökhan, Ingolfson, HanzoHattori, JAnDbot, STSC, Husond, Spahbod, MER-C, CyberAnth, Andonic, Flying tiger, TAnthony, RebelRobot, CarolineBogart, Geniac, Magioladitis, VoABot II, Ka-ru, KConWiki, Cgingold, Seashorewiki, Xhancock, IvoShandor, Arjun01, CommonsDelinker, MapleTree, Leyo, Em Mitchell, J.delanoy, Pharaoh of the Wizards, Filll, Numbo3, Shimaspawn, Davbrck, Yonidebot, Skumarlabot, Neon white, Hodja Nasreddin, Dispenser, LordAnubisBOT, Mrg3105, Drewgupt, NewEnglandYankee, Bernard S. Jansen, MisterBee1966, Hanacy, KylieTastic, Cometstyles, WJBscribe, DH85868993, Beezhive, Squids and Chips, Mokgen, Idioma-bot, Wikieditor06, VolkovBot, Drakheim, C2thek93, Christophenstein, Fences and windows, TXiKiBoT, GimmeBot, Khutuck, RS Archives, Xerxesnine, Plarter, Qxz, Someguy1221, Beyond silence, Leafyplant, Supertask, Cybermaster~enwiki, Mr.NorCal55, Agent of the Reds, Billinghurst, History expert1, WiiRools, W5WMW, Sapphic, Dassiebtekreuz, Doc James, Kiamnomch, Aaron.linderman, EmxBot, Postcardpigs, Someguynobody, Dogen Zenji, SieBot, Milnivri, BotMultichill, Yintan, Cjallen1, Flyer22 Reborn, Oxymoron83, Lightmouse, Hobartimus, MadmanBot, Akarkera, Mygerardromance, Denisarona, Martarius, Beeblebrox, ClueBot, HujiBot, Snigbrook, Theseven7, Blackangel25, Helenabella, Pakaraki, EoGuy, Wraithful, CyrilThePig4, R000t, Chessy999, Robby.is.on, Blanchardb, Masterblooregard, Ktr101, Excirial, Gnome de plume, -Midorihana-, Relata refero, NuclearWarfare, L.tak, JamieS93, CowboySpartan, Kaiba, OG17, Takabeg, Wikimedes, Palindromedairy, Xs10ry, Cmacauley, Riversider2008, Digiweb, Loranchet, Natuut, XLinkBot, BodhisattvaBot, Viking6, Snapperman2, Chemicalnasties, Addbot, Heavenlyblue, Willking1979, DOI bot, Jojhutton, Mabdul, TEh L337 1, Chris19910, Harrymph, Heyitsalexander, Glane23, Favonian, Darkscholar789, Green Squares, Genius1995, Mdnavman, Pmcyclist, Rehman, Tide rolls, Lightbot, Smeagol 17, Luckas-bot, Yobot, MSClaudiu, Guy1890, Mmxx, KamikazeBot, Tonyrex, AnomieBOT, Wikieditoftoday, VX, Jim1138, Karthickbala, KPackard, Materialscientist, Citation bot, Bindiji, Lightroom2008, Prince Ludwig, Taikah, Quebec99, Xqbot, Cloflin, 14160aldora, Joesacnut, Jmundo, Gremi-ch, J04n, Dac28, Tsuchida54, Bellerophon, CalmCalamity, Milbergeralex, Mattis, Oldknock, Ajice, Benzen, Zammy5, FrescoBot, Surv1v4l1st, Boyjf29, Dino Agosto, Teamturnz, Tobby72, MBelzer, Cs32en, Sourjah22, AndresHerutJaim, Mart572, Gire 3pich2005, DivineAlpha, Citation bot 1, Doebler, Recipe For Hate, Loyalist Cannons, YAR08, Tinton5, V.narsikar, MondalorBot, Wickeryby, Full-date unlinking bot, Orionpilot, HelenOnline, Lotje, Huaxia, RjwilmsiBot, Tortuga135, Jackehammond, Angelouss, WildBot, Derim Hunt, Vinnyzz, EmausBot, John of Reading, JohnnyTopQuark, Ajraddatz, Super48paul, Rail88, Ibbn, GoingBatty, Prosweda, 8digits, ZxxZxxZ, Dcirovic, Italia2006, John Cline, Marino ha, Espbuff, Systemofadown44, Akerans, Matthewcgirling, BionicRock, XxDestinyxX, Noodleki, Donner60, XRiamux, Scientific29, Peter Karlsen, Whoop whoop pull up, Will Beback Auto, ClueBot NG, Effective1001, Catlemur, Baseball Watcher, Omegafishes, Frietjes, Charliemattyndjames, ChemicalWarfare2012, O.Koslowski, Widr, Helpful Pixie Bot, Regulov, Qbgeekjtw, BG19bot, Xerxessenior, Jayohen, Cyberpower678, Samuel-two, Mark Arsten, FiveColourMap, Pooptable, Katangais, Mich.kramer, Fotoriety, Wannabemodel, TeamHoBotv, Cyberbot II, Radiochemist, Enterprisey, Dexbot, Hmainsbot1, Mogism, XXzoonamiXX, Lugia2453, Project Osprey, MarchOrDie, Epicgenius, Dratgin, CsDix, Earlgrey T, El Foes, B14709, B. Godden, Ginsuloft, Blurrim, Linnalid, Gooserock, Cjohnson1992, Monkbot, Filedelinkerbot, Madsinpes, Haichina, PersianFire, MRD2014, Chathusha, Julietdeltalima, Angolatake2, Ali Stewart21, CAPTAIN RAJU, Pandalover0009, InternetArchiveBot, 2A02A03F, GreenC bot, SamHolt6, Hanikassah99, Royal Ace22, Mikebone12121212 and Anonymous: 595

• **Cobalt bomb** *Source:* https://en.wikipedia.org/wiki/Cobalt_bomb?oldid=765794630 *Contributors:* Trelvis, Bryan Derksen, The Anome, Fnielsen, Patrick, Dan Koehl, Ehn, Wikiborg, K1Bond007, HarryHenryGebel, Wereon, Vfrickey, Davidjonsson, Ploum's, DocWatson42, Karn, Antandrus, Beland, Kuralyov, Rich Farmbrough, Night Gyr, JustinWick, El C, Martey, Clawson, Apyule, Kjkolb, Ashley Pomeroy, Radical Mallard, Wiccan Quagga, Max rspct, DV8 2XL, Gene Nygaard, Kazvorpal, Killing Vector, Daveydweeb, BillC, Rjwilmsi, Strait, Catsmeat, Nimur, Gwernol, The Rambling Man, YurikBot, Kencaesi, Phmer, RussBot, Xihr, Splash, Hellbus, Shaddack, Brad Rousse, Ino5hiro, RUL3R, Gadget850, Curpsbot-unicodify, Serendipodous, Algae, NetRolller 3D, Scolaire, SmackBot, David Kernow, Arniebuteff, Geoff B, Man with two legs, Dylnuge, Andy M. Wang, Bluebot, Baldghoti, Thumperward, Oli Filth, Sbharris, Colonies Chris, DevSolar, LouScheffer, Crab182, Derek R Bullamore, Springyard, A5b, Curly Turkey, John. A. Parrot, Wizard191, Captainj, Chetvorno, Jafet, Calibanu, SlowSam, Acontorer, Besieged, Gogo Dodo, RomanXNS, Quibik, Robertinventor, NadirAli, Thijs!bot, Kubanczyk, Headbomb, Nick Number, DPdH, Rjmail, Noclevername, Luna Santin, Fru1tbat, Alphachimpbot, LegitimateAndEvenCompelling, MaXiMiUS, MegX, Jrowle, Canonymous, Father Goose, SquidSK, Gundato, Doodledoo, Mschel, Kb1, Captain Infinity, Rod57, Trumpet marietta 45750, Jaimeastorga2000, Smokopilomidanek~enwiki, Cs302b, Tourbillon, Jets5278, Goltz20707, Khutuck, Bearian, StarChaser Tyger, DragonRider13, Antoncampos, Smellersthefeller, Mandsford, HighInBC, Fractalfire, MenoBot, Wwheaton, Apperceptions, Boneyard90, Ahmed91981, John Nevard, Coccyx Bloccyx, Wnt, DumZiBoT, Editorofthewiki, Avoided, Addbot, LaaknorBot, Tide rolls, Frmatt, MileyDavidA, Yobot, Legobot II, PMLawrence, AnomieBOT, KDS4444, Citation bot, Bobbinz, Spectreboy~enwiki, HombreDHojalata, Shadowjams, HJ Mitchell, AM, Antoras, Skiff, RjwilmsiBot, A protohominid, Boundarylayer, Tommy2010, SporkBot, Wingman417, Senator2029, Whoop whoop pull up, ClueBot NG, Petriot333, Ajohnson777, Ramaksoud2000, Bearbin, BG19bot, OliverHargreaves, Jhogatec, Mompson, Datafoxy, Myxomatosis57, Comatmebro, Khazar2, SNAAAAKE!!, Robbru, Dristarg, Parronax, Oranjelo100, Stormmeteo, Abattoir666, Monkbot, Szezerac, SierraNevadan, Nokkenbuer, AggressiveNavel, Madnessgenius, Tiokio, BlusteryBlowers, Fitindia, Bender the Bot and Anonymous: 157

32.8.2 Images

• **File:2inchMortarsPortonDown.jpg** *Source:* https://upload.wikimedia.org/wikipedia/commons/b/bf/2inchMortarsPortonDown.jpg *License:* Public domain *Contributors:* This is photograph PD-CRO-70 from the collections of the Imperial War Museums (collection no. 2000-11-05)

Original artist: Photographer: Porton Down official photographer

- **File:60Co_gamma_spectrum_energy.png** *Source:* https://upload.wikimedia.org/wikipedia/commons/8/88/60Co_gamma_spectrum_energy. png *License:* GFDL *Contributors:* observation made by me at Bonn university, plotted with Origin *Original artist:* Traitor

- **File:8th_Air_Force_psychological_warfare_leaflet.jpg** *Source:* https://upload.wikimedia.org/wikipedia/commons/d/dc/8th_Air_Force_ psychological_warfare_leaflet.jpg *License:* Public domain *Contributors:* Own retouching of U.S. Government propaganda work *Original artist:* U.S. Government

- **File:96602765.lowres.jpeg** *Source:* https://upload.wikimedia.org/wikipedia/commons/3/34/96602765.lowres.jpeg *License:* Public domain *Contributors:* http://imglib.lbl.gov/ImgLib/COLLECTIONS/BERKELEY-LAB/RESEARCH-1930-1990/NUCLEAR-PHYSICS/ TRANSURANIUM-ELEMENTS/index/96602765.html *Original artist:* Berkeley-Laboratory

- **File:A_BCV_(Bulk_Contamination_Vehicle.jpg** *Source:* https://upload.wikimedia.org/wikipedia/commons/e/e2/A_BCV_%28Bulk_ Contamination_Vehicle.jpg *License:* Public domain *Contributors:* IWM H 25575 *Original artist:* Tanner A R (Lt) War Office official photographer

- **File:Activityofuranium233.jpg** *Source:* https://upload.wikimedia.org/wikipedia/commons/2/20/Activityofuranium233.jpg *License:* Public domain *Contributors:* Own work *Original artist:* Knoxgirl

- **File:Activitytotal.jpg** *Source:* https://upload.wikimedia.org/wikipedia/commons/d/d7/Activitytotal.jpg *License:* CC BY-SA 3.0 *Contributors:* Own work *Original artist:* Knoxgirl

- **File:Adriaen_Millaert_-_Portrait_of_Christoph_Bernhardt_von_Galen,_Bishop_of_Munster.jpg** *Source:* https://upload.wikimedia. org/wikipedia/commons/1/1b/Adriaen_Millaert_-_Portrait_of_Christoph_Bernhardt_von_Galen%2C_Bishop_of_Munster.jpg *License:* Public domain *Contributors:* Rijksmuseum Amsterdam *Original artist:* Adriaen Millaert (1633 - 1667/68)

- **File:Aerodynamic_enrichment_nozzle.svg** *Source:* https://upload.wikimedia.org/wikipedia/commons/7/7d/Aerodynamic_enrichment_ nozzle.svg *License:* Public domain *Contributors:* Created by Fastfission in Inkscape. Based on images 26.1.1 (for the labels and general design conventions) and images 26.1.2 (and actual photo of the geometry) from page 34 of chapter 3 of U.S. Department of Energy's Handbook for Notification of Exports to Iraq: Annex 3. *Original artist:* Fastfission (<a href='//commons.wikimedia.org/w/index.php?title=User_talk: Fastfission,,&,,action=edit,,&,,redlink=1' class='new' title='User talk:Fastfission (page does not exist)'>talk</a>)

- **File:Albania_chemweapcanister.jpg** *Source:* https://upload.wikimedia.org/wikipedia/commons/0/04/Albania_chemweapcanister.jpg *License:* Public domain *Contributors:* http://lugar.senate.gov/photos/nunn-lugar.html *Original artist:* US Government

- **File:Alfa_beta_gamma_radiation.svg** *Source:* https://upload.wikimedia.org/wikipedia/commons/d/d6/Alfa_beta_gamma_radiation.svg *License:* CC BY 2.5 *Contributors:* Traced from this PNG image. *Original artist:* User:Stannered

- **File:Ambox_globe_content.svg** *Source:* https://upload.wikimedia.org/wikipedia/commons/b/bd/Ambox_globe_content.svg *License:* Public domain *Contributors:* Own work, using File:Information icon3.svg and File:Earth clip art.svg *Original artist:* penubag

- **File:Americium-241.jpg** *Source:* https://upload.wikimedia.org/wikipedia/commons/1/16/Americium-241.jpg *License:* Public domain *Contributors:* Own work (Original text: *I (Whitepaw (talk)) created this work entirely by myself.) Original artist:* Whitepaw (talk)

- **File:Americium-241_Sample_from_Smoke_Detector.JPG** *Source:* https://upload.wikimedia.org/wikipedia/commons/9/93/ Americium-241_Sample_from_Smoke_Detector.JPG *License:* CC BY-SA 3.0 *Contributors:* Own photo *Original artist:* MedicalReference

- **File:Americium_button_hd.jpg** *Source:* https://upload.wikimedia.org/wikipedia/commons/4/4f/Americium_button_hd.jpg *License:* CC BY 2.0 *Contributors:* https://www.flickr.com/photos/amagill/2939712141 *Original artist:* Andrew Magill

- **File:Anthrax_culture.jpg** *Source:* https://upload.wikimedia.org/wikipedia/commons/4/40/Anthrax_culture.jpg *License:* Public domain *Contributors:* http://www.armymedicine.army.mil/news/photos/fullsize/anthraxculture.cfm *Original artist:* U.S. Army Medical Research Institute of Infectious Diseases photo

- **File:Asterisks_2_(vertical).svg** *Source:* https://upload.wikimedia.org/wikipedia/commons/f/fe/Asterisks_2_%28vertical%29.svg *License:* CC BY-SA 4.0 *Contributors:* Own work *Original artist:* DePiep

- **File:Asterisks_one.svg** *Source:* https://upload.wikimedia.org/wikipedia/commons/4/49/Asterisks_one.svg *License:* CC BY-SA 3.0 *Contributors:* Own work *Original artist:* DePiep

- **File:B-w-scientists.jpg** *Source:* https://upload.wikimedia.org/wikipedia/commons/7/7c/B-w-scientists.jpg *License:* Public domain *Contributors:* http://www.detrick.army.mil/cutting_edge/chapter04.cfm#pic07 *Original artist:* USG

- **File:Bar_magnet.jpg** *Source:* https://upload.wikimedia.org/wikipedia/commons/d/d8/Bar_magnet.jpg *License:* CC-BY-SA-3.0 *Contributors:* ? *Original artist:* ?

- **File:Bedrijfsafval.jpg** *Source:* https://upload.wikimedia.org/wikipedia/commons/3/3a/Bedrijfsafval.jpg *License:* Public domain *Contributors:* No machine-readable source provided. Own work assumed (based on copyright claims). *Original artist:* No machine-readable author provided. Fun4life.nl assumed (based on copyright claims).

- **File:Biohazard_symbol.svg** *Source:* https://upload.wikimedia.org/wikipedia/commons/c/c0/Biohazard_symbol.svg *License:* Public domain *Contributors:* Own work *Original artist:* Silsor

- **File:Blueprint_abomb.JPG** *Source:* https://upload.wikimedia.org/wikipedia/commons/2/26/Blueprint_abomb.JPG *License:* Public domain *Contributors:* U.S. Military or Department of Defense *Original artist:* U.S. Military

- **File:Bluetank.png** *Source:* https://upload.wikimedia.org/wikipedia/commons/5/50/Bluetank.png *License:* Public domain *Contributors:* Own work *Original artist:* LA2

- **File:Bravo_fallout2.png** *Source:* https://upload.wikimedia.org/wikipedia/commons/f/f7/Bravo_fallout2.png *License:* Public domain *Contributors:* Samuel Glasstone and Phillip J. Dolan, eds., *The Effects of Nuclear Weapons*, 3rd. edn. (Washington, D.C.: DOD and DOE, 1977): 437. [Gamma doses are Roentgens from arrival time to 96 hours (4 days) after detonation, outside on land. Glasstone and Dolan mention that because data from the ocean was not obtained in this particular test, *Bravo*, the fallout contours to the north of the islands are uncertain and some other fallout patterns for the same test ascribe the high levels measured on Rongelap to a "hotspot" of the sort measured in the ocean downwind in later tests. *Original artist:* United States Department of Energy

- **File:British_55th_Division_gas_casualties_10_April_1918.jpg** *Source:* https://upload.wikimedia.org/wikipedia/commons/d/dc/British_55th_Division_gas_casualties_10_April_1918.jpg *License:* Public domain *Contributors:* This is photograph Q 11586 from the collections of the Imperial War Museums (collection no. 1900-22) *Original artist:* Thomas Keith Aitken (Second Lieutenant)

- **File:Bundesarchiv_B_145_Bild-F016216-0020A,_Flugzeug_über_der_Halbinsel_Kertsch.jpg** *Source:* https://upload.wikimedia.org/wikipedia/commons/f/fd/Bundesarchiv_B_145_Bild-F016216-0020A%2C_Flugzeug_%C3%BCber_der_Halbinsel_Kertsch.jpg *License:* CC BY-SA 3.0 de *Contributors:* This image was provided to Wikimedia Commons by the German Federal Archive (Deutsches Bundesarchiv) as part of a cooperation project. The German Federal Archive guarantees an authentic representation only using the originals (negative and/or positive), resp. the digitalization of the originals as provided by the Digital Image Archive. *Original artist:* Unknown<a href='https://www.wikidata.org/wiki/Q4233718' title='wikidata:Q4233718'><img alt='wikidata:Q4233718' src='https://upload.wikimedia.org/wikipedia/commons/thumb/f/ff/Wikidata-logo.svg/20px-Wikidata-logo.svg.png' width='20' height='11' srcset='https://upload.wikimedia.org/wikipedia/commons/thumb/f/ff/Wikidata-logo.svg/30px-Wikidata-logo.svg.png 1.5x, https://upload.wikimedia.org/wikipedia/commons/thumb/f/ff/Wikidata-logo.svg/40px-Wikidata-logo.svg.png 2x' data-file-width='1050' data-file-height='590' /></a>

- **File:CBP_X-ray_vehicle_Superbowl.jpg** *Source:* https://upload.wikimedia.org/wikipedia/commons/3/39/CBP_X-ray_vehicle_Superbowl.jpg *License:* Public domain *Contributors:* http://www.cbp.gov/xp/cgov/newsroom/photo_gallery/current_events/superbowl02.xml (file superbowl02.jpg) *Original artist:* Gerald L. Nino, CBP, U.S. Dept. of Homeland Security

- **File:Caesium-137_Gamma_Ray_Spectrum-en.svg** *Source:* https://upload.wikimedia.org/wikipedia/commons/5/5b/Caesium-137_Gamma_Ray_Spectrum-en.svg *License:* Public domain *Contributors:* Caesium-137 Gamma Ray Spectrum-de.svg *Original artist:* Kolbasz

- **File:Checking_Radiation_Exposure_Levels_(FDA_190)_(8227384650).jpg** *Source:* https://upload.wikimedia.org/wikipedia/commons/0/07/Checking_Radiation_Exposure_Levels_%28FDA_190%29_%288227384650%29.jpg *License:* Public domain *Contributors:* Checking Radiation Exposure Levels (FDA 190) *Original artist:* The U.S. Food and Drug Administration

- **File:Chemical_Weapons_Convention_2007.svg** *Source:* https://upload.wikimedia.org/wikipedia/commons/3/37/Chemical_Weapons_Convention_2007.svg *License:* Public domain *Contributors:* Own work This file was derived from: Chemical Weapons Convention 2007.png *Original artist:* Cflm001 (talk)

- **File:Chemical_agent_protection.jpg** *Source:* https://upload.wikimedia.org/wikipedia/commons/1/18/Chemical_agent_protection.jpg *License:* CC-BY-SA-3.0 *Contributors:* Photo taken by myself. User:Johan Elisson. *Original artist:* User:Johan Elisson.

- **File:Cobalt-60.jpg** *Source:* https://upload.wikimedia.org/wikipedia/commons/2/21/Cobalt-60.jpg *License:* CC BY 3.0 *Contributors:* Own work *Original artist:* DMKTirpitz

- **File:Cobalt-60_Irradiator.tif** *Source:* https://upload.wikimedia.org/wikipedia/commons/4/43/Cobalt-60_Irradiator.tif *License:* Public domain *Contributors:* http://picturethis.pnl.gov/picturet.nsf/by+id/AMER-4VXNDP *Original artist:* pnl.gov, US Department of Energy

- **File:Cobalt-60m-decay.svg** *Source:* https://upload.wikimedia.org/wikipedia/commons/0/03/Cobalt-60m-decay.svg *License:* Public domain *Contributors:* Own work *Original artist:* Tubas-en

- **File:Cobalt0001.jpg** *Source:* https://upload.wikimedia.org/wikipedia/commons/f/f4/Cobalt0001.jpg *License:* Public domain *Contributors:* https://www.nal.usda.gov/exhibits/speccoll/items/show/6996 *Original artist:* ?

- **File:Commons-logo.svg** *Source:* https://upload.wikimedia.org/wikipedia/en/4/4a/Commons-logo.svg *License:* PD *Contributors:* ? *Original artist:* ?

- **File:Cosmos_954_-_Recovery_001.jpg** *Source:* https://upload.wikimedia.org/wikipedia/commons/7/7f/Cosmos_954_-_Recovery_001.jpg *License:* Public domain *Contributors:* Operation Morning Light Fact Sheet, DOE/NV1198 *Original artist:* Federal Government of the United States

- **File:Crystal_energy.svg** *Source:* https://upload.wikimedia.org/wikipedia/commons/1/14/Crystal_energy.svg *License:* LGPL *Contributors:* Own work conversion of Image:Crystal_128_energy.png *Original artist:* Dhatfield

- **File:Cs-137-decay.svg** *Source:* https://upload.wikimedia.org/wikipedia/commons/3/3e/Cs-137-decay.svg *License:* Public domain *Contributors:* http://www.springerlink.com/content/rm0v8727k2j02x85/, p. 492, doi:10.1007/978-3-642-00875-7 *Original artist:* Tubas-en

- **File:Cs-137_from_nuclear_tests_vector.svg** *Source:* https://upload.wikimedia.org/wikipedia/commons/3/3c/Cs-137_from_nuclear_tests_vector.svg *License:* Public domain *Contributors:* File:Cs-137 from nuclear tests.png *Original artist:* Harold L. Beck, National Cancer Institute

- **File:Culturaconvencoesgvn.jpg** *Source:* https://upload.wikimedia.org/wikipedia/commons/a/a8/Culturaconvencoesgvn.jpg *License:* Public domain *Contributors:* Own work *Original artist:* Adelano Lázaro

- **File:Dangclass7.svg** *Source:* https://upload.wikimedia.org/wikipedia/commons/b/b4/Dangclass7.svg *License:* Public domain *Contributors:* Own work, based on Dangclass7.png, and Radioactive Trafoil from Radioactive.svg. *Original artist:* User:IRTC1015

- **File:Death_by_haematopoietic_syndrome_of_radiation_sickness-_influence_of_dose_rate.png** *Source:* https://upload.wikimedia.org/wikipedia/commons/f/fb/Death_by_haematopoietic_syndrome_of_radiation_sickness-_influence_of_dose_rate.png *License:* Public domain *Contributors:* self-made from tabular data by Haskin et al. (1997) *Original artist:* Regis Lachaume

- **File:Death_by_haematopoietic_syndrome_of_radiation_sickness-_influence_of_medical_care.png** *Source:* https://upload.wikimedia.org/wikipedia/commons/8/83/Death_by_haematopoietic_syndrome_of_radiation_sickness-_influence_of_medical_care.png *License:* Public domain *Contributors:* self-made from tabular data by Anno et al. (2003) *Original artist:* Régis Lachaume

- **File:HEUraniumC.jpg** *Source:* https://upload.wikimedia.org/wikipedia/commons/d/d8/HEUraniumC.jpg *License:* Public domain *Contributors:* http://web.archive.org/web/20050829231403/http://web.em.doe.gov/takstock/phochp3a.html *Original artist:* ?

- **File:Halflife-sim.gif** *Source:* https://upload.wikimedia.org/wikipedia/commons/3/3f/Halflife-sim.gif *License:* Public domain *Contributors:* Own work *Original artist:* Sbyrnes321

- **File:Hanford_B_Reactor.jpg** *Source:* https://upload.wikimedia.org/wikipedia/commons/e/e5/Hanford_B_Reactor.jpg *License:* Public domain *Contributors:*   ?   *Original artist:*   Unknown<a href='https://www.wikidata.org/wiki/Q4233718' title='wikidata:Q4233718'><img alt='wikidata:Q4233718' src='https://upload.wikimedia.org/wikipedia/commons/thumb/f/ff/Wikidata-logo.svg/20px-Wikidata-logo.svg.png' width='20' height='11' srcset='https://upload.wikimedia.org/wikipedia/commons/thumb/f/ff/Wikidata-logo.svg/30px-Wikidata-logo.svg.png 1.5x, https://upload.wikimedia.org/wikipedia/commons/thumb/f/ff/Wikidata-logo.svg/40px-Wikidata-logo.svg.png 2x' data-file-width='1050' data-file-height='590' /></a>

- **File:Hanford_N_Reactor_adjusted.jpg** *Source:* https://upload.wikimedia.org/wikipedia/commons/d/d3/Hanford_N_Reactor_adjusted.jpg *License:* Public domain *Contributors:* Image N1D0069267. *Original artist:* United States Department of Energy

- **File:Heinkel_He_111_during_the_Battle_of_Britain.jpg** *Source:* https://upload.wikimedia.org/wikipedia/commons/8/82/Heinkel_He_111_during_the_Battle_of_Britain.jpg *License:* Public domain *Contributors:* This is photograph MH6547 from the collections of the Imperial War Museums (collection no. 4700-05) *Original artist:* Unknown<a href='https://www.wikidata.org/wiki/Q4233718' title='wikidata:Q4233718'><img alt='wikidata:Q4233718' src='https://upload.wikimedia.org/wikipedia/commons/thumb/f/ff/Wikidata-logo.svg/20px-Wikidata-logo.svg.png' width='20' height='11' srcset='https://upload.wikimedia.org/wikipedia/commons/thumb/f/ff/Wikidata-logo.svg/30px-Wikidata-logo.svg.png 1.5x, https://upload.wikimedia.org/wikipedia/commons/thumb/f/ff/Wikidata-logo.svg/40px-Wikidata-logo.svg.png 2x' data-file-width='1050' data-file-height='590' /></a>

- **File:High_altitude_EMP.gif** *Source:* https://upload.wikimedia.org/wikipedia/commons/b/b2/High_altitude_EMP.gif *License:* Public domain *Contributors:* (Image composed of the calculations on pages 33 and 36 of Louis W. Seiler, Jr., "A Calculational Model for High Altitude EMP". AIR FORCE INST. OF TECH., WRIGHT-PATTERSON A.F.B., U.S. Government report number AD-A009208, March 1975, available online at http://stinet.dtic.mil/cgi-bin/GetTRDoc?AD=A009208&Location=U2&doc=GetTRDoc.pd *Original artist:* User:Photocopier, from government source.

- **File:Hiroshima_girl.jpg** *Source:* https://upload.wikimedia.org/wikipedia/commons/f/f6/Hiroshima_girl.jpg *License:* Public domain *Contributors:* Transferred from fr.wikipedia *Original artist:* Shunkichi Kikuchi (菊池俊吉). image from Hiroshima Peace Media Center

- **File:Hormesis_dose_response_graph.svg** *Source:* https://upload.wikimedia.org/wikipedia/commons/b/b7/Hormesis_dose_response_graph.svg *License:* Public domain *Contributors:* en:Image:Hormesis dose response graph.png *Original artist:* Composed by User:Stannered

- **File:James_Bryce.jpg** *Source:* https://upload.wikimedia.org/wikipedia/commons/0/00/James_Bryce.jpg *License:* Public domain *Contributors:* [1] *Original artist:* Unknown<a href='https://www.wikidata.org/wiki/Q4233718' title='wikidata:Q4233718'><img alt='wikidata:Q4233718' src='https://upload.wikimedia.org/wikipedia/commons/thumb/f/ff/Wikidata-logo.svg/20px-Wikidata-logo.svg.png' width='20' height='11' srcset='https://upload.wikimedia.org/wikipedia/commons/thumb/f/ff/Wikidata-logo.svg/30px-Wikidata-logo.svg.png 1.5x, https://upload.wikimedia.org/wikipedia/commons/thumb/f/ff/Wikidata-logo.svg/40px-Wikidata-logo.svg.png 2x' data-file-width='1050' data-file-height='590' /></a>. Published by New York Times Company.

- **File:Japanese_Special_Naval_Landing_Forces_in_Battle_of_Shanghai_1937.jpg** *Source:* https://upload.wikimedia.org/wikipedia/commons/3/39/Japanese_Special_Naval_Landing_Forces_in_Battle_of_Shanghai_1937.jpg *License:* Public domain *Contributors:* Brent Jones: *Rising Sun in the East 1937* *Original artist:* Unknown photographer. Ministry of the Navy

- **File:Kepler-solar-system-2.gif** *Source:* https://upload.wikimedia.org/wikipedia/commons/1/1d/Kepler-solar-system-2.gif *License:* Public domain *Contributors:* ? *Original artist:* ?

- **File:LEUPowder.jpg** *Source:* https://upload.wikimedia.org/wikipedia/commons/2/2f/LEUPowder.jpg *License:* Public domain *Contributors:* ? *Original artist:* ?

- **File:Lock-green.svg** *Source:* https://upload.wikimedia.org/wikipedia/commons/6/65/Lock-green.svg *License:* CC0 *Contributors:* en:File: Free-to-read_lock_75.svg *Original artist:* User:Trappist the monk

- **File:Logo_iso_radiation.svg** *Source:* https://upload.wikimedia.org/wikipedia/commons/3/35/Logo_iso_radiation.svg *License:* Public domain *Contributors:* Image:Radiation warning symbol.jpg by User:Yann and User:AnonMoos . For original info see press-release http://www.iaea.org/NewsCenter/News/2007/radiationsymbol.html and PDF file http://www.iaea.org/NewsCenter/News/PDF/newradsymbol.pdf *Original artist:* historicair 19:47, 25 February 2007 (UTC)

- **File:Lyon_Playfair.jpg** *Source:* https://upload.wikimedia.org/wikipedia/commons/9/96/Lyon_Playfair.jpg *License:* Public domain *Contributors:* National Portrait Gallery: NPG x133395 *Original artist:* Lock & Whitfield

- **File:Map_of_Operation_Bodyguard_subordinate_plans.png** *Source:* https://upload.wikimedia.org/wikipedia/commons/d/d6/Map_of_Operation_Bodyguard_subordinate_plans.png *License:* CC BY-SA 3.0 *Contributors:* Own work, based on File:BlankEurope.png & Holt, Thaddeus, The Deceivers: Allied Military Deception in the Second World War (Scribner, New York, 2004) p. 807 - 896 *Original artist:* ErrantX

- **File:Mc-1_gas_bomb.png** *Source:* https://upload.wikimedia.org/wikipedia/commons/8/84/Mc-1_gas_bomb.png *License:* Public domain *Contributors:* ? *Original artist:* ?

- **File:Members_of_the_Ukrainian_Army's_19th_CBRN-Battalion_maintaining_decontamination_skills_in_Support_of_Operation_Iraqi_Freedom_at_Camp_Arifjan,_in_KUWAIT_on_August_3rd_2003.jpg** *Source:* https://upload.wikimedia.org/wikipedia/commons/d/d8/Members_of_the_Ukrainian_Army%E2%80%99s_19th_CBRN-Battalion_maintaining_decontamination_skills_in_Support_of_Operation_Iraqi_Freedom_at_Camp_Arifjan%2C_in_KUWAIT_on_August_3rd_2003.jpg *License:* Public domain *Contributors:* ? *Original artist:* US government

- **File:Merge-arrows.svg** *Source:* https://upload.wikimedia.org/wikipedia/commons/5/52/Merge-arrows.svg *License:* Public domain *Contributors:* ? *Original artist:* ?

- **File:SS-24_silo_destruction.jpg** *Source:* https://upload.wikimedia.org/wikipedia/commons/9/9c/SS-24_silo_destruction.jpg *License:* Public domain *Contributors:* ? *Original artist:* ?

- **File:SanFranHouses06.JPG** *Source:* https://upload.wikimedia.org/wikipedia/commons/e/ec/SanFranHouses06.JPG *License:* Public domain *Contributors:*

  Scanned from the personal collection of en:User:Infrogmation Originally from en.wikipedia; description page is/was here.

  Original artist: ?

- **File:Sarin-2D-skeletal.png** *Source:* https://upload.wikimedia.org/wikipedia/commons/3/36/Sarin-2D-skeletal.png *License:* Public domain *Contributors:* Own work *Original artist:* Ben Mills

- **File:Sasahara.svg** *Source:* https://upload.wikimedia.org/wikipedia/commons/c/cc/Sasahara.svg *License:* CC-BY-SA-3.0 *Contributors:* Transferred from en.wikipedia *Original artist:* Original uploader was JWB at en.wikipedia

- **File:Shiro-ishii.jpg** *Source:* https://upload.wikimedia.org/wikipedia/commons/a/ae/Shiro-ishii.jpg *License:* Public domain *Contributors:* Bulletin of Unit 731(an article not for sale) *Original artist:* Masao Takezawa

- **File:Skull_and_crossbones.svg** *Source:* https://upload.wikimedia.org/wikipedia/commons/5/53/Skull_and_crossbones.svg *License:* Public domain *Contributors:* http://vector4u.com/symbols/skull-and-crossbones-vector-svg/ *Original artist:* Unknown<a href='https://www.wikidata.org/wiki/Q4233718' title='wikidata:Q4233718'><img alt='wikidata:Q4233718' src='https://upload.wikimedia.org/wikipedia/commons/thumb/f/ff/Wikidata-logo.svg/20px-Wikidata-logo.svg.png' width='20' height='11' srcset='https://upload.wikimedia.org/wikipedia/commons/thumb/f/ff/Wikidata-logo.svg/30px-Wikidata-logo.svg.png 1.5x, https://upload.wikimedia.org/wikipedia/commons/thumb/f/ff/Wikidata-logo.svg/40px-Wikidata-logo.svg.png 2x' data-file-width='1050' data-file-height='590' /></a>

- **File:Sound-icon.svg** *Source:* https://upload.wikimedia.org/wikipedia/commons/4/47/Sound-icon.svg *License:* LGPL *Contributors:* Derivative work from Silsor's versio *Original artist:* Crystal SVG icon set

- **File:Stylised_Lithium_Atom.svg** *Source:* https://upload.wikimedia.org/wikipedia/commons/6/6f/Stylised_atom_with_three_Bohr_model_orbits_and_stylised_nucleus.svg *License:* CC-BY-SA-3.0 *Contributors:* based off of Image:Stylised Lithium Atom.png by Halfdan. *Original artist:* SVG by Indolences. Recoloring and ironing out some glitches done by Rainer Klute.

- **File:Symbol_book_class2.svg** *Source:* https://upload.wikimedia.org/wikipedia/commons/8/89/Symbol_book_class2.svg *License:* CC BY-SA 2.5 *Contributors:* Mad by Lokal_Profil by combining: *Original artist:* Lokal_Profil

- **File:Teletherapy_Capsule2.svg** *Source:* https://upload.wikimedia.org/wikipedia/commons/d/d1/Teletherapy_Capsule2.svg *License:* CC BY-SA 3.0 *Contributors:* Own work. This image is based on a similar technical drawing of such a capsule from US patent 3588031 located here. I have changed the viewing angle slightly, changed the degree of completeness of each cylindrical component, and added color, shading, and texture to this version. *Original artist:* KDS444

- **File:Teller-Ulam_device_3D.svg** *Source:* https://upload.wikimedia.org/wikipedia/commons/c/c1/Teller-Ulam_device_3D.svg *License:* Public domain *Contributors:* ? *Original artist:* ?

- **File:Text_document_with_red_question_mark.svg** *Source:* https://upload.wikimedia.org/wikipedia/commons/a/a4/Text_document_with_red_question_mark.svg *License:* Public domain *Contributors:* Created by bdesham with Inkscape; based upon Text-x-generic.svg from the Tango project. *Original artist:* Benjamin D. Esham (bdesham)

- **File:The_Art_of_War-Tangut_script.jpg** *Source:* https://upload.wikimedia.org/wikipedia/commons/3/3a/The_Art_of_War-Tangut_script.jpg *License:* Public domain *Contributors:* 宁夏档案 *Original artist:* Sun Tzu

- **File:Transparent.gif** *Source:* https://upload.wikimedia.org/wikipedia/commons/c/ce/Transparent.gif *License:* Public domain *Contributors:* Own work *Original artist:* Edokter

- **File:Trident_C-4_montage.jpg** *Source:* https://upload.wikimedia.org/wikipedia/commons/7/7f/Trident_C-4_montage.jpg *License:* Public domain *Contributors:* ? *Original artist:* ?

- **File:U.S._Army_loudspeaker_team_in_action_in_Korea.jpg** *Source:* https://upload.wikimedia.org/wikipedia/commons/1/1e/U.S._Army_loudspeaker_team_in_action_in_Korea.jpg *License:* Public domain *Contributors:* http://www.psywarrior.com/KoreanLS005.jpg http://www.psywarrior.com/KW1stLoudspeakerLeaflet.html *Original artist:* U.S. Army

- **File:U.S._Navy_Seabees_assigned_to_Naval_Mobile_Construction_Battalion_1_don_their_MCU-2P_gas_masks_at_the_Naval_Construction_Battalion_Center_in_Gulfport,_Miss.,_Oct_081024-N-LD343-001.jpg** *Source:* https://upload.wikimedia.org/wikipedia/commons/c/cf/U.S._Navy_Seabees_assigned_to_Naval_Mobile_Construction_Battalion_1_don_their_MCU-2P_gas_masks_at_the_Naval_Construction_Battalion_Center_in_Gulfport%2C_Miss.%2C_Oct_081024-N-LD343-001.jpg *License:* Public domain *Contributors:* http://www.defenseimagery.mil/imageRetrieve.action?guid=c64057b4be184cb527c164f3672784fd7ce88ec2&t=2 *Original artist:* MC2 Demetrius Kennon

- **File:US_Army_soldier_hands_out_a_newspaper_to_a_local_Aug_2004.jpg** *Source:* https://upload.wikimedia.org/wikipedia/commons/0/01/US_Army_soldier_hands_out_a_newspaper_to_a_local_Aug_2004.jpg *License:* Public domain *Contributors:* Defense Visual Information Center official site *Original artist:* Jeremiah Johnson

- **File:US_and_USSR_nuclear_stockpiles.svg** *Source:* https://upload.wikimedia.org/wikipedia/commons/b/bb/US_and_USSR_nuclear_stockpiles.svg *License:* Public domain *Contributors:* Own work Source data from: Robert S. Norris and Hans M. Kristensen. "Global nuclear stockpiles, 1945-2006," *Bulletin of the Atomic Scientists* 62, no. 4 (July/August 2006), 64-66. Online at http://thebulletin.metapress.com/content/c4120650912x74k7/fulltext.pdf *Original artist:* Created by User:Fastfission first by mapping the lines using OpenOffice.org's Calc program, then exporting a graph to SVG, and the performing substantial aesthetic modifications in Inkscape.

- **File:US_fallout_exposure.png** *Source:* https://upload.wikimedia.org/wikipedia/commons/3/37/US_fallout_exposure.png *License:* Public domain *Contributors:* Slightly modified (whitespace made transparent, converted to PNG) from Figure 1 in *Study Estimating Thyroid Doses of I-131 Received by Americans From Nevada Atmospheric Nuclear Bomb Test*, National Cancer Institute (1997) [1] *Original artist:* National Cancer Institute

32.8.3 Content license